Tropical Forestry

Tropical Forestry
Volumes Already Published in this Series

Sampling Methods, Remote Sensing and GIS Multiresource Forest Inventory
Authors: Köhl, M., Magnussen, S., Marchetti M.
2006, ISBN: 3-540-32571-9

Tropical Forest Ecology - ***The Basis for Conservation and Management***
Authors: Montagnini, F., Jordan C.
2005, ISBN: 3-540-23797-6

Michael Köhl · Steen Magnussen · Marco Marchetti

Sampling Methods, Remote Sensing and GIS Multiresource Forest Inventory

With 113 Figures, 5 in Color and 27 Tables

Springer

Professor Dr. Michael Köhl
University of Hamburg
Department of Wood Science
Section World Forestry
Leuschnerstr. 91
D-21031 Hamburg
Germany

Dr. Steen Magnussen
Canadian Forest Service
506 W. Burnside Rd
Victoria
BC V8Z 1M5
Canada

Professor Dr. Marco Marchetti
University of Molise
Dept. STAT – Science and Technologies
for Land and Environment
Via Mazzini, 8
86170 Isernia
Italy

ISSN: 1614-9785

ISBN 978-3-642-06898-0 e-ISBN 978-3-540-32572-7

Springer-Verlag is a part of Springer Science+Business Media

springer.com

Editor: Dr. Dieter Czeschlik, Heidelberg
Desk Editor: Anette Lindqvist, Heidelberg
Cover Design: Design & Production, Heidelberg

Author Profiles

Michael Köhl is Professor of World Forestry at the University of Hamburg, Germany, and head of the Institute for World Forestry, at the Federal Research Center for Forestry and Forest Products. His main research interests are in the field of quantitative methods for natural resources assessments and forest planning, including statistical design and analysis of complex sample surveys, remote sensing, information systems, and geographic information systems. He has authored more than 150 scientific papers. He is former leader of the International Union of Forest Research Organizations' (IUFRO) unit 4.02.00 "Forest resources inventory and monitoring," leader of the Team of Specialists for the UN-FAO/Economic Commission for Europe (ECE) Temperate and Boreal Forest Resources Assessment (TBFRA) 2000, coordinator for the Ministerial Conference on the Protection of Forests in Europe (MCPFE) – an advisory group on the Improvement of the Pan-European Indicators for Sustainable Forest Management – and lead author of the UN Intergovernmental Panel on Climate Change (IPCC) *Good practice guidance for land use, land-use change and forestry*. He is currently Associate Editor of the *Schweizerische Forstzeitung* and Subject Editor of the *European Journal of Forest Research*. He is an honorary member of the Finnish Society of Forest Science, a member of the Scientific Advisory Board of the European Forest Institute, and an associate member of the Italian Academy of Forest Sciences.

Marco Marchetti is Full Professor of Forest Inventory and of Forest Conservation at the University of Molise (Italy). He has a research record of 20 years in the domain of geographic information systems and remote sensing for forest management and forest assessment. He is the principal investigator in

several national and international projects. His research activities span the domain of development and application of new technologies. His current research focus is on multiresource inventory and forest mapping, forest management and land planning, fire monitoring, and biodiversity assessment. His accomplishments have been communicated in more than 100 publications. Marco is on the editorial boards of national and international journals. He is former deputy of IUFRO unit 8.07.01 "Key factors and ecological functions for forest biodiversity." Currently he is leader of IUFRO unit 4.02.00 "Forest resources inventory and monitoring," and is a member of the Team of Specialists for the UN-FAO/ECE Forest Resources Assessment. Marco is an ordinary member of the Italian Academy of Forest Sciences and a scientific advisor of the Italian Remote Sensing Association.

Steen Magnussen is a senior research scientist with the Canadian Forest Service (Natural Resources Canada). His research covers statistics and biometrics of multiresource forest inventories, spatial analysis, and classification and accuracy of forest inventory information. Steen is a member of the American Statistical Association and the International Association of Statistical Computing. He is currently Associate Editor of the *European Journal of Forest Research.* Similar positions were held with *Forest Science,* the *Canadian Journal of Forest Research,* and the *Forestry Chronicle.* Steen holds a Ph.D. from the University of Göttingen (Germany) and a M.Sc. in Forestry from the Agricultural University of Copenhagen (Denmark) His research has been published in more than 100 peer-reviewed scientific journals and in over 40 articles and technical reports.

Acknowledgements

Several people have assisted in the preparation of this book. We thank Andre Iost, Bernhard Kenter, Raul Köhler, Aljoscha Requardt, Thomas Riedel, Katja Oehmichen, Konstantin Olschofsky, and Dr. Wolfgang Stümer, all Ph.D. (former) students at the University of Hamburg, and Ugo Chiavetta, Marco Ottaviano, and Fabio Lombardi, Ph.D. students at the University of Molise, for their patience, input, and help with the first rudiments of this book.

Special appreciation goes to Infoterra GmbH, Friedrichshafen, Germany, GAF 17G, Munich, Germany, the German Aerospace Center, Bonn, and the EC Joint Research Center, Ispra, Italy, for granting us permission to use material from their research projects. Otto Ueli Bräker, Markus Keller, and Andreas Zingg at the Swiss Federal Institute for Forest, Snow, and Landscape Research, Birmensdorf, kindly supported us with material and graphs. We thank Alexander Korotkov, UN-ECE Timber Section, for allowing us to use material from a special study on reporting nonwood goods and services prepared for the UN-ECE.

Thanks are also due to Melanie Wöbb for her diligence and professionalism in handling the manuscript and her talent for making graphs and figures. She never ceased to remind us about missing quotations and incomplete text fragments, forever exhorting us to comply to a timetable.

We are indebted to Anette Lindqvist from Springer, Heidelberg, for not losing her patience with us. Despite numerous delays and missed deadlines she never gave up on us and continued her support of this book.

Finally, special thanks go to our families. Three wives and seven kids became entangled by this book at one point or another. They braved adverse side effects, such as limited availability, absentmindedness, or mental absence of the authors. It is our luck that they are still with us. Michaela, Nancy, and Elena: this book is for you – but you do not have to read it.

Preface

When we talk about forests, we talk about 30% of our planet's land surface area. In 2000 there was an estimated 3 870×10^6 ha of forest worldwide, of which 5% is in forest plantations and 95% in natural[1] forests (FAO 2003). Forests are not only a source for timber; they also generate significant nonwood goods and services, mitigate climate change, conserve biological diversity, provide protection from natural hazards, and not least: provide recreational areas for an increasingly urbanized world population. The availability of timber and nonwood goods and services is waning as deforestation and degradation of tropical forests continue. While forest area has stabilized or is slightly increasing in the boreal and temperate regions, the annual loss of forest area in the tropics and subtropics is decreasing. Between 1990 and 2000 the annual rate of deforestation was estimated to be 14.6×10^6 ha (approximately 0.38%) and took place mainly in tropical and subtropical forests (FAO 2003). The net annual rate of change is about 9.4×10^6 ha (0.2%).

Maintaining and enhancing forest areas and the vitality of forest ecosystems is a widely accepted political goal, which is often opposed by conflicting demands of various stakeholders. Solutions to conflicts of this nature require actions at different scales ranging from managing demands of local communities to resolutions of transboundary problems such as global climate change (Jackson and Ingeles 1998; Mayers and Bass 2004; Sliggers and Kakabeeke 2005). Decisions about political measures as well as local management issues will not be effective unless they rest on reliable, timely, and readily available information. Forest inventories offer a tool to provide objective and reliable information about the multiple functions of forest ecosystems and their potential to satisfy various demands.

There is always a direct relation between the quality of information available and the cost involved in obtaining it. The complexity, diversity, and wide spatial extension of forests preclude a 100% assessment in most cases. An alternative to a complete enumeration is sampling, which is the process of obtaining information by assessing only a proportion of and drawing inference for the

[1] In the FAO terminology "natural" includes both managed and unmanaged forests.

whole. Where spatial information is to be provided, remote sensing offers a suite of methods.

This book is intended to be a primer on multiresource forest inventories, with special reference to tropical and subtropical forests. The focus is on sustainable forest management, which requires an assessment of both the current state and changes over time. The information needs to be satisfied by forest inventories cover a wide range, which extends far beyond the forests' productive function and timber supply. Nonwood goods and services, environmental functions – such as mitigating climate change – biodiversity, watershed protection, protective functions, or recreation are related issues.

Besides the diversity of topics, the size of the area for which information is required is to be considered when designing and implementing a forest inventory. Local assessments require different approaches from regional, national, or multinational assessments. While field assessments may be a sufficient data source for inventorying and monitoring small areas, extensive inventories for large areas may involve the combination of different data sources for reasons of cost-effectiveness. Thus, remote sensing has become a prominent tool for multiscale forest resources assessments (Franklin 2001; Wulder and Franklin 2003).

Today's information needs about the forest resource often touch on areas outside the forests as well. For example, information on the accessibility of forest areas, road network inside and outside forests, wildlife habitats at the edge of and in close proximity to a forest, and the protective function of forests. The forest is part of a larger landscape and its function and services can only be fully appreciated in an integrated multidiciplinary approach to forest inventory. The increasing availability of georeferenced data in digital format and the widespread availability of powerful geographic information systems (GIS) have greatly facilitated this integration and paved the way for cross-cutting spatial analyses of inventory information.

The short annotation above portends to the diversity of methods and approaches needed to carry out a multiresource forest inventory. It would be far beyond the scope of this book to give an overarching collection of available methods for forest resources assessments. Our intent is to give an introduction to and overview of basic concepts, which can be easily adapted for real-world situations.

M. Köhl
S. Magnussen
M. Marchetti

May 2006

Contents

List of Abbreviations

AIC	Appreciation–influence–control
ALI	Advanced Land Imager
APA	American Pulpwood Association
ASTER	Advanced Spaceborne Thermal Emission and Reflection Radiometer
ATO	African Timber Organization
AVHRR	Advanced Very High Resolution Radiometer
AVIRIS	Airborne Visible Infrared Imaging Spectrometer
BLUP	Best linear unbiased prediction
CASI	Compact Airborne Spectrographic Imager
CBD	Convention on Biological Diversity
CCAD	Central American Commission for Environment and Development, Comisión Centroamericana de Ambiente y Desarrollo
CFI	Continuous forest inventory
CIFOR	Center for International Forestry Research
CIR	Color infrared
DBH	Diameter at breast height
DCMI	Dublin Core Metadata Initiative
DEM	Digital elevation model
DTM	Digital terrain model
EB	Empirical Bayesian
EBULP	Empirical best linear unbiased predictor
ECE	UN Economic Commission for Europe
EO	Earth observation/Earth Observer
ETM+	Enhanced Thematic Mapper Plus
EUNIS	European Nature Information System
FAO	Food and Agriculture Organization
FIFS	Forest industries feasibility study
FIS	Forest information systems
FMU	Forest management unit
fpc	Correction factor for the sampling fraction

FRCC	Fire regime condition class
FSC	Forest Stewardship Council
GCP	Ground control point
GIS	Geographic information systems
GPS	Global Positioning System
HH	Hansen–Hurwitz
HT	Tree height
ICP	International Cooperative Program
IGBP	International Geosphere and Biosphere Program
IPCC	Intergovernmental Panel on Climate Change
IRS	Indian Remote Sensing Satellite
IS	Importance sampling
ISRO	Indian Space Research Organization
ITTO	International Tropical Timber Organization
IUCN	International Union for the Conservation of Nature and Natural Resources (since 1990, the name "World Conservation Union" has been used)
IUFRO	International Union of Forest Research Organizations
ITTO	International Tropical Timber Organization
LCE	Linear contrast enhancement
MCAR	Missing completely at random
MCPFE	Ministerial Conference on the Protection of Forests in Europe
MERIS	Medium-Resolution Imaging Spectrometer
MIMICS	Michigan microwave canopy scattering model
MODIS	Moderate-Resolution Imaging Spectroradiometer
MS	Multispectral Scanner
MSE	Mean square error
MSR	Multiphase sampling with regression estimators
MSS	Multispectral Scanner
NALC	North American Land Characterization
NCDCDS	US National Committee for Digital Cartographic Data Standards
NDVI	Normalized-difference vegetation index
NFI	National Forest Inventory
NN	Nearest neighbor
NOAA	National Oceanic and Atmospheric Administration
NWGS	Nonwood goods and services
OLS	Ordinary least squares
PA	Population area
PCA	Principal component analysis
PLA	Participators learning and action
PPS	Probability proportional to size

PRA	Participators rural assessment
RBS	Randomized branch sampling
RGB	Red–green–blue
RRA	Rapid rural assessment
SAR	Synthetic aperture radar
SF	Spatial filtering
SFM	Sustainable forest management
SPOT	Système Pour l'Observation de la Terre
SPR	Sampling with partial replacement
SRS	Simple random sampling
TARA	Tarapoto Process
TBFRA	Temperate and Boreal Forest Resources Assessment
TM	Thematic Mapper
UNCED	United Nations Conference on Environment and Development
UNEP	United Nations Environment Program
UNFCCC	United Nations Framework Convention on Climate Change
UTM	Universal Transverse Mercator
VI	Vegetation index
ZOPP	Objective-oriented project planning

Forest Inventories — an Overview 1

1.1 Focus

Forest inventory is a process for obtaining information on the quality and quantity of forest resources and forms the foundation of forest planning and forest policy. While early concepts of sustainable forest management and forest inventory focused on timber production (Hartig 1795; Cotta 1804), modern forest inventory concepts support a holistic view of forest ecosystems addressing not only timber production but also the multiple functions of forests as well as the need to understand the functioning mechanisms of forest ecosystems (von Gadow et al. 2002; Corona et al. 2003).

Forest resources assessment facilitates a multifaceted analysis and study of forests not only as an important source of subsistence, employment, revenue earnings, and raw materials to a number of industries but also for their vital role in ecological balance, environmental stability, biodiversity conservation, food security, and sustainable development of countries and the entire biosphere. Forests have to be managed judiciously not only for environmental protection and other services but also for various products and industrial raw material. In some parts of the world biological resources are being depleted faster than they can regenerate. Following the 1992 United Nations Conference on Environment and Development (UNCED) conference in Rio de Janeiro considerable progress has been made in the area of sustainable forest management. Among others, the International Tropical Timber Organization (ITTO) and the Forest Stewardship Council (FSC) developed criteria and indicators for sustainable forest management and certification. The Kyoto Protocol of the United Nations Framework Convention on Climate Change (UNFCCC) describes measures to mitigate greenhouse gasses and addresses in Article 3.3 in particular the impact of deforestation and afforestation on global climate change. The Convention on Biological Diversity (CBD) that was ratified in 1994 deals with the protection and maintenance of biodiversity.

Forest resources assessments have their focus on the provision of information, which has several implications:

1. The information provided has to satisfy user needs. An inventory is generally not carried out for the needs of a single stakeholder; multiple issues of forests have to be covered. The objectives must be defined by those who require the data to be collected. All groups of users of inventory results should be involved in this phase of planning. Very often, the number of those interested in the inventory results increases after an inventory has already begun or after the findings have been published, so the data collected usually fail to satisfy all demands for information. Before defining the objectives it is advisable to make an inquiry not only among forest authorities but also among private forest owners, the wood-processing industry, land-use planning and environmental protection agencies, consumers of secondary forest products, wildlife organizations, and other potentially interested parties, thus enabling them to articulate their particular needs for information. In addition, this approach increases the possibility of finding partners who will make a financial contribution.
2. The information obtained by an inventory is typically presented in maps and statistical estimates. The basic concept of any statistical presentation is to summarize the population of interest and extract the facts important for potential users. This is generally done by presenting statistical parameters such as mean values, totals, or ratios and percentages. In addition, information on the variability or diversity of a population is an important ecological issue.
3. The information has to be objective. All parts of the population should be covered by the inventory; no part should be intentionally excluded. Data must be assessed in a nonsubjective way. Objective information requires the objective assessment of data. When information is gathered by some form of sampling, only application of a statistical design with known selection probabilities for any population element can ensure the integrity of the information-gathering process (inventory).
4. The information has to be reliable. The results of any sampling survey are always estimates rather than true values and are thus subject to a certain degree of uncertainty, as only part of the population is assessed. The uncertainty can be reduced through an optimal combination of sampling design and sample size in order to increase the precision of the estimates and to reduce sampling errors. The measurements themselves may be subject to error caused by, for example, inappropriate measurement devices, poor training, or subjective interpretation of measurement rules and definitions. Investments in improved instruments and the provision of intensive

training of field crews usually generate a handsome payback in the form of an increase in the quality and accuracy of data. It is necessary to specify the degree of precision and accuracy (see Chap. 3.4) to which the results should attain. This must be decided by the prospective users, though it is often difficult for administrators to think in terms of sampling error.

5. The information must be assessed in a cost-efficient way. Once forest managers and decision-makers have provided a rough definition of the objectives, several alternative inventory designs should be investigated. Alternatives can be based on different sampling design, sampling intensities, or data sources. Comparison of these alternatives allows assessment of the cost—benefit relationship and the final definition and weighing of the objectives.
6. The results of an inventory should be intuitively clear for potential users. Users are normally not very familiar with sampling statistics and thus the results should not require a Ph.D. in statistics for any immediate and basic interpretation. Users will have confidence only in information that they can understand. The inventory design should be documented and give advice for the impartial interpretation of data. As sample-based results are always subject to sampling errors, it is necessary to accompany any statistical estimate with estimates of sampling error or confidence interval.
7. Forest inventories should "only" present information in statistical and mapped format. It is beyond the mandate of a forest inventory to interpret results. However, forest inventory specialists should give advice for the interpretation of data. This restraint is also intended as a safeguard for the integrity of the inventory process.
8. In inventories on successive occasions, terms and definitions should not be changed unless it can be argued that the benefits outweigh the problems introduced by a change. When terms and definitions are changed between assessments one cannot distinguished between true change and change due to change in definitions.
9. Planning of a forest inventory is a complex task that involves expertise from many fields; thus, experts from silviculture, forest management planning, economics, policy, ecology, or timber products need to be consulted at an early stage.

1.2 Objectives

The main elements of an inventory depend very much upon the inventory objective; thus the objectives of an inventory have to be laid down in a very early phase of inventory planning. The exact definition is a joint action by the inventory designer and the potential user group. It is a very laborious undertaking and consequently too little effort is often committed; yet without it, one may lose sight of the real objectives with the risk of accepting inappropriate methods and procedures. The formulation of the objectives constitutes not only the basis for the design of the sampling methods but also an instrument for checking success once the survey is complete.

Four specific guidelines should be considered when determining inventory objectives (FAO 1998):

1. Objectives need to be determined jointly by the people who will use the results, including forest managers, planners, and decision-makers, as well as by inventory specialists. Inventory objectives should not be determined by inventory specialists alone.
2. Not all inventory objectives have the same level of importance. Some have higher priority than others and it is the objectives having highest priority that should determine the inventory design and the presentation of results.
3. Inventory objectives should reflect the physical effort that will be required to conduct an inventory, the organization, estimated costs and time, the existing knowledge of resources, the availability of specific aspects of inventory technologies, and institutional capability. All have a direct bearing upon the implementation of an inventory. An overriding consideration is that an inventory must be practicable and achievable. The value of an incomplete inventory that lacks important information and thus limits the possibility to establish causal relationships could be zero or close to zero.
4. All objectives should be *SMART*
   - *S*pecific
     - Well defined.
     - They are clear to anyone who has a basic knowledge of the project.
   - *M*easurable: They provide quantifiable measures of achievement and variance from set objectives.
   - *A*greed upon: There is agreement between the users and the project team on what the objectives should be.
   - *R*ealistic: Looking at the resources, knowledge, and time available, can the objective be accomplished?
   - *T*ime-framed
     - How much time is needed to accomplish the objective?
     - Having too much time can affect the project performance.

The information requirements from forest owners, policy planners, the scientific community, and society in general concerning forest resources have been growing steadily since the 1950s when the main focus was on information about timber supply (Table 1.1). The multiple functions of forests, biomass,

Table 1.1. Increase in information needs about forest lands in the USA (after Lund and Smith 1997)

					Nonforest lands,
					habitats, old
					growth and
					primary forests
				Ecosystems,	Ecosystems,
				biodiversity,	biodiversity,
				NWGS	NWGS
			Global	Global	Global
			warming	warming	warming
		Biomass	Biomass	Biomass	Biomass
	Multiple	Multiple	Multiple	Multiple	Multiple
	resources	resources	resources	resources	resources
Timber	Timber	Timber	Timber	Timber	Timber
1950s	1960s	1970s	1980s	1990s	2000+

NWGS nonwood goods and services

global warming, biodiversity, and nonwood goods and services have since gained prominence (Lund and Smith 1997).

The thematic scope of forest inventories can vary considerably. It is advisable to review global initiatives and obligations in order to get a broad view on potential information topics to be covered by a forest inventory. UNCED criteria and indicators for sustainable forest management have been formulated through several international, national, and nongovernmental processes. These include the Pan-European (or Helsinki) process (for European forests), the Montreal Process (for temperate and boreal forests), the Tarapoto Proposal of Criteria and Indicators for Sustainability of the Amazon Forest, the United Nations Environment Program (UNEP)/Food and Agriculture Organization (FAO) Expert Meeting on Criteria and Indicators for Sustainable Forest Management in Dry-Zone Africa, or the Lepaterique Process of Central America. The ITTO, the Tarapoto Process (TARA), the Center for International Forestry Research (CIFOR), the African Timber Organization (ATO), and the Central American Commission for Environment and Development (CCAD) developed systems of criteria and indicators for sustainable forest management which cover administrative, economic, legal, social, technical, and scientific issues which affect natural forests and plantations. The criteria define the essential factors of forest management against which forest sustainability may be assessed. Each criterion relates to a key management factor which may be described by one or more qualitative, quantitative, or descriptive indicators. Through measurement and monitoring of selected indicators, the effects of forest management action, or inaction, can be assessed and evaluated and action adjusted to ensure that forest management objectives are more likely to be achieved. Table 1.2 summarizes the criteria

Table 1.2. Criteria and indicators for sustainable management (after FAO 1998)

1. **Extent of forest resources and global carbon cycles**
   Area of forest cover
   Wood-growing stock
   Successional stage
   Age structure
   Rate of conversion of forest to other use
2. **Forest ecosystem health and vitality external influence**
   Deposition of air pollutants
   Damage by wind erosion
   Forest vitality indicators
   Incidence of defoliators
   Reproductive health
   Forest influence indicators
   Insect/disease damage
   Fire and storm damage
   Wild-animal damage
   Anthropogenic influence indicators
   Competition from introduction of nonnative plants
   Nutrient balance and acidity
   Trends in crop yields
3. **Biological diversity in forest ecosystem**
   Ecosystem indicators
   Distribution of forest ecosystems
   Extent of protected areas
   Habitat suitability
   Forest fragmentation
   Area cleared annually of endemic species
   Area and percentage of forest lands with fundamental ecological changes
   Forest fire control and prevention measures
   Species indicators
   Number of forest-dependent species
   Number of forest-dependent species at risk
   Reliance on natural regeneration
   Resources exploitation systems used
   Measures for in situ conservation of species at risk
   Genetic indicators
   Number of forest-dependent species with reduced range
4. **Productive functions of forests**
   Percentage of forests/other wooded lands managed according to management plans
   Growing stock
   Wood production
   Production of nonwood forest products
   Annual balance between growth and removal of wood products
   Level of diversification of sustainable forest production
   Degree of utilization of environmentally friendly technologies

Table 1.2. (continued)

5. Protective functions of forests
Soil conditions
Water conditions
Management for soil protection
Watershed management
Areas managed for scenic and amenity purposes
Areas and percentage of forest lands managed for environmental protection
Infrastructure density by FMU category
6. Socioeconomic functions and conditions
Indicators for economic benefits
Value of wood products
Value of nonwood products
Value from primary and secondary industries
Value from biomass energy
Economic profitability of SFM
Efficiency and competitiveness of forest products production, processing, and diversification
Degree of private and nonprivate involvement in SFM
Local community information and reference mechanisms in SFM
Indicators for the distribution of benefits
Employment generation/conditions
Recreation and tourism
Forest-dependent communities
Impact on the economic use of forests on the availability of forests for local people
Quality of life of local populations
Average per capita income in different forest sector activities
Gender-focused participation rate in SFM
7. Political, legal, and institutional framework
Legal framework that ensures participation by local governments and private landowners
Technical and regulatory standards of management plans
Cadastral updating of the FMU
Percentage of investment on forest management for forest research
Rate of investment on the FMU level activities: regeneration, protection, etc.
Technical, human, and financial resources

FMU forest management unit, SFM sustainable forest management

and indicators identified by the processes and initiatives and should facilitate the definition of inventory objectives.

As not all objectives have the same importance, the priority of inventory objectives has to be assessed before designing the inventory. Before a final decision on the inventory objectives, all issues that could constrain an implementation of the inventory should be listed and considered. Issues include cost limits, availability of staff, presentation of the findings, the schedule, or the population for which estimates should be given.

The listing of inventory objectives should not be confused with the list of attributes to be assessed. Based on the objectives, the attributes for field assessments, remote-sensing imagery, or other data sources have to be derived. The attributes have to be defined in detail on the basis of data type, desired error, and units of reference.

1. *Populations for which estimates are to be presented.* The term population is used to describe that aggregate from which the samples are to be taken and for which valid conclusions are to be drawn. In forest inventories it can be relatively difficult to define the population, as the exact borders and the surface area of the region to be surveyed must first be determined. Here, maps, land-use classification, and remote-sensing imagery can be of great assistance.
2. *Data type.* For each type of information required the measurement scale must be defined. The results may be presented in one of three ways:
   (a) Graphically – through maps, charts, diagrams, etc.
   (b) Descriptively, or qualitatively – for example, forest type, stage of development
   (c) Quantitatively – for example, stem count, total standing volume, or mean increment

   The simplest method of measurement is to group observations into qualitative classes. Here, a general system of classification is based on given characteristics and takes only the major manifestations of these into consideration.

   The allocation of data into different classes of equal standing is termed nominal scaling as opposed to ordinal scaling, which reveals a ranked order. Nominal scaling may, for instance, involve classification according to tree species, administrative unit, or soil type. Examples of ordinal scaling are the stand tree development class, or the stem quality class. Mathematical operations for nominal as well as ordinal scales are limited.

   Where quantitative data are to be furnished, any given parameter (i.e., population characteristic) is allocated a number reflecting the intensity of a certain characteristic. Where the measurements are linearly related to values and the zero point is arbitrarily set, we have an interval scaling. An example is the measurement of temperature: on the Fahrenheit scale the freezing point of water is 32°F, and water boils at 212°F (at standard pressure); on the Celsius scale the equivalent temperatures are 0 and 100°C, respectively. Interval scaling allows conclusions about the differences between observations made at different scales.

A further system for scaling quantitative data is the ratio or relative scaling. Here, a true zero point exists, and the relationships between numerical measurements can be directly interpreted as the relationship between the dimensions of objects. For instance, a tree 20-m tall is twice as tall as one only 10-m tall. Such a statement cannot be made for two temperatures. Examples in which relative scaling can be applied are weight, length, and time.

Whenever quantitative data are considered, it must be specified whether overall values (e.g., total standing crop of the inventory area) or ratios (per tree, unit area, or other units of reference) are to be given.

3. *Error Limits.* The results of any sampling survey are always subject to a certain degree of uncertainty, as only part of the population is measured and the measurements themselves may be subject to error. The uncertainty can be reduced through a careful choice of sampling strategy (design and estimators), taking more measurements, or employing better instruments. Most improvements impose additional costs. Consequently, it is necessary to specify the degree of precision to which the results should attain.

   In deciding the limits of error, the two components (1) sampling error and (2) bias must be taken into consideration just as much as the chosen significance level for confidence intervals. The desired error must be related to the population, be it the entire inventory area or only a part of it. The given population and the desired error strongly influence the intensity of the survey.

1.3 A Typology of Forest Inventories

Forest inventories can be differentiated according to combination and emphasis of different data categories, i.e., the inventory objectives and the size of the area to be surveyed.

Global forest inventories are conducted to determine forest resources at the global level. This usually means the compilation of results from separate national inventories. Thanks to advances in remote-sensing techniques, satellite data can now be used to determine the distribution of forest vegetation throughout the world (Itten et al. 1985; Kuswaha 1990; FAO 2003). Global inventories were conducted by the FAO in 1946, 1953, 1958, and 1963 (World Forest Inventory) and in 1990 and 2000 (Global Forest Resources Assessment). Much discussion and readjustment of the data catalogues from national inventories was necessary before the data could be compiled to give a global picture (UN-ECE/FAO 2000; FAO 2001).

Individually designed national inventories do not permit monitoring of either the environment or the development of resources on a global or continental scale. Joint or multinational coordinated surveys allow the collection of data for large, cohesive areas, for instance, the UNEP/FAO Project on Monitoring Tropical Forest Cover or the annual UN Economic Commission for Europe International Cooperative Program (ECE-ICP) reports on forest condition in Europe, which began in the mid-1980s.

National forest inventories are already being conducted in many countries. In western Europe, almost every country has its own national inventory (EC 1997), some of which, especially those in Scandinavia, have been running for many years. Timber volume is usually employed as a key parameter, though information on the distribution of forested areas, the condition of the forest, and productivity is also collected (Lund et al. 1987; Lund 1998). The nonwood functions of the forest are receiving increased attention. Ideally, national inventories should be planned as permanent surveys and conducted by a specific organization with a permanent staff. Information obtained through national inventories is mainly applied in questions of national forest policy (Clement 1988; Lund 1998).

Land-use inventories record not only forest resources but also the distribution of other types of land use. Aerial photographs and satellite data are of especial importance here. The value of a forest inventory can be considerably increased by extending it to give a land-use inventory. Where noncommercial forms of vegetation, such as swamps, barren areas, or maquis, are recorded in addition to the various types of agricultural use, areas potentially suited for forestry can be identified. In Africa, FAO introduced an integrated system of nomenclature for agricultural and forest areas (AFRICOVER).[1]

Regional inventories register only a part of the national forested area and usually cover some hundreds of thousands up to two million hectares. Similarly to national inventories, they are intended to provide a general picture of the situation regarding forestry (Pellico-Netto 1979).

Reconnaissance inventories aim at furnishing a rough outline of the forest conditions. As well as the location and extent of forested areas, they may aim to register access, species composition, tree dimensions, the distribution of various forest types, and a crude assessment of timber quality (Touber et al. 1989). Through the employment of aerial photography and the restriction of field surveys to the minimum, reconnaissance inventories can be conducted at little cost. They frequently serve the preparation of a more intensive forest inventory. Data on the degree of variation and time-and-motion studies conducted during a reconnaissance inventory facilitate the planning of the definitive inventory.

Exploitation surveys or logging plan surveys are conducted in forests to provide a basis for the planning of programs for timber harvesting. The main focus is on determining the standing crop, classified according to species, dimensions, and assortment, and describing the accessibility of the area concerned. Little or no attention is paid to increment or ecological conditions.

[1] http://www.africover.org

Where the economic potential of establishing a wood-processing industry is to be examined, *a forest industries feasibility study* (FIFS) is standard practice. A FIFS comprises the collection of data not only on the forest resources as such but also on the situation regarding demand and marketing, potential sites for processing plants, the job market, sources of water and power, transport possibilities, and existing industries. For further details see Higgins et al. (1973), Philip (1976), Lanly (1977) and Staepelaere and Ginsburger (1978). As the establishment of a timber industry is only worthwhile where there is a steady supply of raw material, it is necessary to determine the forest resources in considerably more detail than is usual in exploitation surveys. In particular, the sustained yield of exploitable timber must be computed.

Working plan surveys are the most intensive type of forest inventory. The preparation of working plans for intensively managed but restricted areas requires relatively detailed information. Usually the data are computed on a stand-by-stand basis for each species. Information on increment, detailed forest maps, data on the quality of the various sites are just as necessary as details on topography, ownership, and access.

Forest condition inventories report on the symptoms of diseases and stress (water, nutrient, competition, air pollution, climate change). In central Europe and North America forest condition surveys are conducted annually in order to track the course of development of different types of damage (UN-ECE 1998, 2004).

In addition to the types of inventories just described, special surveys are sometimes conducted, for instance, to determine regeneration, available biomass, or carbon sequestration.

Forest inventories may also be classified in terms of time. Static inventories may be conducted simply to determine conditions at a given point in time, and do not require consideration of possible subsequent inventories – a fact which considerably simplifies their planning. Nevertheless, the additional expense of permanently marking the sample plots is often a worthwhile investment.

The various types of inventory are not distinct types. In practice they will often overlap. Neither is their nomenclature static; increasing demands on the forest and forest information will invariably stimulate the development of new types of inventory.

1.4 Inventory Planning

Lund (1998) outlined the steps needed to develop and implement a forest inventory (Fig. 1.1). As the planning of an inventory may take a long time, usually involves experts in different fields, and requires networking between different tasks, it is advisable to base the entire inventory on a sound project

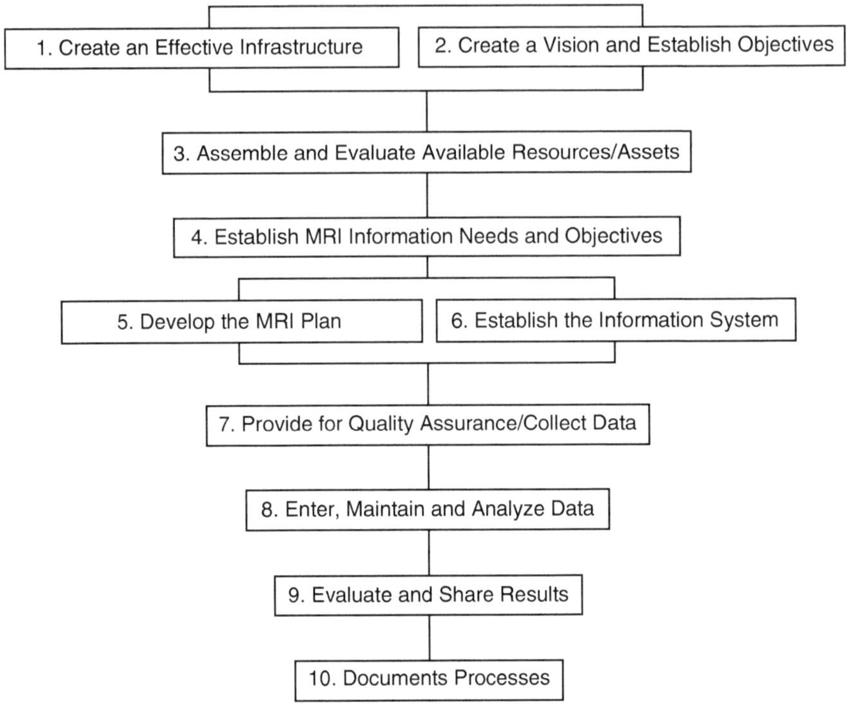

Fig. 1.1. Steps in implementing a forest inventory (after Lund 1998)

management concept (Burke 2003). Good project management deals with three factors: time, cost, and performance. Projects are successful if they are completed on time, within budget, and meet performance requirements. A multitude of components in any large project needs to be controlled. A large toolkit of techniques, methodologies, and tools has been developed for this purpose. They provide the tools for managing different components involved in a project: planning and scheduling, developing a product, managing financial and capital resources, and monitoring progress. However, the success of a project will always rest on the abilities of a project manager and the team members. A project life cycle includes the four phases (1) study phase, (2) design phase, (3) development phase, and (4) operation phase.

The following overview presents the separate steps in the four project management phases and is intended to serve as a checklist for inventory planning:

1. Study phase
   (a) User needs.
   (b) Initial investigation.

(c) Formulation of the inventory objectives:
   - Necessity of the inventory, information needed.
   - Potential users of the results.
   - Formulation of the inventory objectives.
   - Priorities of the objectives.

(d) Determination of the administrative and logistic situation:
   - Bodies responsible for the execution.
   - Budget (available funds, bodies providing funds, financial administration, time available).
   - Legal basis (right of access to privately owned forest, labor laws, protection of private forest owners from information leaks).
   - Available information (maps, aerial photographs, data from previous forest inventories and other types of survey, scientific studies in the inventory area, general details on the forest. Data on variation, description of the terrain, accessibility, and climatic conditions).
   - Potential use of aerial photography and remote sensing imagery.
   - Possibilities for recruiting qualified staff.
   - Available equipment (vehicles, computers and software, measuring instruments, tents).
   - Responsible bodies (staff management, financial administration, monitoring of data security, data release, dissemination of data, definition of inventory objectives and methods, execution of field surveys, data evaluation, formulation and release of the final results, publication, additional analyses).

(e) User review.

(f) Study phase report.

2. Design phase
   (a) General system review
   (b) Compilation of the data catalogue and stipulations for measurements
      - Listing of all variables to be analyzed (depending on inventory objectives)
      - Definition of qualitative data
      - Instructions for measurement of quantitative data
   (c) Inventory design
      - Description of the design to be employed
      - Sampling methods
      - Description of the sampling units, especially their form, size, number, and distribution
      - Computation of the necessary sample size for each inventory level, survey intensity
      - Description of inventory levels (aerial photographic survey, interpretation of satellite data, field surveys, questionnaires)

   - Map construction
   - Estimation of areas
   - Description of statistical methods for evaluation, estimation procedures, correlations to be applied, and the computed parameters
   - Methods of volume determination (e.g., volume functions, points of measurement on the tree, volume inside or outside bark)
   - Determination of regeneration conditions
   - Determination of timber quality
   - Description of road and transport networks
   (d) Data base and/or information system design
   (e) Control requirements
   (f) Software selection
   (g) Equipment selection/acquisition
   (h) Staff recruitment
   (i) Field manual
   (j) Plans for work progress
   (k) Design phase report
3. Development phase
   (a) Implementation planning
   (b) Computer program design
   (c) User review
   (d) Equipment acquisition and installation
   (e) Field tests/pilot survey
   (f) Computer program testing
   (g) System testing
   (h) Reference manual preparation
   (i) Personnel training
   (j) Changeover plan preparation
   (k) Development phase report preparation
   (l) User acceptance review
4. Operation phase
   (a) Interpretation of aerial photographs and/or remote sensing data
   - Instruments (interpretation instruments, computers, software)
   - Organizations, staff, competence, duties
   - Documentation and backup of the results
   (b) Field surveys
   - Organization, central coordination
   - Communication between field survey teams and central coordinators
   - Recording and delivery of data
   - Training of field staff (localization of sample plot centers, assessments on sample plots, use of instruments)
   - Check cruises

(c) Data evaluation
   - Digitalization of data
   - Checking and correction of data
   - Data analysis
   - Operating, system management, data security
(d) Final report
   - Preparation (output format printed, Web-based, or CD-ROM)
   - Approval for release
   - Reproduction.
   - Dissemination
(e) Performance evaluation

Forest Mensuration 2

2.1 Introduction

Remote sensing and field assessments are the main data sources for forest inventories. While remote sensing provides spatially explicit data for large areas at considerable costs, the number of attributes that can be extracted from remote-sensing imagery is rather limited; thus, field assessments continue to be an essential component of forest inventories.

Many aspects dealing with the in situ assessment of trees and forests are dealt with in textbooks such as Prodan (1965), Loetsch et al. (1973), Hush et al. (2003a, b), and Avery and Burkhardt (2001). An assessment of the productive function has been the traditional focal point of forest inventories. In recent decades information on the multiple functions of forests, especially nonwood goods and services, has gained significant importance (Bachman et al. 1998; Corona et al. 2003).

In assessing forest functions it is essential to distinguish between the actual existence of a function and the potential of a forest to provide a function. Brassel (1995) described the problem of integrating the assessment of nonproductive forest functions in a national forest inventory program. He gives a set of attributes that provide information on the functional potentials of forests and that can be directly recorded in the field:

- Wood production: standing volume, increment, drain, accessibility, structure, stage of development, age, size of stand
- Biodiversity: number of woody species, especially trees and shrubs
- Nature protection: forests and forest edges as habitats for flora and fauna
- Pasturage: traces of other usage, game damage
- Recreational functions: evidence of human influence and overutilization
- Protective functions: evidence of surface erosions, stand density, surface (bare soil, long grass, litter) as indicators of possible avalanches, evidence of rockfall

A number of forest functions cannot be assessed from within the forest itself but only in its environment. Brassel (1995) gives the following examples:

- Protection against avalanches
- Mitigation of floodings
- A safeguard against winds
- Improvement of quality of life by, for example, providing recreation grounds, lowering of noise, screening from visual affronts, or filtering of dust
- Provision of drinking water
- Maintenance and enhancement of landscapes and their ecological functions
- Supply of nontimber products, especially food
- Combating global climate change by providing storages and sinks for carbon

2.2 Area Information

Information about the area and area proportion presents a major result of forest inventories for the following reasons:

1. Information about the forested area itself is of major interest for, for example, forest policy, management planning, and nature conservation. Detailed accounts of subareas are frequently requested for location-specific management purposes, i.e., the absolute size and proportions of forest types, ownership categories, or age classes.
2. Many attributes are presented in terms of unit area, such as the growing stock per hectare. Area-related ratios serve to standardize and facilitate comparisons over time and between different units of reference.
3. Quantifying forest area changes is especially important in regions with strong land-use changes and forest area dynamics.

Information on area can be expressed in two different ways: (1) the total forest area and (2) the proportion of forest area within a given region.

2.2.1 Forest Area Definitions

There exists no unique concept about what qualifies as a forest. The hyperdictionary[1] provides a definition of a forest that represents two different aspects:

1. The trees and other plants in a large densely wooded area
2. Land that is covered with trees and shrubs

[1]http://www.hyperdictionary. com/dictionary/forest

When we talk about forest area we relate to the area covered with trees and shrubs. National legislations often have a legal definition of forest, but they are generally not applicable to forest resource assessments. Forest is a qualitative attribute of an area which cannot be measured directly. Instead forest is defined by a set of quantifiable and measurable attributes. In order to increase the reliability of forest area assessments, forest area definitions are based upon attributes that can easily be measured, such as crown density, size or width of the forested patch, tree height, or productivity. For those attributes threshold values are specified and whenever a patch qualifies for the selected set of attributes the patch is considered to be forest land.

A key question is, where to draw the borderline between trees inside a forest and trees outside forests. Figure 2.1 shows an example from the Swiss Alps, where trees grow close to the timberline. A predominant characteristic of forests close to natural timberlines is that tree density is gradually lowered towards the treeline. A forest area definition uses quantitative criteria to define which areas stocked by trees qualify as forests and which areas do not.

The set of attributes selected as well as the specified threshold values vary in individual forest area definitions. Table 2.1 shows some forest area definitions used by international organizations.

Forest area definitions may also contain specifications about the allowed or disallowed use of forests and forest types. As an example, the forest area definition as used by the Food and Agriculture Organization (FAO) for the global *Forest resources assessment 2000* is presented, and reads as follows (FAO 2001):

Fig. 2.1. Trees close to the timberline in the Swiss Alps

Table 2.1. Forest area definitions

UNESCO (UNESCO 1973)	Closed forest: trees 5 m or taller with crowns interlocking Woodland: trees 5 m or taller with crowns not usually touching but with 40% or more canopy cover
FAO *Forest resources assessment 1990* (FAO 1995)	Forest (developing countries): 10% crown cover of trees and/ or bamboos Forest (developed countries): tree crown cover (stand density) of more than 20% of the area Closed forest (tropical countries): tree crown greater than 40%
FAO *Forest resources assessment 2000* (FAO 2001)	10% crown cover of trees, minimum area 0.5 ha

"Land with tree crown cover (or equivalent stocking level) of more than 10 percent and area of more than 0.5 ha. The trees should be able to reach a minimum height of 5 m at maturity in situ. May consist *either* of closed forest formations where trees of various storeys and undergrowth cover a high proportion of the ground; or of open forest formations with a continuous vegetation cover in which tree crown cover exceeds 10 percent. Young natural stands and all plantations established for forestry purposes which have yet to reach a crown density of 10 percent or tree height of 5 m are included under forest, as are areas normally forming part of the forest area which are temporarily unstocked as a result of human intervention or natural causes but which are expected to revert to forest.

Includes: Forest nurseries and seed orchards that constitute an integral part of the forest; forest roads, cleared tracts, firebreaks and other small open areas within the forest; forest in national parks, nature reserves and other protected areas such as those of special environmental, scientific, historical, cultural or spiritual interest; windbreaks and shelterbelts of trees with an area of more than 0.5 ha and a width of more than 20 m. Rubberwood plantations and cork oak stands are included.

Excludes: Land predominantly used for agricultural practices."

Table 2.2 summarizes quantitative criteria used in selected countries to assess their forest area. The quantitative criteria found in definitions for forest area are crown cover[2] (5–30 %), width of the stand (9–50 m) and minimum area (0.01–2 ha). These quantitative criteria are measured either on aerial photographs or in sample plots on the ground. Scandinavian countries also require a minimum potential increment of at least 1 m^3/ha/year for the site to be considered as forest area. Australia has no national standards except for the minimum crown cover, which is 20%.

Köhl et al. (2000) studied the effect of different national forest area definitions on the estimated size of forest area in a simulation study. Differences in the spatial distribution of trees and forested patches – as they can be found in

[2]Proportion of area covered by the vertical projection of tree crowns

Table 2.2. Forest area definitions in selected countries

Country	Minimum width (m)	Minimum crown cover	Minimum area (ha)	Minimum production (m^3/ha/year)
Austria	10	30%	0.05	–
Finland	–	–	0.25	1
France	15	500 stems/ha or 10%	0.05	–
Germany	10	–	0.1	–
Greece	30	10%	0.5	–
Iceland	–	–	0.25	–
Ireland	40	20%	0.5	4
Italy	20	20%	0.2	–
Portugal	15	10%	0.2	–
Spain	20	5%	0.2	–
Sweden	–	–	0.25	1
Switzerland	25–50[a]	100–20%	–	–
UK	20	20%	0.25	–
Australia	–	20%	–	–
New Zealand[b]	–	20%	0.5	–
Japan	–	30%	0.3	–
USA	40	10%	0.40	1.4

[a]Depending on crown cover
[b]Definition for forest and other wooded land

the Mediterranean, the central and the Nordic regions of Europe – were simulated in computer-generated forest/nonforest maps. The computer-generated forests were used to simulate the impact of exchanging one national definition with another in a number of European countries. This approach allowed estimation of the effect of different national forest area definitions in absolute terms. For example, in Spain (total forest area 259,840 km^2) adopting the forest area definitions of Sweden and Finland would lead to a forest area that is more than 12,000 km^2 larger than if the area was delineated according to the Spanish national definition. In contrast, the definition of the UK would result in a forest area that is roughly 19,000 km^2 smaller than the figure reported according to the Spanish forest area definition. Depending on the chosen definition, the range of the Spanish forest area could vary from about 240,000 km^2 (reference UK) to 274,000 km^2 (reference Luxembourg).

A universal forest definition remains elusive. Sample-based estimates of forest area must therefore carefully consider the existing forest definition and take competing definitions into account to ensure that estimates of forest areas can be obtained for more than one definition. In the early phase of planning a forest inventory it should be considered whether only areas currently supporting forest vegetation are to be surveyed or whether former forest areas and sites suitable for reforestation or afforestation should be included (Pancel 1984; Weaver and Birdsey 1986).

2.2.2
Assessment of Forest Area

Area information may be obtained from maps, aerial photographs, digital remote-sensing data, or terrestrial surveys. Information on the size of an area can be provided by two techniques: (1) the measurement of an area or (2) the estimation of an area by sampling.

Areas may be measured by means of area calculations, weighing, planimetry, or counting the number of cells in a square or parallel grid that covers the area of interest. Area measurements can be very time consuming and impractical for large-scale inventories. Especially where large-scale inventories are carried out, areas of concern should not be delineated on the basis of the aerial photographs. Where wall-to-wall maps are available in digital format geographic information systems offer an efficient technique for calculating areas (Chou 1997).

The reliability of area estimates provided by measurement techniques depends on the errors given by the method, the instruments applied, the process of execution, and the particularity of the staff involved. Knowledge of the positional error of a line (boundary) and the process involved in delineation of objects of interest allows a model-based estimation of the error of a polygon drawn by this process. Magnussen (1994) provides a model for estimating the area of a forest stand when the stand is delineated by photographic interpretation. Næsset (1998, 1999) also quantifies the error and estimates its impact on volume estimation.

In forest inventories, area estimation by means of sampling is often preferable to measuring techniques. There are two possibilities: point sampling and parallel lines or "transects."

Point sampling is based on the concept of random points: a point is chosen at random from the possible locations and a value is assigned to the random point, for example, it is either forest or nonforest. In practical applications point sampling is realized by applying dot grids, where the dots are considered to be a realization of a random process vis-à-vis the map. The localization of one point corresponds to a Bernoulli experiment with the possible values of nonforest and forest. The binomial distribution describes the probability of all possible outcomes of the random sampling design completely. In the estimation of the total area it is helpful to note that the number of sample locations with forest (or nonforest) is asymptotically (when both sample size n and population size N go toward infinity) a realization of a Poisson process with a density λ where each point represents on average an area of $1/\lambda$. An estimator for the total forest area A_w is $A_w = N/\lambda$.

If n_w out of n random points are found in the forest, $\lambda = n_w/n$ is an unbiased estimator of the true forest density (viz., proportion p). The forest area proportion is estimated according to Cochran (1977) by

$$\hat{\lambda} = \hat{p} = \frac{n_w}{n},$$

$$v(\hat{p}) = s_p^2 \approx \frac{\hat{p}\hat{q}}{n},$$

$$\hat{s}_p = \sqrt{v(\hat{p})},$$

where $(\hat{p})$ is the forest area proportion $\hat{q} = 1 - (\hat{p})$ is the proportion of nonforest area, $v(\hat{p})$ is the variance of $(\hat{p})$, s_p is the standard error of $\hat{p}$, n is the number of dots on the dot grid inside the area A of interest, and n_w is the number of forest dots on the dot grid.

The area proportion always has to be seen in relation to the total area (Fig. 2.2).

In practical applications the total forest area A_w is often estimated by multiplying the (known) total area A by the estimated forest area proportion $(\hat{p})$:

$$\hat{A}_w = n_w \frac{A}{n} = A \times \hat{p},$$

with the variance $v(A_w)$ and the standard error $s(A_w)$

$$v(A_w) = A^2 s_p^2,$$

$$s(A_w) = \sqrt{v(A_w)}.$$

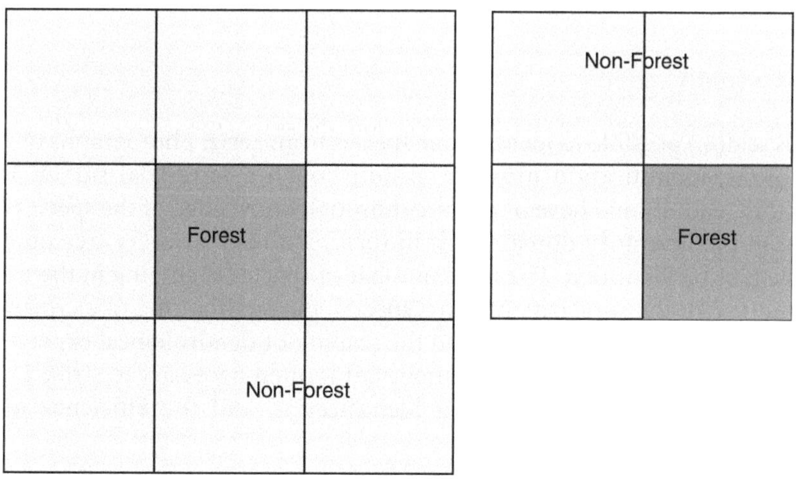

Fig. 2.2. Area proportion assessed by dot grids

If these estimation equations are used for a systematic dot grid, as opposed to a random sampling of dot locations, the standard error is generally overestimated. The form and the spatial distribution pattern of the forest areas influence the amount of overestimation. Kleinn (1991, pp. 26–27) shows by using systematically distributed points that the difference between the sampling error of the area estimate and the true sampling error depends on the forest distribution. It is moderately low for small-scale fragmented forest areas. Nevertheless, experience shows that the binomial distribution gives acceptable results for large inventory areas, as long as the forest patches are small compared with the sample grid and as long as they are irregularly distributed (Trachsler et al. 1980).

Sampling techniques for area estimation may be based on maps, aerial photographs, digital classification, or field surveys. Where maps and aerial photographs are used, it must be ensured not only that the latest data are verified but it must also be recognized that precision will depend on the scale at which the sampling is done. As a general rule, the larger the scale, the better is the accuracy. Forested areas are often only found on maps when they exceed a certain area. Small patches of forest area may consequently not appear, which leads to underestimation of the total forest area. Only maps with a high-dimensional stability should be employed, or even better, recent aerial photographs, if available.

2.3 Tree Information and Information for Characterizing the Growing Stock

2.3.1 Species Identification

As it is seldom possible to identify tree species from aerial photographs of tropical forests, identification must be made through a terrestrial survey. Every inventory team should have or acquire thorough knowledge of the species to be found in the inventory domain; this in turn requires that every tree on every sample plot be identified. The large number of species occurring in the tropics (Brünig 1973), the similarity of some species, the fact that the tree crowns are often invisible from the ground, and the scarcity of dendrological experts render this a very difficult task to accomplish. It is often necessary to strike a compromise between precise botanical identification and the efficiency of the terrestrial surveys (Noack 1971).

Every field survey team should include at least one "tree finder." This may be an experienced forester – though the services of such are often difficult to obtain – or one of the local inhabitants, who often possess an astonishing knowledge of the species occurring in their neighborhood. Potential deficits in the expertise of the selected tree finder should not be underestimated: basic training courses should be provided to all and a continuous checking of survey results is necessary.

Tree finders are often only familiar with the local name of a tree species. Consequently it is necessary to compile a list of local and botanical species names. One species may have several local names, while one local name may be applied to a number of different species. Local names may vary with the gender or age of a tree. Consistency of records may be enhanced by making extensive use of the same tree finder over longer periods of time.

In addition, an identification key based on morphological and anatomical characteristics of a species, such as bark, roots, leaves, and fruits, can be of great value (Ella and Escobin 1993; Luxmi-Chauhan et al. 1995; Salang and Sugawa 1997; Batalha et al. 1998; Ella and Pitarguen 1998). Subsidiary methods have been described to support species identification, such as chlorophyll fluorescence (Tyystjarvi et al. 1998) subepidermal features (Luxmi-Chauhan et al. 1998), comparative DNA amplification fingerprinting (Böhm et al. 1993), or allozymes (Rajora and Zsuffa 1991).

It is by no means always possible to survey every species, yet no species should be ignored whether commercially important or not. In any case, the list of identifiable species should be supplemented by a catalogue illustrating those that were not identifiable. The catalogue leaves a legacy for future inventories and a possible delayed identification.

In summary, the quality of species identifications may be improved through various measures:

- Keeping a small number of permanently employed tree finders on the staff.
- Conducting conscientious, repeat intensive training courses throughout the survey period.
- Compiling a list of local and botanical species names.
- Constructing a simple identification key, at least for the most important species and those easily confused.
- Checking of inventory records on the sampling plots by particularly well qualified staff. The findings of such checks should be presented to field crews.
- Systematic checking by dendrologists, including cross-checking of parts from randomly selected trees from a botanical laboratory. The frequency of such checks should be high at the beginning of an inventory.
- Employing the services of a specialist who can, in case of doubt, identify the species from various samples.

2.3.2
Diameter at Breast Height and Upper-Stem Diameters

Diameter at breast height (DBH) is the most important measurable parameter of a tree in the context of a forest inventory. Of all measurements on a tree, this is one of the easiest to take and should consequently be recorded for all trees on each sampling site. It is measured not only in forest inventories but in almost all studies related to forestry and forest resources, and is thus a valuable measure for comparisons. The breast height is selected as it is the most convenient point to measure a diameter along the stem. In countries using the metric system, DBH is taken at 1.3 m above ground level, in the USA and Canada at 1.37 m, and in Japan at 1.25 m.

DBH serves as a basis for computing other tree parameters, such as the basal area or volume. Through a summation of the cross-sectional area at breast height of individual trees, which is a function of DBH, the basal area of a stand can be computed, which in combination with the number of trees reflects the density of the stock.

DBH measurements can be utilized to estimate diameter distributions (de Carvalho 1981), which present the number of trees per diameter class (Fig. 2.3). Diameter distributions form an essential part of inventory results, as they allow the determination of stand structure, stage of development, and in some cases even the silvicultural approach to stand management.

A formal model describing the frequency distribution of DBH values or basal areas in a stand is often of great utility in forest management. Yields and economic values of thinning and harvest operations and often growth depend

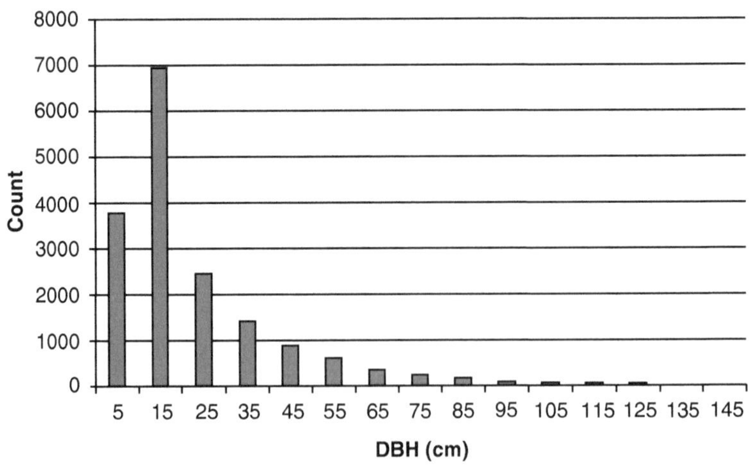

Fig. 2.3. Diameter distribution of a selective cut forest (Schallenberg Rauchgrat, Switzerland). *DBH* diameter at breast height

critically on the size distribution of trees. Models are commonly obtained by fitting an observed diameter distribution to a statistical density distribution of choice. Distributions such as the Weibull, Johnson, or beta have traditionally been popular (Magnussen 1986; von Gadow 1987; Gove and Patil 1998; Cao 2004). According to Pretzsch (2001), the three-parameter Weibull function will often provide a reasonable fit to a diameter distribution. The function of the three-parameter Weibull is

$$F(x) = 1 - e^{-[(x-a)/b]^c},$$

where a is the lower limit of the distribution, b the scale parameter, and c is a shape parameter ($\{x, a, b, c\}>0$).

According to the standardized system of the International Union of Forest Research Organizations (IUFRO 1959), the following labels are used to signify the various diameter classes:

d the DBH over bark,
c the circumference of the stem at breast height over bark,
g the basal area at breast height over bark,
d_u, c_u, g_u the under-bark values of d, c, and g, respectively,
d_b the diameter over bark at buttress height,
d_{st} the diameter over bark at stump height,
d_x the diameter over bark at x meters height, for example,
d_7 is the diameter at 7-m height, and
$d_{0.xh}$ the diameter over bark at 0.x of tree height, for example, $d_{0.1}$ is the diameter at one tenth of tree height

The most widely used instruments for measuring DBH are calipers and diameter tapes. The great girth of trees in tropical forests renders diameter tapes preferable to calipers, as they provide more consistent measurements. In order to measure tree diameter, the diameter tape should be wound around the trunk in the horizontal plane and pulled tight with no bend, wrinkle, kink, or buckle anywhere along the tape. Moss, lichens, and loose bark should be removed prior to measurement.

Other measuring instruments are the Biltmore stick and Bitterlich's sector fork (Fig. 2.4). Subjective errors may occur with either of these, which make them more suitable for rough measurements rather than for precise measurements. A detailed description of these instruments can be found in Hush et al. (1982).

Readings should be accurate to the nearest millimeter. Rounding-off to the next centimeter or allocation to general classes should not be carried out during the actual measurement in the forest. Rounding-off of the DBH does not allow precise determination of growth in diameter. DBH is a continuous parameter but may be transformed into a discontinuous one if registered in classes. A tree is then only considered to display an increase in diameter if the latest measurement exceeds the threshold of the class to which the tree was previously assigned. On permanent sampling plots established for the purpose of quantifying increment,

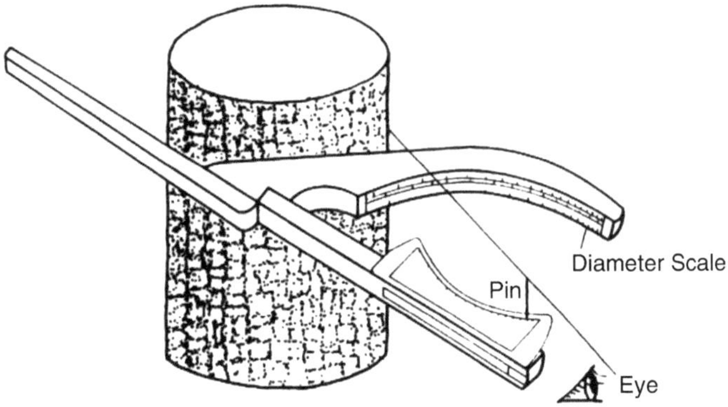

Fig. 2.4. Bitterlich's sector fork

the DBH must be measured to the nearest millimeter. The exact height and, where callipers are used, the direction of the calliper arms must be recorded.

Considerable bias occurs when the cross-sectional area is calculated from class observations of DBH, as

$$\text{cDBH} = \text{DBH} + \text{rounding error}$$

and for measures related to basal area the bias can be obtained from

$$E\,(\text{cDBH}^2) = E\,(\text{DBH} + \text{rounding error})^2,$$

where E denotes an expectation (across diameters).

Errors in DBH measurements may occur in measuring with calipers or diameter tapes. Errors arise from the tree, the instrument, and the operator. For noncircular cross sections the tape is always positively biased. The bias of a DBH measurement obtained with a calliper depends on the location of the measurement and can be smaller or larger than the bias of a tape measurement. The accuracy of diameter measurements of trees with irregular shapes by callipers can be improved by the following measurement rules:

- Measure the maximum and minimum diameters
- Measure the maximum diameter and the one at right angles to it
- Measure any two diameters at right angles to each other

The instrumental error of cloth and fiberglass tapes is due to stretching of the tape; tapes made of steel are much less prone to stretching. Worn-out callipers, where the angle of the calliper arms is no longer 90° or where the arms are not simultaneously in a horizontal plane, can result in substantial errors.

The main sources of operator error with both tapes and callipers are caused by (1) incorrect location of the point of measurement (i.e., above or below 1.3

m above ground), (2) incorrect tension (tape) and pressure (calliper) during the measurement, and (3) incorrect reading of the scale of the instrument. The last of these occurs more often than expected when a tape shows the reading of circumference on one side and the diameter on the other.

In tropical rain forests and mangrove swamps, trees often have buttress or stilt roots with root collars extending far up the stem. If this is the case then the diameter is measured above the point at which the thickening ceases, and just like DBH is taken as a basis for the determination of volume and stem form. The instructions for field surveys should include clear definitions of the positions at which the base diameter of trees with forked or branched trunks or those growing on slopes is to be measured (Fig. 2.5). Some forest inventories require additional measurements of upper diameters as parameters for the estimation of stem form or volume, or the computation of volume functions.

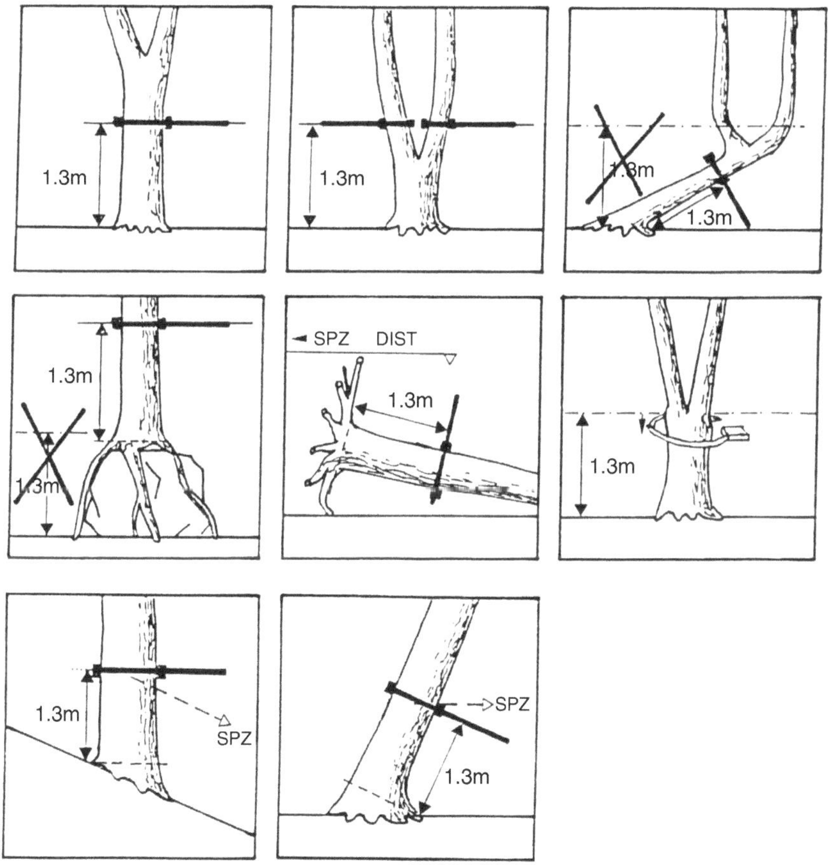

Fig. 2.5. Measurement of diameter at breast height (from Zingg 1988)

Optical measurements of stem diameter are made with a dendrometer. A dendrometer is an instrument for measuring the diameter and height above ground of points along the stem. There are four types of optical dendrometers for the measurement of upper-stem diameters of standing trees:

1. Optical forks
2. Optical callipers
3. Fixed-base range finder
4. Fixed-angle range finder

Optical forks works on the basis of an angle subtended by two vertical sides of a stem. The point of intersection of the two lines (the vertex) is the observer's eye. The geometry is shown in Fig. 2.6.

The use and the design of optical forks depend on whether the angle is constant or variable. If the angle is constant, the distance from the tree must be varied; if the angle is variable, the distance must be constant. Where the angle is combined with the angle of inclination, the measurement and inclusion of the latter in further calculations can be dispensed with.

The best-known types of optical forks are the mirror Relaskop and the Tele-Relaskop (. In both cases the angle of inclination is automatically corrected for – the angle stays the same but its confounding impact on the measurement is eliminated. As the angle can be varied, it is possible to measure the diameter at several different heights from a single location. The measurement of the diameter at various positions along the stem with a mirror Relaskop is illustrated in Fig. 2.7. Readings are in tachymeter units. At a distance of 20 m, one unit corresponds to 10-cm width. For instance, if the measurements shown in Fig. 2.7 had been taken every 20 m along the stem, the first reading of 6 would correspond to a diameter of 60 cm; the second reading of 4.6 to 46 cm; the third of 1.3 to 13 cm, and the fourth of 0.5 to 5 cm.

The precision of measurement of the diameter at higher levels depends on the distance from the tree and the visibility; where the view angle is smaller than 2% and larger than 10% the precision of the measurements is affected. Specifically for the measurement of large-diameter trees the wide-scale Relaskop has been developed. For precision measurements, the Tele-Relaskop, with its magnification of ×8 is to be recommended (Fig. 2.8). Any kind of Relaskop should always be mounted on a tripod.

Optical callipers comprise a scale with two pentaprisms attached and a sighting device. One pentaprism is in a fixed position, the other can be moved along the scale. The observer looks through the fixed prism at one side of the tree (as seen in cross section). The movable prism is moved along the scale until the two sides of the trunk appear to coincide (Fig. 2.9). As the measurement is based on

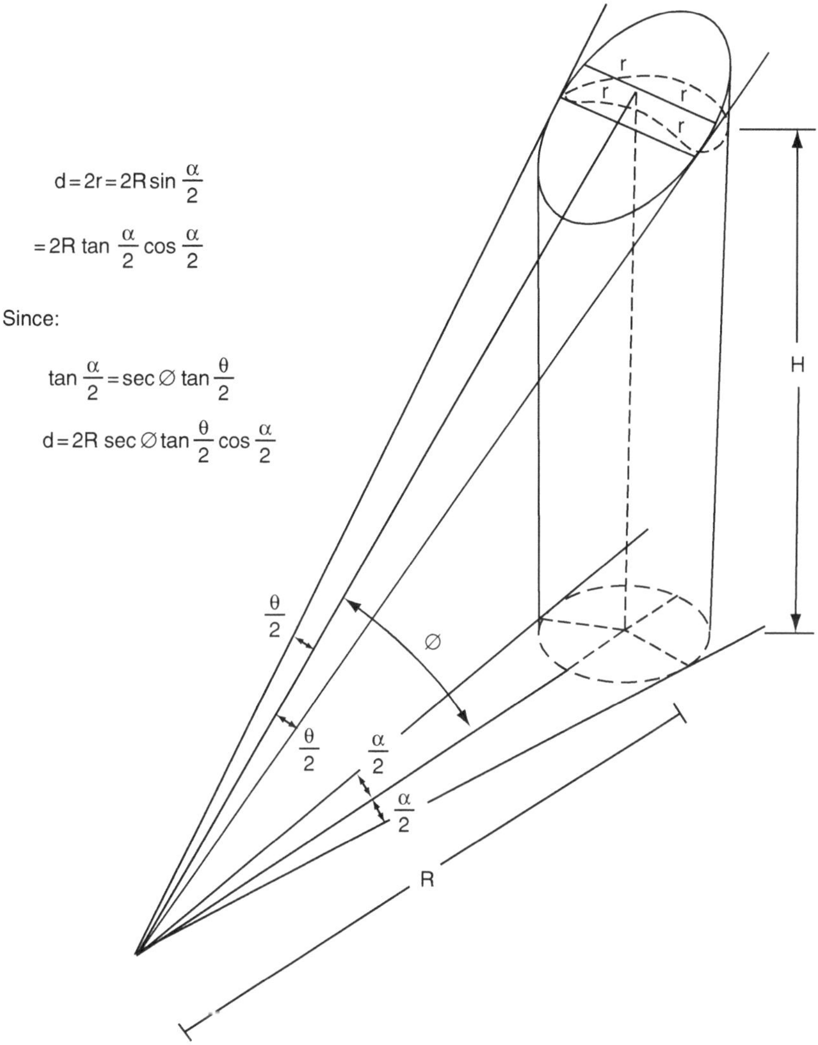

Fig. 2.6. Geometry of optical forks

the principle of parallelism of lines of sight, correction of the inclination is unnecessary. The diameter is read off from the position of the moveable prism on the scale. The best-known optical calliper is the Wheeler pentaprism. If a clinometer is attached, the height at which the various diameter measurements were taken is determined simultaneously.

Range finders do not use parallel lines of sight, but allow determination of the stem diameter through the relationship between a base and the angle of sight.

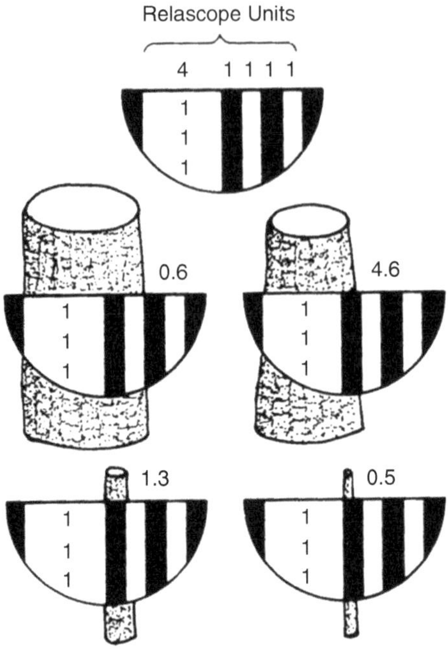

Fig. 2.7. Measurement of upper-stem diameters with the Relaskop

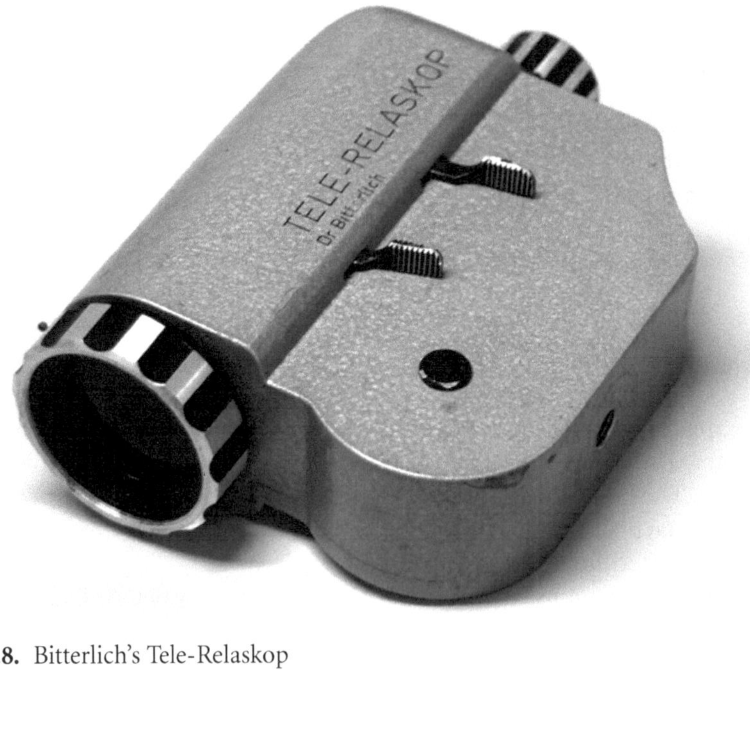

Fig. 2.8. Bitterlich's Tele-Relaskop

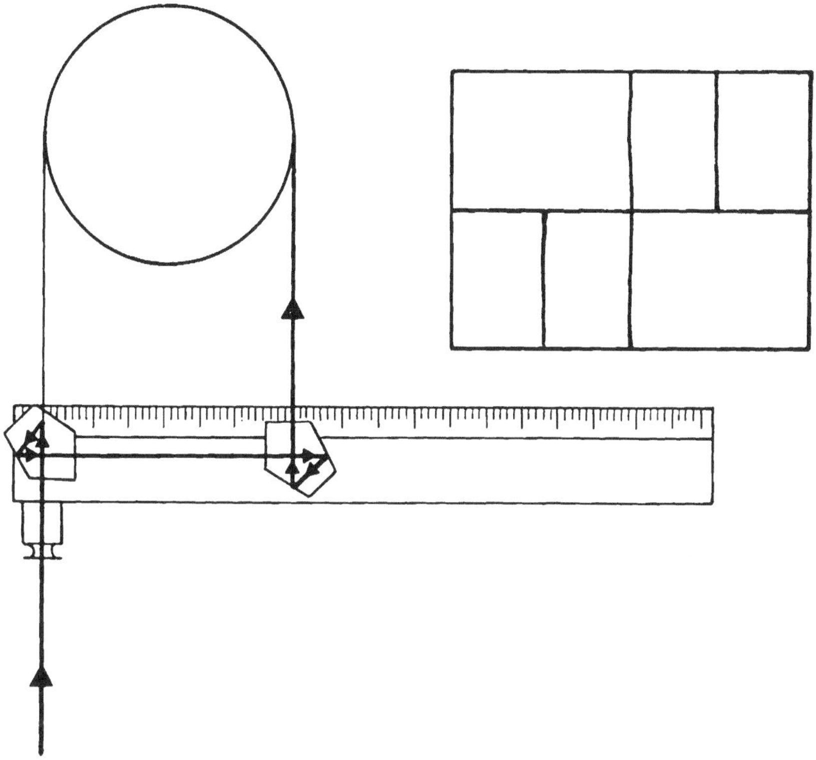

Fig. 2.9. Optical calipers

Fixed-base range finders, such as the Barr & Stroud dendrometer, are mounted on a fixed base, and the angle of measurement is varied. With fixed-angle range finders (e.g., the Breithaupt TODIS) the angle of the optics is predetermined and the baseline is varied. Range finders are particularly suitable for precision measurements and in conditions of poor visibility are far superior to other optical dendrometers.

Stem diameter above 1.3 m can theoretically be determined with instruments normally employed for measuring DBH, as long as these are fixed on a pole or tripod. The only practicable instrument of this type is the Finnish calliper, comprising a parabolic measuring scale attached to a telescopic arm. It has been employed in national inventories in Scandinavia, Germany, and Switzerland. This device, however, should not be used for measuring diameters at heights above 10 m (Fig. 2.10).

Older textbooks suggest that the best way to measure the upper diameter of a tree is simply to climb up and use a measuring tape or calliper (Fig. 2.11), but

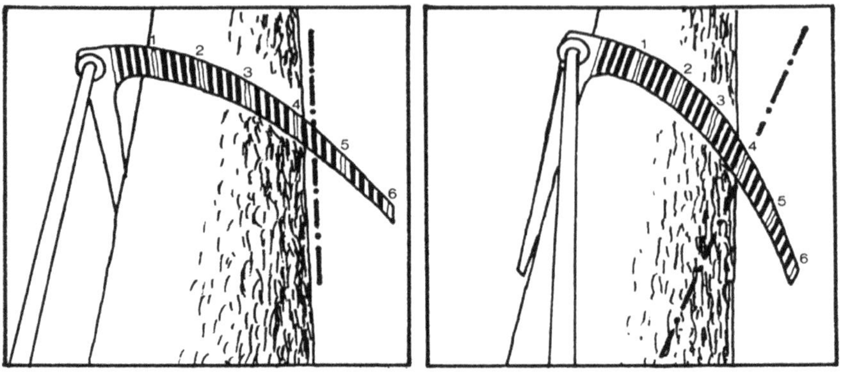

Fig. 2.10. Finnish calliper

Fig. 2.11. Measurement of upper-stem diameters with callipers (by courtesy of Andreas Zingg, WSL, Switzerland)

this method is not only dangerous but also rarely achieves set standards of accuracy. Climbing should be limited to a few exceptional cases.

2.3.3 Cross-Sectional Area Measurement

The cross-sectional area of a tree is that area it covers in cross section at breast height. A synonym for cross-sectional area is basal area. The total cross-sectional area of all trees or classes of trees per unit area (e.g., per hectare) is an important parameter for determining stock density. It is generally assumed that the cross-sectional area, g, is circular and can thus be directly computed either from the diameter, d, or the circumference, c, according to the following equations:

$$g=(\pi/4)d^2$$

and

$$g=c^2/(4\pi),$$

where d is the diameter at 1.3 m and c is the circumference at 1.3 m.

The cross-sectional area of a stand is measured in square meters or square feet. Unfortunately, the cross-sectional area of a tree rarely has an outline that is a perfect circle. Rather, the assumed circular outline is an approximation. Some species, for example, *Tectonia grandes*, have a very irregular stem form. Matérn (1956) suggests that deviations from the circular can be described through (1) the convex deficit (Fig. 2.12) and (2) the isoperimetric deficit. The convex deficit is the difference between the area as determined with a measuring tape and the actual cross-sectional area of the stem. If the cross-sectional area of the stem is not circular, the circumference is greater than that of a circle of the same area.

Diameter measurements by tape and by callipers affect the calculation of cross-sectional area for trees with irregular shapes in different ways. Wherever measurements by *tape* and equations for computing cross-sectional area assuming a circular form are applied, the computed area is always greater than the true cross-sectional area. All areas of convex form deviating from the circular display an isoperimetric deficit.

Where a *calliper* is used for measuring noncircular trees, the error may be smaller or larger and positive or negative compared with tape measurements. Actual tree cross sections are neither circular nor elliptical. To control bias arising from the assumption of form it is advisable to measure two diameters perpendicular to each or to measure the smallest (d_1) and largest (d_2) diameters and calculate the geometric mean, $d_g=\sqrt{d_1 d_2}$.

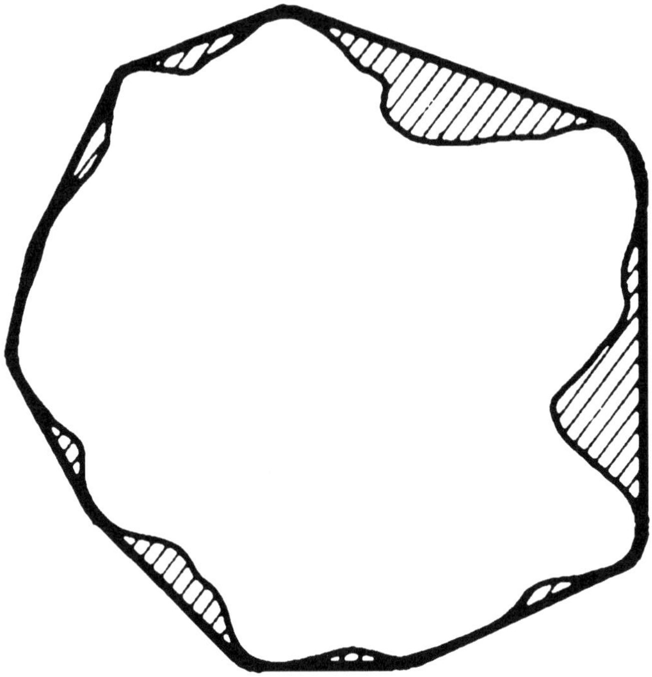

Fig. 2.12. Cross section of a stem demonstrating the convex deficit

2.3.4 Height

Tree height is another tree attribute which can be measured directly. Tree height is a parameter in volume and increment functions and, in combination with stand age, serves as a measure of site quality. The term "height" is relatively vague and must be defined according to inventory objectives. The most common definitions are (Fig. 2.13):

- *Total height.* This is the vertical distance between the base of the tree stem (ground level) and the topmost tip of the tree. However, practical problems arise in measuring the topmost tip of trees, for example, in tropical forests where the tip of a tree is seldom visible from the ground or where trees have drooping tops (e.g., hemlocks)
- *Bole length.* This is the distance along the stem between the stem base at ground level and the base of the (live) crown. The base of the crown is the point where the lowest living branch is attached to the stem. Stipulations for measuring lopsided crowns or crowns with dead branches must be laid down in the survey instructions.

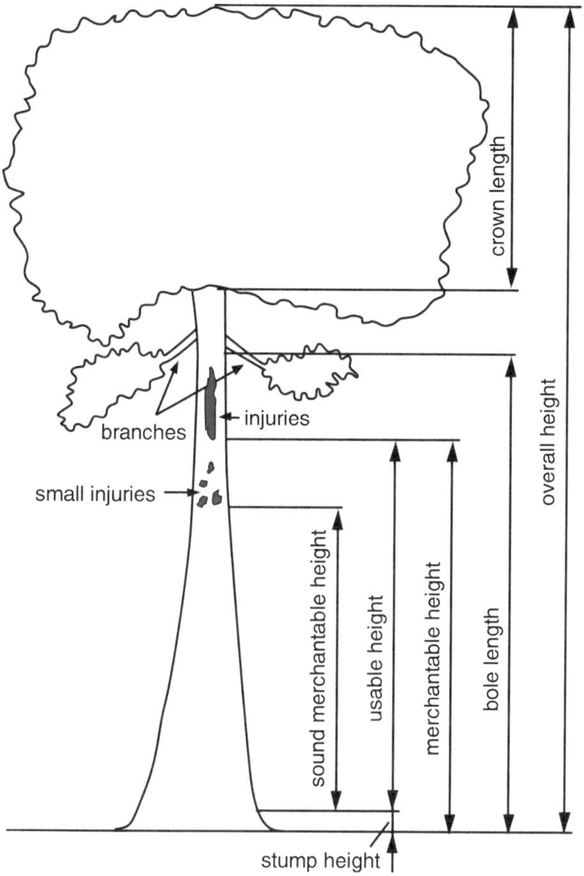

Fig. 2.13. Definitions of tree heights

- *Merchantable height.* This is the distance between the base of the stem at ground level and upper end of the last merchantable section of the stem. This point is defined by product-specific minimum-diameter standards, or on the basis of qualitative features such as branches, irregular stem form, or stem injury. Its determination in field surveys is liable to subjective assessment errors.
- *Stump height.* This is the distance from the ground to the point at which the stem has been or will normally be cut. In computing volume functions, it is taken as a constant based on average practice.
- *Usable length.* This is the merchantable height minus the stump height.
- *Sound merchantable length.* This is the usable length minus the length of stem displaying injury.
- *Crown length.* This is the distance from the crown base to the tip of the tree.

Only in exceptional cases can tree height be measured directly. There are many instruments for measuring heights. They are known as hypsometers, altimeters, or clinometers. Their use is based on one of two principles :

1. The relationship between similar triangles (geometric principle) (Fig. 2.14)
2. The determination of angles of inclination (trigonometric principle) (Fig. 2.15)

Measurements using the geometric principle are based on the equation

$$\frac{A'C'}{AC} = \frac{A'B'}{AB},$$

where AB corresponds to the tree height. Instruments applying the geometric principle use fixed distances of A′B′, A′C′ and AC, where A′B′ and A′C′ are given on the instrument and AC is set up by some reference fixed at the tree. Examples of instruments applying the geometric principle are the Christen, Merritt, and JAL altimeters. They all include a scale about 30-cm long. With the Christen altimeter, the visual image of the tree or part of the tree to be measured must be fit exactly between the upper and lower ends of the scale. The height or length of a tree, a stem, or a stem section is then determined on the basis of a fixed reference length on the stem. Where the upper diameter is measured with a Finnish calliper, the telescopic tube can be employed as a reference length (Fig. 2.16).

The advantages of these instruments are they are relatively simple in construction, the distance from the tree need not be measured, only one reading

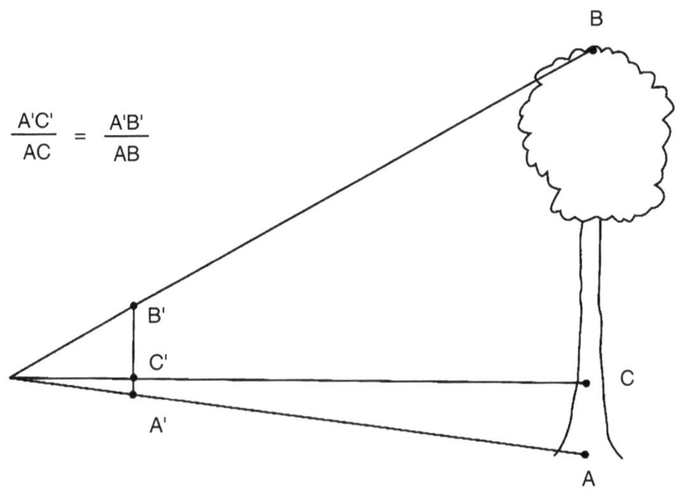

Fig. 2.14. Height measurements: geometric principle

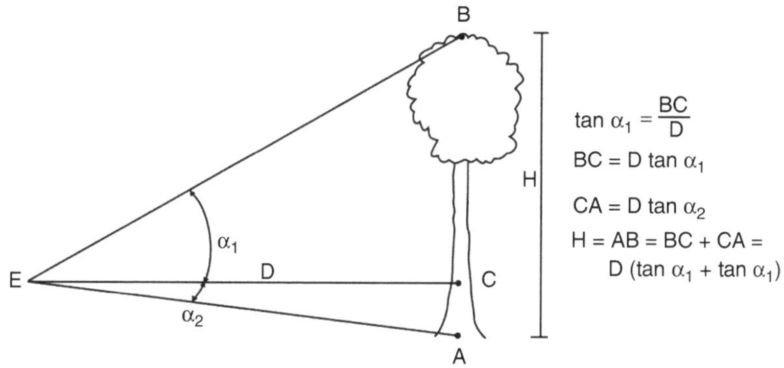

Fig. 2.15. Height measurements: trigonometric principle

is necessary, and the measurement is not affected by the inclination of the terrain.

The trigonometric principle is illustrated in Fig. 2.15. One measurement is made at the tree tip, another at the stem base. The two angles α_1 and α_2 are read, and the distance of the observer from the tree, D, is measured. The tree height is then determined from these three known variables according to

$$BC = D \tan\alpha_1$$

and

$$CA = D \tan\alpha_2.$$

As the tree height AB is given by BC+CA, it follows that

$$AB = D\,(\tan\alpha_1 + \tan\alpha_2).$$

Examples of instruments using the trigonometric principle are the Abney level, the Haga altimeter, and the Blume Leiss and Suunto clinometers (Fig. 2.17). A modern instrument for measuring tree height is the Vertex altimeter (Fig. 2.18), which uses ultrasonic signals to obtain the exact distance from the observer to a tree. The height is calculated trigonomically through the distance and angle. The Vertex altimeter can be used for measuring height, distances, angle, inclination, and current air temperature. The Bitterlich Relaskop can also be used.

Systematic errors in the measurement of tree height are especially likely when trees are leaning or have large crowns (Fig. 2.19). Deciduous trees often pose a special problem when the crown has no distinct tip.

As the measurement of tree height is relatively time consuming, it can usually only be undertaken on a subsample. The height of nonmeasured trees is often based on predictions obtained via a model estimated from the subsample

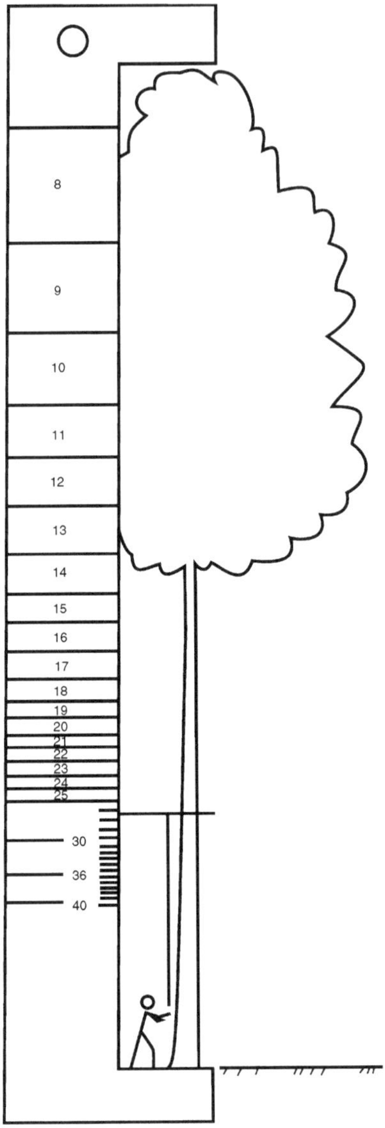

Fig. 2.16. Height measurements with the Christen altimeter (from Schmid-Haas et al. 1978)

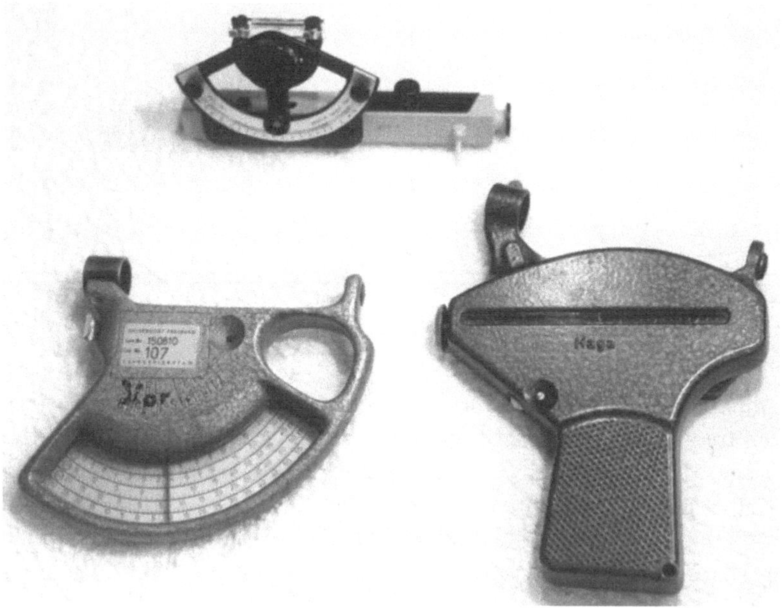

Fig. 2.17. Altimeters using the trigonometric principle

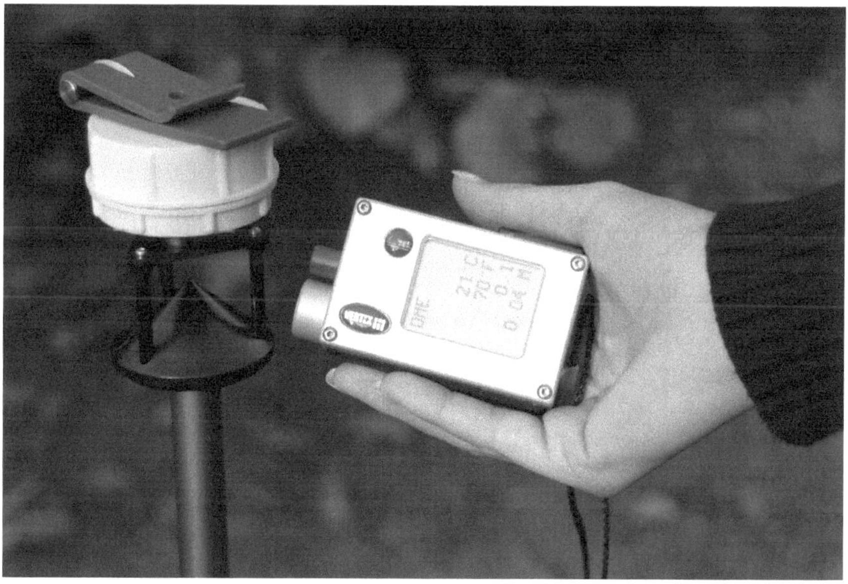

Fig. 2.18. The Vertex altimeter

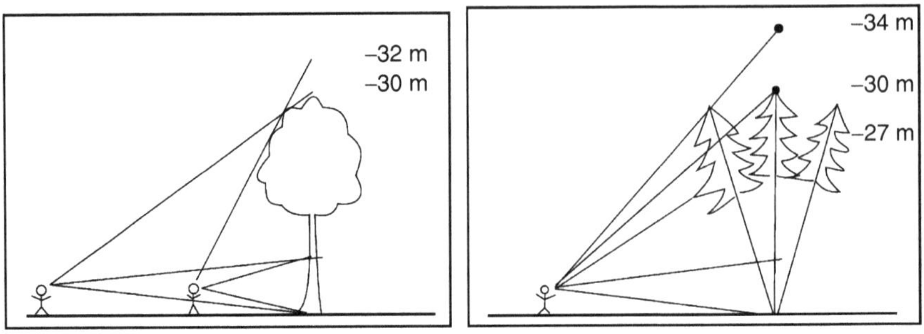

Fig. 2.19. Sources of errors in height measurements (from Schmid-Haas et al. 1978)

of heights. The most common model is a height curve, showing the relationship between tree height and DBH (Loetsch et al. 1973; Hush et al. 1982). In managed forests the relationship between tree height and age is used to compute the yield potential of a stand or site (Prodan 1965).

It is often observed that the total production of a stand is proportional to its height. Within a climatic region and conditional on this height, the total production appears to be independent of stand age, site, and silvicultural treatment. The apparent conformity of this relationship is the basis of what has become known as Eichorn's Law (Eichhorn 1904). Different definitions are used for presenting the stand height:

- Mean stand height, $\bar{h} = \left(\sum_{i=1}^{N} h_j\right) / N$
- Height of the mean cross sectional area of a tree, h_g
- Lorey's mean height, $h_L = \left(\sum_{i=1}^{N} g_i h_i\right) / G$
- Height of dominant trees, h_{dom} The arithmetic mean of the height of the 100 (h_{100}) or 200 (h_{200}) trees with the largest DBHs

The following functions are often used to describe the general development of height over time in homogeneous monospecific single-age stands:

- Bertalanffy:

  $h = a\,[\,(1 - b)\,e^{ct}\,]^{1/(1-m)}$,

  where a is the maximum height (horizontal asymptote), t is age, and c and m are coefficients to be determined
- Schumacher:

  $h = æ^{-b/t}$,

with a horizontal asymptote at h=a and a point of inversion at t=$b/2$

- Lundquist:

  $h = \text{ae},^{(-b/t)c},$

  with a horizontal asymptote at h=a, and a point of inversion at $t = (bc)^{1/c} / (c+1)$, and where b and c are coefficients to be determined
- Decourt:

  $h = (t^2 / a)\ [\ (t^2 + b)\ (t + c)\ ],$

  with a horizontal asymptote at h=$1/a$, and where b and c are coefficients to be determined

Stand height curves are the graphical display of height above DBH and show the dependency of the tree height on tree diameter. Stand height curves are shifted "upwards" with increasing age., but the shift is not uniform across all diameter classes since trees with the same diameter may belong to different stand layers with contrasting height growth profiles (Fig. 2.20).

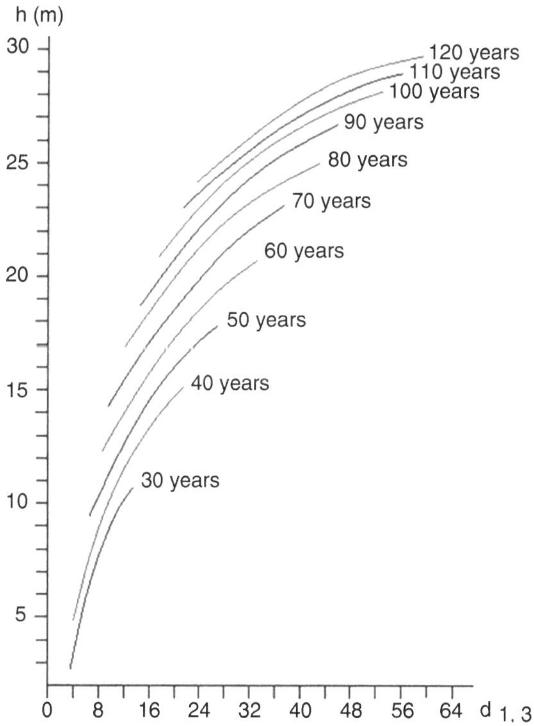

Fig. 2.20. Stand height curves in spruce stands (after Prodan 1965)

Table 2.3. Functions for stand height curves in homogeneous stands

$h = ad^b$
$h = a(1 - e^{-bd})$
$h = a + b\ln d$
$h = 1.3 + d^2/(a + bd)^2$
$h = 1.3 + a[d/(1 + d)]^2$
$h = 1.3 + ae^{-b/d}$
$h = 1.3 + a + bd + cd^2$

a, b, and c are coefficients to be estimated, d is the diameter at breast height

Table 2.3 presents commonly used functions for stand height curves in homogeneous stands.

2.3.5 Bark Thickness

The diameter of standing trees naturally includes twice the thickness of the bark. For commercial purposes, however, only the volume of the actual timber is important. To calculate this, twice the thickness of the bark must be subtracted from the overall diameter. Instruments for measuring bark thickness are the Swedish bark gauge and the bark hammer (Fig. 2.21).

Bark measurements should be conducted with great care. With a bark thickness of, for instance, 10 mm, a measurement error of ±1 mm leads to an error of ±10%. Determination of bark thickness through coring is not to be recommended, as this incurs the loss of loose bark tissues and a compression of the bark itself, both of which can lead to pronounced bias in the recorded thickness values.

Stem diameter under bark (d_u) is often a linear function of stem diameter over bark ($d_{1.3}$). If we have an estimate of this relationship then we can obtain an estimate of bark thickness by predicting d_u from $d_{1.3}$ and then take the difference as twice the thickness of the bark. The linear relationship has the basic form $d_u = a + bd_{1.3}$. Values of a and b for three tree species are given as follows (after Loetsch et al. 1973):

- *Tectona grandis*: $d_u = 4.962 + 0.003d_{1.3}$
- *Dipterocarpus alatus*: $d_u = 4.356 + 0.0065d_{1.3}$
- *Shorea albiba*: $d_u = 2.170 + 0.0079d_{1.3}$

The proportion of bark varies with species, age, buttressing, site, tree size, and position on the stem (Santander Flores and Albertin 1974; Smith 1979).

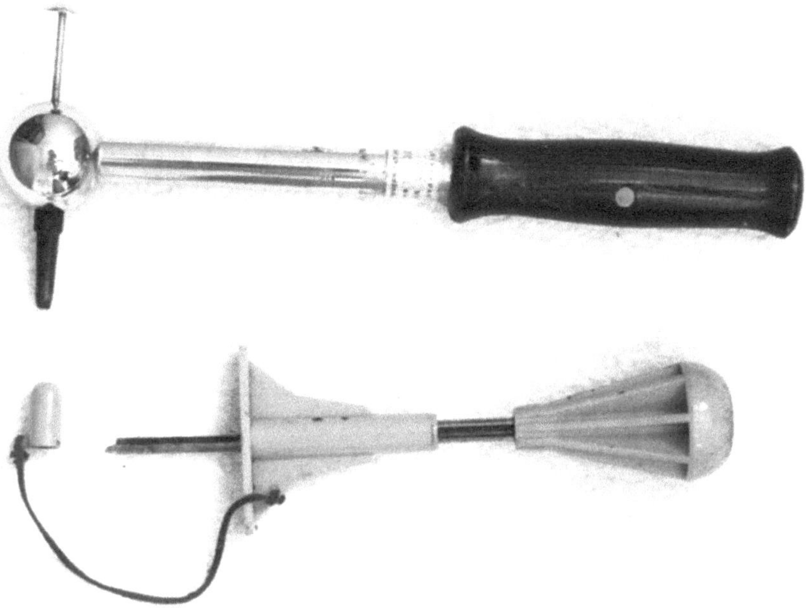

Fig. 2.21. Instruments for measuring bark thickness

2.3.6 Tree Form

The form of a tree is captured by the diminution of the diameter from the base to the tip. The diminution in diameter varies with species, site, stock density, or age. Tree form determines the utilization of timber and is a major component for the estimation of the volume of a tree.

Tree form can be described in terms of theoretical solids. The simplest solids approximating the various stem sections are the cylinder, the paraboloid, the cone, and the neiloid.

The form factor f is the relationship between the stem volume and the volume of a geometric figure – usually a cylinder – with the same length (height) and cross section g as the stem. The volume is given by

$$v = ghf;$$

thus, the form factor for a cylinder is less than 1 and it compensates for the overestimation that would be incurred if one assumed a cylindrical stem form.

The form factor cannot be measured directly. An alternative way for describing stem form is to calculate the form quotient, which is the ratio of an upper-stem diameter to a reference diameter. If the DBH is taken as a reference diameter, the resulting form quotient is termed the breast height form quotient. It is not a true expression of the stem form, as the form quotients for trees with the same stem form but differing dimensions vary. Consequently, breast height form quotients are not directly comparable.

Hohenadl (1936) suggested that the diameter at one tenth, $d_{0.9}$, of the tree height be taken as the reference measurement. This allows computation of what is termed the natural form function, which in turn permits the direct comparison of stem forms independent of tree dimensions. Hohenadl (1936) divided the stem into five sections and derived a form quotient for each section:

$$\eta_{0.9} = \frac{d_{0.9}}{d_{0.9}} = 1; \ \eta_{0.7} = \frac{d_{0.7}}{d_{0.9}}; \ \eta_{0.5} = \frac{d_{0.5}}{d_{0.9}}; \ \eta_{0.3} = \frac{d_{0.3}}{d_{0.9}}; \ \eta_{0.1} = \frac{d_{0.1}}{d_{0.9}}.$$

The tree form quotient according to Hohenadl can then be computed from

$$\lambda_{0.9} = 0.2\ (1 + \eta_{0.7}^{2} + \eta_{0.5}^{2} + \eta_{0.3}^{2} + \eta_{0.1}^{2})$$

and used for the immediate calculation of stem volume:

$$V = \frac{\Pi}{4} d_{0.9}^{2} h \lambda_{0.9}.$$

Analogue series can be calculated with DBH as a reference diameter.

An alternative way to describe tree form is through a taper function. Taper functions express the decrease in diameter from top to base. Scatter plots of stem diameter against stem height visualize the taper of a tree in a graphical format and allow for fitting the data to a function that describes the trend in the data. The result is a taper curve (function). With the advent of powerful computers, a variety of polynomial curves have been fitted to the profile of individual trees. For example, Fries and Matérn (1965) fitted a polynomial up to 58th power. Kozak (Kozak and Smith 1966; Kozak et al. 1969) fitted a ratio of two second-order polynomials and found it acceptable. Species vary greatly in branching habits, basal swell, and stem form, and variation within a species is also important; hence, no single taper model is equally suitable to all species and all sizes of trees. Rather, specific taper functions are needed on a species-specific and a regional basis (Czaplewski 1989; Czaplewski et al. 1989). Intensive studies have been undertaken in order to develop mathematical models that describe the stem taper (e.g., Sterba 1980; Reed and Green 1984; Bruce and Max 1990). Recently flexible trigonometric and mixture modeling (i.e., the taper is assumed to be a mixture of several standard geometric shape-curves) has advanced our arsenal of taper functions to be considered in a specific application (Sharma and Zhang 2004).

The first step towards obtaining a taper equation is to convert paired measurements of stem diameter and stem height into relative heights and diameters, for example, the height of a measurement point divided by the total height, or the diameter at a measurement point divided by DBH. The result is a set of dimensionless numbers, where most of the correlation between size and variation is eliminated. Raw values of height and diameter often show large changes in volumes of trees of different size with only minor differences in taper (shape).

Taper functions are used to estimate stem diameters for any height and vice versa. Where a taper function is available, it can be used in the inventory to compute volumes for various heights or sections. It is desirable to have unbiased estimates of both upper-stem diameters d and their squared counterparts d^2. Estimates of d are used to derive assortments or products from a stem and estimates, whereas d^2 is used to estimate volume.

Typical taper functions for relative diameter have an inflection point at about 20% of the height (from below), a point where the second derivative of the taper curve changes sign. This change cannot be seen at stems but emerges visually when the taper profiles are plotted with vertical scales compressed to 1/50 and 1/100 of the horizontal scale.

Many trees show a substantial difference between the shape of butt swell and the shape of the upper part of the stem, which is difficult to capture in a single function. Therefore, many taper functions are segmented polynomials. One part describes, for example, the butt, while the other describes the upper stem. The lower joint point is at the butt height or between the butt height and the inflection point. Segmented curves need to be kept smooth and continuous at the joint point, which can be realized by conditioning submodels to be connected at these points and smoothed by conditioning their first-order and higher-order derivatives (Max and Burkhart 1976).

Max and Burkhart (1976) developed a segmented polynomial model that has consistently performed well. The from of the model is

$$rd^2 = b_1\left(\frac{h}{H} - 1\right) + b_2\left(\frac{h^2}{H^2} - 1\right) + b_3\left(a_1 - \frac{h}{H}\right)^2 I_1 + b_4\left(a_2 - \frac{h}{H}\right)^2 I_2,$$

where rd is the relative diameter (i.e., the ratio of an actual diameter at a measuring point above DBH to DBH, b_i are regression coefficients, $i=\{1,2,3,4\}$, h is the height above ground estimated, H is the total tree height, a_i are join point parameters estimated from sample data, $i=\{1,2\}$, and

$$I_i = \begin{cases} 1, \text{ for } h/H < a_i \\ 0, \text{ for } h/H > a_i \end{cases}.$$

2.3.7 Volume

Volume is among the most important attributes determined in forest inventory; the sampling design of forest management inventories is commonly optimized for volume. Procedures for volume determination are manifold and cannot be discussed in detail here. For those interested, textbooks written by Prodan (1965) or Hush et al. (1982) provide extensive information. Methods and problems of volume determination are outlined in Cunia (1979) and Kublin (1987). These historic references are still the most comprehensive and they answer all basic questions associated with volume estimation in practice. Computation of volume functions requires a solid basis in regression analysis. The books by Draper and Smith (1981) and Myers (1986) give an excellent undergraduate-level introduction to the topic.

The selection of the method to be used for volume determination is one of the most important decisions in the planning of an inventory. Volume estimation must be unbiased. Biased volume estimates may invalidate all main results of the inventory (Cunia 1981; Gertner and Köhl 1992).

Existing volume functions offer convenience and expediency, but they should be carefully checked before they are accepted for use. Existing volume functions have often been derived from data representing only a small region or a nonspecific population. The verification of existing volume functions before their application in a forest inventory is essential (Kaufmann 1991).

It is of the utmost importance that the volume to be quantified in the inventory is unambiguously defined. To this end, it is necessary to stipulate from the beginning:

- The minimum and maximum diameter of the trees to be surveyed
- To what minimum diameter of the upper stem the volume is to be calculated (this gives – together with the maximum and minimum diameter – the defined spectrum of tree forms)
- Whether the volume is to be determined inside or outside bark
- Whether branches, roots, and stump are to be included or not
- Whether damaged stem sections are to be excluded or not (this includes a definition of damage and guidelines for delineating the extent of damage)
- Whether the gross or only the merchantable volume is to be determined
- Whether the term "usable volume" refers to the volume without logging loss or loss through processing at the sawmill

FAO/UN Economic Commission for Europe (ECE) proposed a definition of standing volume to be applied in forest inventories (p. 393 in UN-ECE/FAO 2000,):

Volume of standing trees, living or dead, above-stump measured overbark to top (0 cm). Includes all trees with diameter over 0 cm (d.b.h.)

Includes: Tops of stems, large branches; dead trees lying on the ground which can still be used for fibre or fuel.

Excludes: Small branches, twigs and foliage.

Volume may be expressed in several different units. It is advisable, not least to allow international comparison of results, to use the metric system (cubic meters). Volume is usually determined in three steps:

1. Computation of volume functions for individual trees (preliminary study)
2. Measurement of individual trees (field surveys)
3. Determination of individual tree volume and compilation for data analysis

Besides methods for determining the volume of individual trees there are also methods for estimating the volume of whole stands. The following remarks are restricted to volume determination of individual trees as forest inventories almost exclusively employ methods for the determination of such volumes. The single tree volumes are accumulated on a plot basis to give figures for larger units of reference.

Before the actual field survey is conducted, the method of volume estimation for single trees must be determined, preferably through a preliminary study. The aim is to derive valid (ideally unbiased with minimum prediction error) volume functions for single trees, species groups, vegetation types, or particular sites. A decision has to be made whether global or stratified models are to be used (Lappi 1991).

In the past, volume tables were often constructed by means of graphical methods, and were often subjectively biased. As statistical errors cannot be computed from such tables, they should not be used in inventories. The construction of volume tables is not discussed here.

To obtain representative data for the computation of volume functions, the volume of a sufficient number of sample trees must be measured by sections for which the assumption of a simple geometric shape is appropriate. This task is considerably easier if the sample trees can be felled. For reasons of cost or time, however, this is seldom possible, so it is usually necessary to take a sufficient number of measurements of diameters at upper levels on standing trees.

Ideally, four to ten sections or diameters should be measured, the measurements in the lower portion of the stem being taken at smaller intervals than those in the upper portion. The measurements can be used to estimate a specific geometric shape of the tree stem or to approximate the trend in diameter as a function of height by means of polynomials, spline functions, or form functions. Examples of the application of spline and form functions may be found in Kublin (1987) und Schmid (1971). If none of these methods are used, the volume of individual stem sections has to be calculated based on an assumption about the geometric shape of each section.

Given g_b as the basal area of the lower cross section, g_u as that of the upper cross section, and g_m as that in the middle of a stem of length L, the volume of a section can be calculated from the following three equations and provides an approximation to the volume of a cone frustrum:

1. Smalian's formula: $V = \frac{g_u + g_b}{2} L$
2. Huber's formula: $V = g_m L$
3. Newton's formula: $V = \frac{g_u + 4g_m + g_b}{6} L$

The accuracy of the approximated volumes is best for Newton's formula. Huber's and Smalian's equations give reliable estimates only for sections that in fact have the shape of a paraboloid; in other situations they will give biased results. For Huber's equation the bias is negative for conoids and paraboloids and positive for the fuller shapes of third-degree paraboloids. In all cases the bias of Smalian's equation is opposite and twice that of Huber's equation. For many applications it is preferable to derive volume by integrating a cubic spline curve fitted through data pairs of height and diameter measurements (Goulding 1979).

The total volume can be computed by summing the volume of individual sections. This approach is called the sectional method. When measuring sectional stem diameters, DBH and tree height (or length) should be recorded at the same time, as they are normally used as input parameters in volume functions. These measurements are used as independent variables (predictors) in regression functions with volume as the dependent variable. These regression functions can be derived through standard computer programs such as SAS, SPSS, SPLUS, or STATA.

The most important predictor of volume is usually DBH. Depending on the desired precision and availability of additional predictors, a measurement of height and an upper diameter can also be included if they significantly reduce the volume prediction error. The upper diameter serves as a surrogate of the stem form.

Volume functions can be divided into three classes. The first class has DBH as the sole predictor. Functions of this class may be termed *local volume functions.* Their validity is obviously limited. They do not account for the potential impact that age, site conditions, stem form, and growth history can have on the volume of a stem with a given DBH. The most common local volume functions are:

- $V=b_0+b_1d_{1.3}^2$
- $\log V=b_0+b_1\log d_{1.3}$
- $V= b_0+b_1d_{1.3} +b_2d_{1.3}$

The second class has DBH and a height measurement as predictors. These functions have a wider range of validity than functions in the first class. They are frequently referred to as *regional volume functions.* Regional volume functions are considerably more accurate than local volume functions. Examples of regional volume functions are:

- $V=b_1 d_{1.3}{}^2 h$
- $V=b_0+b_1 d_{1.3}{}^2 h$
- $V=b_0+b_1 d_{1.3}{}^2+b_2 d_{1.3}{}^2 h+b_3 h$
- $\log V=b_0+b_1 \log d_{1.3}{}^2+b_2 \log h$
- $\log V=b_0+b_1 \log d_{1.3}+b_2 \log^2 d_{1.3}+b_3 \log h+b_4 \log^2 h$

The third class, which can be termed *'large-scale' volume functions*, includes an upper-stem diameter d_x as a surrogate for stem form. Large-scale functions have been developed specifically for scientific studies and national inventories. Examples are:

- $V=b_0+b_1 d_{1.3}\, d_x h$
- $V=\pi/4(b_0\, d_{1.3}{}^2 h+b_1 d_{1.3} d_x h+b_2 h^2)$

The sample trees to be measured for the computation of volume functions must be very carefully selected. Their geographical distribution should be based on random sampling. For logistic reasons, sample trees are often concentrated at only a few sites. This, however, usually means that only a part of the total range of variability over the whole inventory area is captured. Further, sample trees are often selected in the vicinity of roads or in large forest clearings because of easier access. Data from trees in such locations, however, often lead to a distortion of the inventory results, as the growth conditions differ greatly from those in the rest of the inventory area.

The ideal number of sample trees lies between 50 and 100 for local volume functions and between several hundred and several thousand for large-scale functions. The optimum depends on a number of factors, such as the range of diameter, tree dimensions, variation in site conditions, or stand type. The number of trees needed to derive regression estimates can be calculated according to the level of significance, the power, and the R^2 values. Software such as nQuery Advisor[3] facilitates the calculation of sample sizes. Figure 2.22 presents the required sample sizes for different R^2 levels and for k normally distributed covariates. The significance level of $\alpha=0.05$ and a power of 80% were chosen. The results in Fig. 2.22 have to be interpreted in the following way (e.g., R^2 value of 0.2):

When the sample size is 113, the multiple linear regression test of $R^2=0$ ($\alpha=0.050$) for four normally distributed covariates will have 80% power to detect an R^2 of 0.1000.

Sample sizes estimated from generic formulae can only serve as rough guidelines. The actual problem is very complex (Dees 1988; Shieh 2001).

[3]http://www.statsol.ie/nquery/nquery.htm

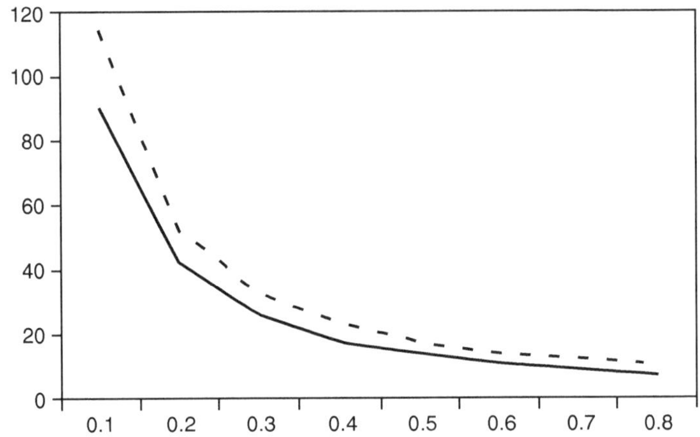

Fig. 2.22. Sample size estimation for regression analysis (n over R^2) (*solid curve* two covariates, *dashed curve* four covariates)

Suffice to say that one should always be prepared to acquire additional data if the predictive precision of an estimated function falls short of a set target.

It is essential to select sample trees covering the entire range of the independent variables ($d_{1.3}$, $d_{0.h}$). Simple random sampling of trees for volume measurement will rarely satisfy this requirement. Most sample trees would be clustered around the mean value of the predictors. To get sample trees from the entire range of predictor values the principle of random selection must be modified in order to ensure that trees of smaller and larger sizes are included in the sample.

The precision of an estimated volume function is determined in one of two distinctly different ways. In the first approach, about one third of the data is withheld, for validation purposes, from the estimation process. The volume function estimated from a random subsample of about two thirds of the collected data is then tested on the data withheld from the estimation process. The predicted volume is compared to the measured volume and an estimate of the prediction error and prediction variance is obtained by standard techniques. Once satisfied that the function meets specifications, a new set of functions are computed from all measured trees. It would be a waste of valuable information if one were to leave about one third of the data used for no other purpose than estimating precision. The precision of predictions derived from the complete sample is almost always better than that of predictions derived from a smaller sample.

The second approach to estimation of the prediction error is via cross-validation or some form of resampling. In a cross-validation the data from one tree are withheld from the estimation and a prediction is made for this tree from the estimated volume function. The error of prediction is noted. This process is repeated for all sample trees. The standard error of prediction is obtained directly from the individual error estimates. In a resampling approach, a model-based prediction error is drawn for each measured tree from a presumed distribution of errors and a new data set is generated as the sum of a prediction and a random prediction error. A new "bootstrap" model is obtained from these new data in the same way as the original volume function and a new bootstrap prediction is made in turn for all trees. The bootstrap prediction error is obtained for each tree and the prediction variance is computed by standard techniques. This process is repeated a large number of times (say 200) and the average prediction variance is calculated (Miller 1974; Li and Schreuder 1985; Magnussen and Burgess 1996). It is important to note that these methods rely on the assumption of independently sampled observations (trees). When more than one tree is selected per sample location, the assumption of independence may be questionable. In the case of multiple sample tree per location (cluster sampling) it is a safer practice to treat all trees from a sample location as a unit in the cross-validation or bootstrap procedure.

2.3.8 Weight and Wood Density

Weight may also be used as a measure of production. Weight is increasingly used for the scaling of traditional forest products and, for several reasons, seems to be a sensible measure of quantification. It takes a great deal of time and work to estimate the volume of branches, twigs, or any other irregularly shaped parts of a tree, and the results are subject to considerable error, whereas weighing is easy and direct. Many forest products, such as pulp, charcoal, particle boards, and other wood composites, are conveniently quantified in terms of weight. It would therefore be logical and attractive to define the raw products in the same units.

The approximate weight of a tree can be obtained on the basis of stem volume measurements. Hush (1962), Taras and Clark (1977), and Crow (1978), have derived equations relating the weight of various tree components to DBH and tree height.

Gross weight of wood includes both the dry weight of wood and the moisture content. The moisture content of a living tree stem varies according to Kollmann and Côté (1968) between:

- Parts of the tree
- Heartwood and sapwood
- Species
- Tree size within a species
- Growing location of the tree
- Time of day
- Time of year

If in addition to gross weight moisture content is also measured, the oven-dry weight can be determined. When gross weight has been measured small samples of wood are selected and dried in an oven at 105°C for 24 h. After this time the water content of the wood samples should have gone (Panshin and Zeeuw 1980). In order to ensure this a subsample of the oven-dried wood samples should be weighed and again dried for 12 h to ensure that there has been no further weight reduction. It is important that the samples are weighed as soon as they come out of the oven, as dried wood is hygroscopic and will reabsorb moisture from the surrounding air and thus regain weight.

If large differences between the moisture content of the different parts of the stem and branches can be found it may be advisable to calculate the dry weight separately for tree components such as stem, branches, or twigs and then sum up the components' weights to obtain the total dry weight.

In addition to the moisture content the wood density of trees can be calculated. Wood density is defined as mass divided by volume at a specific moisture content:

$$Wood\ density = \frac{mass}{volume}.$$

The units of wood density are kilograms per cubic meter or grams per cubic centimeter. Published or otherwise available estimates of wood density allow the calculation of the mass of a tree or its components for a given volume or, conversely, the volume of a tree stem with a given weight. This concept can be applied to convert the volume that has been measured on bigger parts of the sample trees into weight. It can also be used to convert the weight of the biomass into volume.

2.3.9 Biomass

Woody biomass is a renewable resource. If sufficient time is afforded a new crop of biomass will replace one that was harvested When managed sustainably the supply of biomass is safeguarded for future generations. Biomass is, on one hand, an important forest product that can provide a significant amount of energy for local and regional consumption but, on the other hand, biomass is also a major player in the global biochemical cycles, especially in the carbon cycle. Since about 50% of the forest biomass is carbon (Gifford 2000), forest biomass provides a good estimate of the carbon pools in forests.

Various definitions exist for woody biomass. Biomass can be described as the plant material being produced by or resulting from photosynthesis. Schreuder et al. (1992) define biomass as the amount of living matter per unit area or the volume of the forest habitat. In the scope of the ECE/FAO global

Forest resources assessment 2000 (FAO 2001) the attribute "above-stump woody biomass" is defined as

The mass of the woody part (stem, bark, branches, twigs) of trees, alive or dead, shrubs and bushes, excluding stumps and roots.

"Woody biomass" is defined as

The mass of the woody parts (wood, bark, branches, twigs, stumps and roots) of trees, alive and dead, shrubs and bushes, measured to a minimum diameter of 0 mm (d.b.h.)

According to the *Forest resources assessment 2000* definition woody biomass includes the above-stump woody biomass plus the biomass of stumps and roots but excludes foliage. Biomass is often divided into different components in order to quantify the availability of woody material for different utilization, for example, sawn timber, construction wood, feed, or bioenergy.

The biomass of trees, on a unit area basis (Young et al. 1964; Lodhiyal and Lodhiyal 2003), may be taken as a measure of site productivity. Biomass of components such as roots, branches, stem, bark, top sections, and even needle or leaf mass can be interpreted as indicators of stand vigor, stand health, and the need for silvicultural treatment. Biomass is expressed in kilograms of dry weight. Surveys of studies on the biomass production of various tree species throughout the world can be found in Young (1976), Pardé (1980), Cannell (1982), Auclair (1983), Cervantes et al. (1998), Fearnside et al. (1999), Nelson et al. (1999), Parresol (1999), Komiyama et al. (2000), Cordero and Kanninen (2002), Cairns et al. (2003), Laclau (2003), and Muller-Landau (2004).

The direct assessment of the woody biomass of a tree is done by a destructive process. A complete quantification of a tree requires the felling of the tree, separating the woody parts into different assortments, and assessing the total weight of the assortments. As the sampling is destructive this method cannot be used for monitoring tree growth by permanent assessments. This procedure can be very time consuming and – especially for large trees – can entail substantial assessment errors owing to the difficult logistics of handling large amounts and volumes of woody material during the entire process of determination.

As an alternative to the direct determination of biomass, Brown (1997) suggested an "expansion" of the volume of a bole (over bark) to biomass by means of a volume-weighted average wood density factor. The method assumes the availability of precise volume estimates. Applications should be limited to cases where this assumption is justified.

In extensive inventories a direct assessment of woody biomass becomes impractical. One alternative is to derive biomass functions with tree attributes such as diameters and tree height or crown attributes, moisture content, and wood density as independent predictors of biomass. Frost (1990) and Mushove (1994) derived single tree biomass functions for *Brachystegia spiciformis* in

Zimbabwe, with height and DBH as independent variables. Grundy (1995) derived biomass functions for *Brachystegia spiciformis* and *Julbernardia globiflora* in Miombo Woodlands, Zimbabwe, which provide the total biomass per hectare, i.e., they are not single-tree functions. Velle (1995) suggested in the scope of a national biomass survey in Uganda three different types of biomass functions, which reflect different growth conditions and phenotypes of trees. He used DBH, tree height, and crown radius as predictors. Montagu et al. (2002) developed an allometric equation for *Eucalyptus pilularis* in New South Wales, Australia, to calculate aboveground biomass.

$$\text{Aboveground biomass (kg)} = 1.021^{a} \times e(\ln\text{DBH} \times 2.589 - 2.733),$$

where lnDBH is the natural logarithm of tree DBH (centimeters) and a is a bias correction factor based on Baskerville (1965)

The Intergovernmental Panel on Climate Change (IPCC) has default values for estimating root biomass when the aboveground biomass is known (IPCC 2004)[4]. These values were used in the carbon calculator to estimate the belowground component of the total tree biomass.

Suitable equations (functions) for predicting biomass of trees or stands may not be available for a particular forest inventory. The equations must then be obtained from a sample of trees selected specifically for this purpose. Sample trees have to be selected from the population. The selection procedure for sample trees is driven by three requirements (Cunia 1979):

1. The sample trees should be representative for the population of interest
2. The selection procedure should allow a valid regression analysis
3. The selection procedure should be cost-efficient.

The trees to be selected for a destructive biomass assessment should not be selected randomly but should be selected in a way that the individuals cover the whole size range. On the other hand, the trees should not be selected subjectively. A subjective selection invariably results in a bias. For example, trees with a low number of branches should not be selected preferentially just because they reduce the assessment effort of crew members. Objective selection can be achieved by stratified random sampling in the following way. All diameters of trees in a plot or stand are measured. Then the trees are ranked by diameter. The smallest trees are deleted from the list until the number of remaining trees can be divided by 3. The remaining trees are then divided into three classes: "small", "medium," and "large." From each class an equal number of trees is randomly selected (Steward et al. 1992).

According to Steward et al. (1992) at least 12 trees should be taken for each species or species group. In a national biomass survey in Uganda, roughly 3,000

[4]See their website at http://www.ipcc.ch

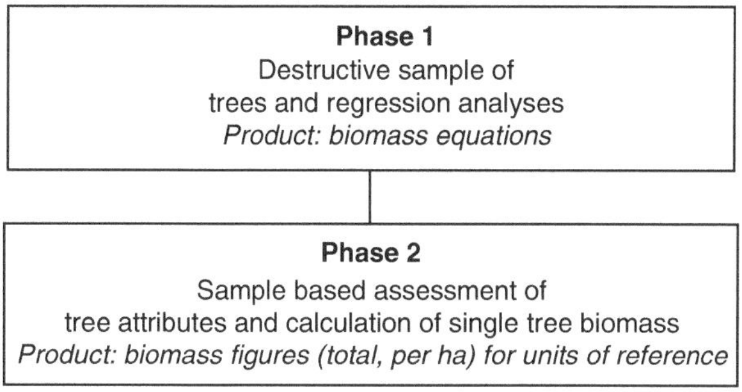

Fig. 2.23. Phases of biomass assessments

trees were destructively sampled in order to derive biomass functions for 123 species (Velle 1995).

Extensive biomass assessments generally follow a two-phase procedure (Fig. 2.23). In the first (experimental) phase, a small sample of trees is taken and destructively measured. Based on this, fairly small sample regression analyses are carried out and final biomass equations that give the single-tree biomass as a function of easy-to-measure tree parameters (e.g., diameter at 30-cm, 1.3-m, and/or 7-m height, crown diameter, or total tree height) are selected

In the second phase, a large set of sample trees is assessed. For all second-phase trees the tree parameters serving as predictors of biomass are measured Once these parameters have been assessed, they are used to predict the biomass of individual trees through the biomass function. Individual tree biomass predictions are then aggregated by statistical algorithms to estimates of interest. As this phase does not require any destructive sampling it is easily integrated into an already-existing sampling design and extended to cover applications in monitoring.

In order to increase the efficiency of single-tree biomass estimates, Valentine et al. (1984) introduced to forestry a sample-based method for biomass assessment of single trees that provides unbiased estimates of volume, dry or fresh weight, and other components such as weight of fruits or mineral content. The method consists of two sampling phases: (1) employment of randomized branch sampling (RBS) to construct a random path through the tree, and (2) application of importance sampling (IS) to select the location of sample discs to be extracted and weighed.

RBS was pioneered by Jessen (1942) to estimate the amount of fruit on trees. Valentine and Hilton (1977) used RBS to estimate the foliar area and mass and leaf count on oak. Furness (1976) used RBS to estimate the number of insect eggs and the larval population on trees. Gregoire and Valentine (1996) applied RBS and IS to assess the stem length and surface (bark) area of tropical tree species. IS has been applied as a method for the estimation of bole volume (Wiant et al. 1989; Schreuder et al. 1992; Valentine et al. 1992; Gregoire et al.

1995; van Deusen and Baldwin 1993) and for quantifying fuel wood (de Gier 1991).

In RBS a tree is considered to be the entire stem system that develops from a single bud. A branch segment or segment is the part of a branch between two nodes. A path is the sequence of connected branch segments. At each node of branching a decision about which branch to follow has to be made. The result of these decisions is a random path through the entire tree. The choice of which branch to follow is decided with probability proportional to size (PPS), where "size" is a measure of the biomass above the node of each branch considered for selection. In other words, the probability of selecting a given branch is proportional to the estimated biomass supported by the branch. A suitable and effective measure of size is the product of the length of the branch to the end of the path, l, and its diameter squared, d^2, at the node of branching. The selection probabilities assigned to branches at a node must sum to 1 (Valentine et al. 1984). In the lexicon of RBS no operational distinction is made between branch, stem, and twig. The total selection probability of a random branch path is the product of all the nodal selection probabilities along the chosen path. Whole tree estimates of biomass are then obtained as Horwitz–Thompson estimators (a sample value is expanded by the inverse of its selection probability) (Thompson 1992).

2.3.10 Quantification of Timber Quality

Even in intensively managed forests, a single stand is rarely stocked exclusively with sound, defect-free timber. Tropical virgin forests, in particular, often have a high proportion of trees with defects. The type of defect affects the merchantable volume and the value of a tree. In tropical forests, the ratio of gross volume to net volume can be as high as 10:8. Consequently, many inventories require a quantification of the timber quality in order to facilitate an estimation of net volume or volume loss due to defects.

The definition of quality (quality grade) depends on the use for which the timber is intended. Uses like lumber, poles, ties, shingles, fuel wood, logs, or veneer each come with a set of quality requirements. The quality is usually defined according to the grade of the highest value. Quality is almost always determined on the basis of discrete variables on an ordinal scale. There are two types of grading systems:

1. Absolute grades: here the quality classes are based on tree dimensions. This is termed dimension grading and may be done on the basis of, for instance, minimal length or minimum top diameter.
2. Relative grades: the classification is based on the most valuable product that can be obtained from the stem.

Quality may be expressed in terms of (1) merchantable value, (2) quality index – a figure representing the percentage of the value of a class in comparison to a reference value – or (3) the volume distribution of quality classes of the product concerned.

Quality is determined on the grounds of dimensions and defects. Parameters such as the shape of the trunk, branch diameter and branch length, number of branches, angle between branch and trunk, and number of secondary branches can be used to estimate the potential assortments on standing trees (Borner et al. 2003). Makela and Makinen (2003) used a process-based growth model to generate 3D sawlogs. Ruschel et al. (2003) studied the market value of timer for different species in Brazil.

Defects are differentiated into technical and internal defects (Fig. 2.24). Indicators of technical defects are forks, leans, sweep and crook, spiral grain limbs, and knots, which indicate the presence of reaction wood or decay or may impair the technical use of the timber. Internal defects may be indicated by the presence of conks, suspect scars, fire scars, lightning scars, dead or broken tops, forks or pronounced crooks, and logging and other scars due to harvesting.

Internal defects may be detected by means of microwaves, X-ray and γ-ray computed tomography, or magnetic resonance. Detailed descriptions of these techniques are given by Bucur (1985) and Hailey and Morris (1988).

The allocation of a tree to a given quality class is difficult, as a whole tree can be assigned several qualities and the mixture of these varies from tree to tree.

(a) (b)

Fig. 2.24. Examples of stem defects (from *top left* to *bottom right*: unveiled stem, frost scar, dead branch, bifurcation, scars from raisin extraction, epiphytes)

Fig. 2.24. (Continued)

Further, there are a number of commercial usages, each of which is based on different quality classes (Palmer 1975; Hemmila and Sipi 2004; Zell et al. 2004).Continuous changes in the timber market, new technologies, and the advent of new products with overlapping specifications force the provider or seller of forest products to constantly adapt definitions of quality classes. For this reason, a quality classification system used in forest inventories must be as flexible as possible and based on a set of generic rules that permits a later reclassification against a revised or new set of criteria.

Information on the wood quality and utilization of tropical wood can be found in Brazier (1981), and on the detection of decay in Panzer (1975).

2.3.11 Age

Where management reports of tree planting or sowing are available tree age can easily be determined. In temperate zones, information on tree age can be achieved through counting annual rings. Seasonal patterns in the number of cells, cell size, and cell shape laid down in the wood make it relatively easy to discern annual rings in the wood. For trees with an undisturbed record of growth the number of rings from the pith to the inner bark is equal to the number of years passed since the tree reached the height at which the ring counting is done.

Several difficulties can arise in the determination of tree age. Seedlings and young trees may have endured suppression for periods of time which can vary greatly in length, from 1 year to several decades, where young trees are suppressed. Increased illumination caused by the opening of the canopy results in rapid growth in height until the tree reaches the intermediate canopy. During this phase, increase in diameter remains relatively modest. Only after the tree has developed a full crown does increase in diameter begin to accelerate. For trees with this growth pattern an age determination based on size would be not only be extremely difficult but in all likelihood also biased. Consequently age is often limited to indicate the period of time a tree has been part of the upper canopy layer and how long it has taken to grow from one diameter class into the next.

In tropical forests an estimate of age can be provided by repeatedly measuring the same trees (Vanclay 1996). Reliable and useful estimates require selection of a sufficient number of trees in each size class. The diameter is measured accurately to the nearest millimeter with a tape. It has proved useful to mark the level at which the diameter was last measured with paint, so as to minimize measurement errors. The measurements should always be conducted at the same time of year and should cover a period of at least 5 years.

It is widely thought that cambial growth in tropical trees is continuous and generally not sensitive to climatic variation (Richards 1996); tropical trees should not produce reliable ring chronologies. Worbes (1989, 1992) and Devall et al. (1995) provide a historic perspective of this viewpoint. A corollary to this viewpoint has been that owing to seasonal variation in temperature and soil water, many woody temperate species interrupt cambial activity and add a morphologically distinct xylem layer, which results in a discontinuity between annual growth increments. Even though the seasonal variation is not distinct most tropical locations show intra-annual variation with respect to humidity, rainfall and solar radiation (Richards 1996; Enquist and Leffler 2001). Eckstein et al. (1981) discuss the problems of dendrochronology and the prospects for tree ring analyses. Tropical tree ring chronologies have been described among others by Pumijumnong et al. (1995) for teak (*Tectona grandis)* in northern Thailand, Vetter and Botosso (1989) for Amazonian trees, Devall et al. (1995) for equatorial laurel (*Cordia alliodora*), barrigon (*Pseudobombax septenatum*), and chirimoya *(Annona sraguei)* in central Panama, and Palmer and Murphy (1993) for teak in Java. Boninsegna et al. (1989) analyzed wood samples of 13 tree species from three sites in Argentina for the occurrence and formation of tree rings. Well-defined annual tree rings were found in *Cedrela fissilis, Parapiptadenia rigida, Cordia trichomata,* and *Chorisia* spp. Stuiver et al. (1981) and Worbes and Junk (1989) used radioactive and stable isotopes in estimating age. Bomb-produced ^{14}C seemed to be the most promising method for dating between 1955 and the present.

More aspects of age determination can be found in Borman and Berlyn (1981), Mariaux (1981), Singh (1981), and del Valle (1986).

2.3.12 Growth and Increment

Growth is the dimensional increase of any organic system, while increment is the dimensional increase within a given period of time. The main process behind tree growth is photosynthesis. The growth and increment generated by photosynthesis depend on environmental factors, growth history, and the state of the living tissues in various plant organs, such as crown, root, and stem. During a life cycle, which generally exceeds 100 years, trees are exposed to a long list of different impacts that each can influence the health and vigor of a tree and even its shape and structure. Growth (increment) is the net result of this complex web of interacting processes. The main external growth determinants are soils, nutrient supply, water supply, atmospheric conditions, climate, competition, and pollution levels.

Growth and increment of single trees is mainly assessed for the following attributes:

- Tree height (h)
- DBH ($d_{1.3}$)
- Basal area (g)
- From factor (f)
- Height-to-DBH ratio
- Volume (V)
- Value
- Needle/leaf area

The growth of a tree is characterized by short-term and medium-term variations; however, growth trends common to either single trees or stands can be summarized in a set of general observations.

Growth is understood as the dimension of an attribute under study reached at a given point in time. The growth curve (Fig. 2.25) is the graphical rendition or a mathematical formulation of the realized dimension as a function of time (t), $y=f(t)$.

A generic example for the growth of tree height is the function

$$h = ae^{-b/t},$$

where the asymptote a and the relative growth rate b are constants that have to be estimated for a particular application.

The growth curve of any attribute in an organic system often shows the shape of an asymmetric "S" (sigmoid). The curve starts with an exponential rate of increase, which represents the period of unconstrained growth. During this period the acceleration in increment is positive. After a point of inflection (zero acceleration), which is caused by one or more constraints in external or internal factors, the growth rate decreases. Finally the curve asymptotically approaches a maximum value (which is rarely, if ever, observed in practice).

To quantify the growth of a stand is a much more complex challenge, as it is sum of the growth of the trees that make the stand. Stand growth can be quantified by:

- The sum of attribute values per unit area (e.g., number of stems per hectare or volume per hectare)
- Mean values (e.g., mean tree height, mean diameter, mean volume)

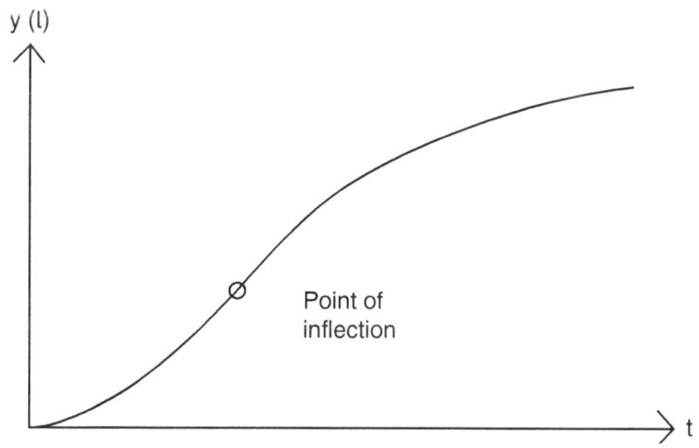

Fig. 2.25. Growth curve

Additional stand-level parameters and characteristics must be quantified as well, for example:

- The frequency distribution of DBH, tree volumes, or tree heights
- Density (e.g., number of stems per hectare)
- Tree species composition
- Vertical structure

Forest management can have a significant impact on both the shape of a growth curve (of trees and stands) and the amount of growth harvested during thinning and harvest operations. It is therefore important to distinguished between:

- The growth of remaining/residual trees in a stand
- The growth extracted during forest management operations (e.g., by thinning)
- The total growth, which is the sum of remaining and extracted stand

Increment is the dimensional increase within a given time interval:

Increment=$\Delta y/\Delta t$.

Depending on the selected time interval the following types of increment can be given:

- The *acceleration of growth* is obtained for the hypothetical situation where the time interval is infinitely small ($\Delta t \to 0$). It is the first derivative (inclination) of the growth curve (Fig. 2.26).
- *Current annual increment*, i, is obtained for a time interval of 1 year.
- *Periodic annual increment*, $i_{t_1-t_2}$, is normally calculated for time intervals of 2–20 years (Fig. 2.27). For longer periods the periodic annual increment is the average annual increment and it is approximately equal to the current annual increment halfway through the period in question, given that the increment is approximately a linear function of time during the period in question.
$$i_{t_1-t_2} = \frac{\Delta y}{\Delta t} = \frac{y_{t_2} - y_{t_1}}{t_2 - t_1}$$
- *Mean annual increment*, i_{0-t}, is related to the time interval from age 0 to age t (Fig. 2.28):
$$i_{0-t_1} = \frac{y_{t_1} - y_0}{t_1}.$$

Increment curves represent the acceleration of growth in relation to time and have different shapes (e.g., asymptotical, bell-shaped, or negative exponential) (Fig. 2.29). Several rules can be derived:

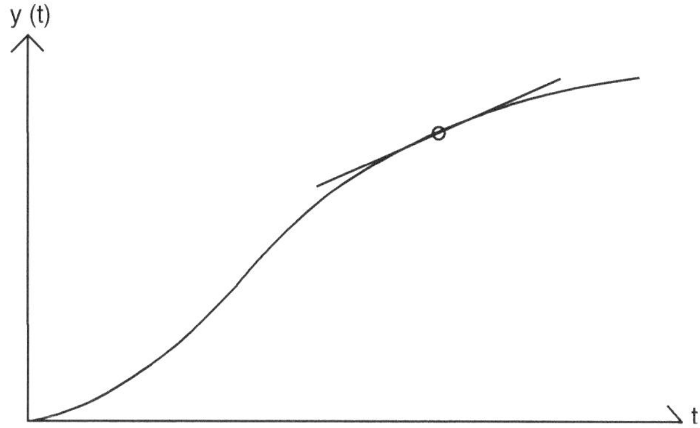

Fig. 2.26. Acceleration of growth (i.e., the first derivative of the growth curve)

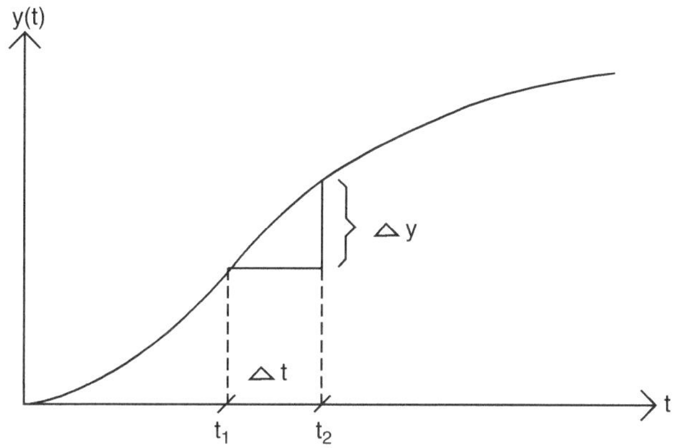

Fig. 2.27. Periodic annual increment

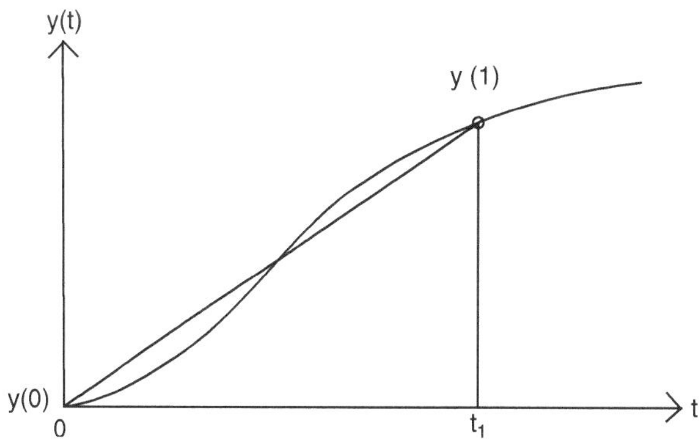

Fig. 2.28. Mean annual increment

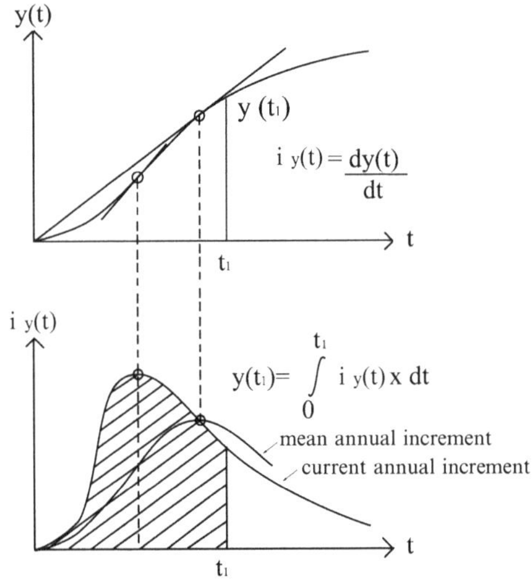

Fig. 2.29. Relationship growth curve, mean annual increment and current annual increment

- Current annual increment always culminates earlier and at a higher level than does mean annual increment.
- Before reaching its maximum the mean annual increment is smaller than the current annual increment.
- Acceleration of growth is the first derivative of the growth curve. Consequently, increment reaches a maximum when the growth curve shows its first point of inflection.
- The tangent to the growth curve that goes through the origin at t=0 touches the growth curve at a point t where the mean annual increment is at a maximum.

In tropical forests, increment can be determined from repeated measurements on permanent sample plots. In natural forest the growth curves of individual trees might substantially differ from a sigmoid shape, which is typical for trees growing in even-aged stands. The concurrence by other trees varies significantly during the lifetime of trees growing in natural forests, so periods of suppression may be followed by periods of release.

In temperate and boreal forests diameter increment can be measured by increment boring. By means of increment borers (Fig. 2.30) an increment core is extracted from the stem. From the increment core the related radial increment is obtained by measuring the corresponding width of the tree rings. However, in years with extreme climatic variations, tree rings might be missing or more than one ring might be generated. These situations lead to overestimation or underestimation of growth. As tree age, viz., time periods, can also be determined from

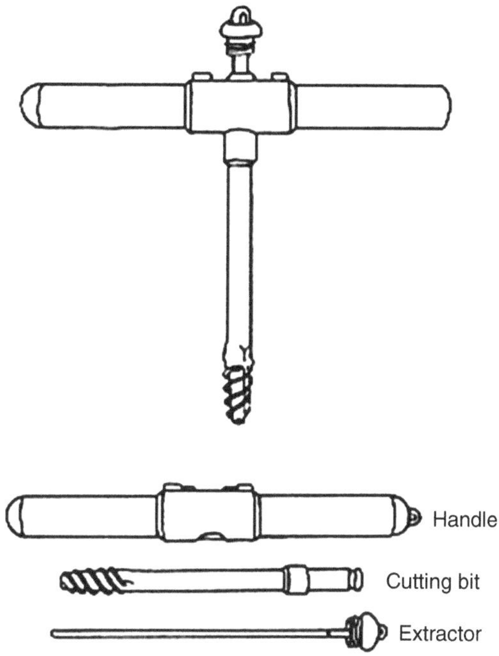

Fig. 2.30. Increment borer

the increment core, it is possible to calculate current annual and periodic annual increment for the tree in question. Where the stem center is hit by the increment core, one can also obtain an estimate of the mean annual increment. Where the core does not hit the center or pith, the estimation of growth is subject to errors.

The apparent growth of a forest stand is the net outcome of several growth components, some of which depend on thresholds and limits of observation. Stand growth between two points in time is conveniently decomposed into survivor growth, ingrowth, mortality, and cut. Ingrowth is the number or volume of trees growing to measurable size between the two points in time. Mortality is the number or volume of trees dying from natural causes between the two points in time, cut is the volume or number of trees felled between the two points in time, and survivor growth is the growth of trees present at both points in time. On the basis of these definitions, Beers (1962) presented growth terms for continuous forest inventory analyses. They are given for two different approaches. One approach deals with pooled volume records. Tree volumes at the beginning and at the end of the time period totaled with no attempt to pair successive volumes of individual trees. In the second approach, successive tree volumes are paired to determine the growth contribution of individual trees. The second approach is called "the tree level approach." The growth equations of Beers (1962) are given in Table 2.4.

Growth equations for volume totals and individual tree growth figures (tree level approach) differ but, fortunately, lead to the same result. Although Beers

Table 2.4. Equations for growth estimation (after Beers 1962)

Type of growth	Formula if using	
	Volume totals	Individual tree growth figures
Gross growth of initial volume	$=V_2+M+C\text{-}I\text{–}V$	$=V_{S2}+I+M+C\text{–}I\text{–}V_{S1}\text{–}M\text{–}C$ $=V_{S2}\text{–}V_{S1}=G_S$
Gross growth	$=V_2+M+C\text{–}V_1$	$=V_{S2}+I+M+C\text{–}V_{S1}\text{–}M\text{–}C=G_S+I$
Net growth of initial volume	$=V_2+C\text{–}I\text{–}V_1$	$=V_{S2}+M+C\text{–}I\text{–}V_{S1}\text{–}M\text{–}C=G_S\text{–}M$
Net growth	$=V_2+C\text{–}V_1$	$=V_{S2}+M+C\text{–}V_{S1}\text{–}M\text{–}C=G_S+I\text{–}M$
Net increase	$=V_2\text{–}V_1$	$=V_{S2}+I\text{–}V_{S1}\text{–}M\text{–}C=G_S+I\text{–}M\text{–}C$

V_1 the volume of trees measured on the first occasion, V_2 the volume of trees measured on the second occasion, V_{S1} the initial volume of survivor trees, V_{S2} the final volume of survivor trees, G_S survivor growth, M the initial volume of trees dying during the period between inventories, C the initial volume of trees cut during the period between inventories, I the volume of trees at the second inventory that were below the measurable size on the first occasion

stressed volume growth, the terms in the growth equations are equally appropriate for a different growth attribute, such as basal area growth.

If stump inventories are carried out to assess cut, the determination of the felling date has to be done with great care. The volume of felled trees can be calculated by two regression functions, one giving the DBH, $d_{1.3}$, as a function of the diameter of the stump, d_s, and a second function giving the volume of the stem, V, as a function of DBH:

$$\hat{d}_{1.3}=f(d_s),$$

$$\hat{V}=f'(d_{1.3}).$$

As estimates of growth depend on their definition (type of growth), it becomes essential to specify the definition of growth when results from independent studies are compared. The definitions according to Beers (1962) should be applied whenever possible. When comparing growth estimates over time or between entities of interest (population, strata, species, age class, etc.) it is advisable to examine how the growth estimates were obtained. The following questions can help you to identify potential problems of analysis and interpretation:

- Has there been a change of methods or definitions during the course of time?
- Are results of successive inventories comparable?
- Have seasonal effects (effect of time of assessment during the year) been handled correctly and documented?
- How was bark increment estimated?

While tree growth has been intensively studied in even-aged temperate and boreal forests (Burschel and Huss 2003; Pretzsch 2001), studies on tropical forests are comparatively rare. The following selection indicates the diversity of topics treated when growth of tropical forests is studied. Singh (1989) discusses the various analytical techniques for determining increment. Lieberman et al. (1988) describe growth in height in the palm *Welfia georgii* in a tropical wet forest in Costa Rica, while Wan Razali (1988) illustrates the derivation of growth models for regenerated mixed tropical forests with an example from Malaysia. Lugo et al. (1988) investigated mean annual biomass increment in seven tropical plantation species. Rai (1975, 1978, 1983) studied basal area, volume, and diameter increment in various species in India, while Rai and Sarma (1987) present the periodic annual increment for different DBH classes and the mean annual increment calculated from DBH measurements made over 5–21 years at five locations in India. Peralta et al. (1987) present the results of observations on permanent plots in Costa Rica. Uriarte et al. (2004) present a likelihood-based regression method that was developed to analyze the effects of neighborhood competitive interactions and hurricane damage on tree growth and survival.

2.3.13 Density

The density of stocking is one of the factors included in the basic definition of forest; hence, "forest," as a form of land use can only be distinguished from treed vegetation when the trees occupy a minimum area or when their density exceeds a minimum standard. Density may be expressed in terms of stem count per unit area, volume per unit area, or biomass per unit area. One important attribute used in ecological studies is crown density. Crown density is the proportion of ground occupied by an (orthogonal) projection of the crowns onto a flat terrain surface (Whitmore 1989). In exploited areas, the crown density may provide useful information about harvest practices. As such it is a informative attribute to consider in many forest inventories. Aerial photographs constitute a convenient medium for the provision of crown density estimates.

In undisturbed tropical forests, the trees tend to form a closed canopy. In these cases the canopy density would be 1.0. Consequently, stem count per unit area rather than crown density provides a more informative estimate of density.

Sampling in Forest Surveys

3

3.1 Introduction

Inductive reasoning based on assessing a part of a whole is very much part of our daily life. For example, when cooking a pot of spaghetti we infer the *al* dente quality of all noodles by checking just a few. Where "the whole" is diverse, complex, and extensive it might be risky to draw conclusions from only one or a few instances. Sampling is an example of inductive logic by which conclusions are inferred on the basis of a limited number of instances. A sample is a subset of a population, which itself is the entire set of elements for which estimates about specific characteristics are to be obtained.

In the context of forest resource assessments the collection of information by means of a complete enumeration is only feasible in exceptional situations. An alternative to complete enumeration is a sample survey, which serves as the basis for estimates or inference for the underlying population. The process of selecting a sample from a population is called sampling.

The first part of this chapter presents some basic terms and concepts, while the second part describes some sampling procedures important for forest resource assessments. For further reading the textbooks of Kish (1965), Cochran (1977), Sukhatme et al. (1984), de Vries (1986), Särndal et al. (1992), Thompson (1992), Schreuder et al. (1993), and Shiver and Borders (1996) are recommended.

3.2 Basic Concepts

3.2.1 Population, Samples, and Estimators

A population comprises all elements from which the sample is to be taken. It may be defined very simply, for instance, the trees of a forest stand or the participants of a workshop. The definition of the population must be unique and allow an operational and comprehensible decision on whether a questionable element belongs to the population or not.

The population from which the sample is taken is termed the sampled population and must match the target population for which estimates are desired. Only then can representative conclusions for a target population be drawn. The population can be either infinite or finite depending on the definition. A population defined as the forest in a given region may comprise an infinite number of spatial locations but only a finite number of trees. We also need a temporal definition of a population. Few populations remain constant over time; most undergo changes due to birth, mortality, emigration and immigration processes. In the example before, the number of trees in a forest is likely to change over time. When a population is finite and countable it is customary to denote the population size by N, where N is a positive integer. For example, for a population composed of a single forest stand with 200 trees N=200.

A sample consists of a number of sampling units (or simply units) selected from the population by some design. The population is also uniquely defined in terms of these units as the union of all possible samples of such units. Sample units can be either unique discrete nonoverlapping units or arbitrarily sized units located at random in the population (Williams and Eriksson 2002). In the former case we view the population as a finite set of unique units that completely tesselate the population. The tesselated paradigm ensures that every population element can be clearly allocated to one unique unit. Examples of sample units are single trees, sample plots, or districts. In the second case we view the population as composed of an infinite set of possible locations for our sampling units. A sample location provides attribute information that is representative of the sampled locations.

For finite countable populations the N individual units of a population are identifiable, if they can be uniquely labeled from 1 to N and the label of each unit is known (Schreuder et al. 1993). While it may be relatively easy to identify and label N trees in a forest stand, the issue of identifiability quickly becomes an insurmountable logistic obstacle when the population of the trees in a large forested area. In extensive forest surveys the construction of an exhaustive list of sampling units, called the *sampling frame*, is often one of the major practical problems (Särndal et al. 1992). Without a complete sampling frame one must adopt the point paradigm for a definition of the population. Populations defined by the point paradigm are often less than intuitively clear.

A forest inventory sampling frame is often assembled from a unique description of what qualifies as forest (i.e., the forest area definition). Forest area definitions utilize quantitative criteria such as crown density, minimum patch size, or minimum patch width to facilitate a forest/nonforest decision in order to construct the sampling frame. Care must be given to the definitions to ensure that the qualifying population is indeed the population of interest. Even minor changes in a definition may lead to substantial and often surprising changes to the qualifying population. The following example illustrates this phenomenon. A survey of two treed areas with 25 sample plot locations in each area (circles) arranged in a regular grid found that 13 sample locations were covered by a tree canopy in the first area while only four were covered in the second area (Fig. 3.1). Accordingly the sample-based estimates of the crown density in each area are 13/25 (52%) and 4/25 (16%), respectively. A forest qualifying threshold of 10% would result in a classification of both areas as forest. In contrast, only the first area would qualify as forest if the defining threshold was raised to 30%.

Each sample unit and population element possess a series of attributes of interest. The attribute may be intrinsic or derived. The cellulose content would be an example of an intrinsic property. A market value, on the other hand, is an example of a derived attribute – an attribute that can only be obtained via another attribute or several other attributes. Natural resource attribute values often exhibit a considerable variation between units (elements). An attribute may or may not be measurable or quantifiable. Instead we may define countable or measurable variables linked to the attribute of interest. Knowledge of and information on these variables are used for inference about the attributes of interest. For example, the attribute of interest may simply be the "trees" in a forest with the element attribute being "tree." To characterize this attribute beyond a mere count of trees we may choose to measure variables such as tree height and stem diameter at breast height, identify the tree species, and assess crown form. The number of variables to include will depend on what is needed to be known about the attribute of interest.

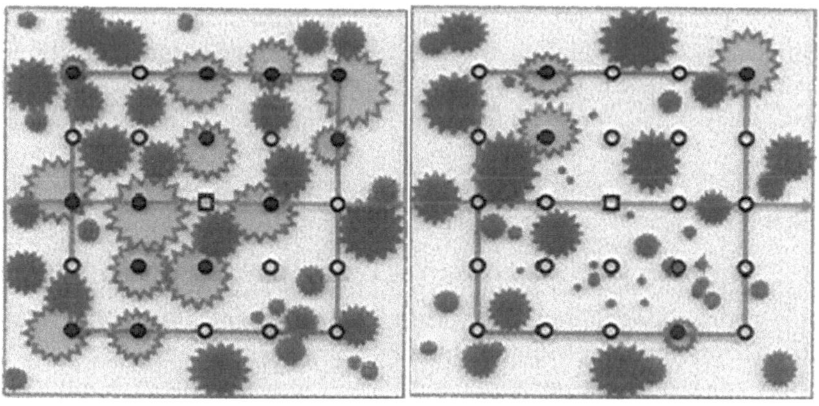

Fig.3.1. Two treed areas with 25 sample locations on a regular grid (*circles*). Sample locations under a tree canopy cover are indicated by *filled circles* (13 in the *leftmost area* and four in the *rightmost area*). (Courtesy of Markus Keller, WSL, Switzerland)

In order to characterize the attributes (variables) of a population certain parameters are employed. When the parameters relate to all units/elements in a population they are called population parameters. The aim of surveys is to estimate population parameters or the functions of one or more of them. The value of a parameter derived from a sample is called an estimate. The formula for calculating this estimate is called an estimator. Parameters include aggregates (e.g., total volume, total area) and averages (e.g., mean tree height) of values associated with each population element or unit. Ratios of pairs of population parameters (e.g., volume per hectare as the ratio of total volume and total area), counts (e.g., number of trees), and proportions (e.g., proportion of forest area with a specific attribute) are further examples of population parameters.

Estimates of population parameters are obtained via estimators. The estimators treated in this book are either design-based or model-based estimators (Särndal et al. 1992; Gregoire 1998; Little 2004). The underlying principle behind a design-based estimator is that the population from which samples are taken is considered as a fixed entity. The random selection of units/elements to include in the sample is the only source of stochastic variation (sampling error). Model-based estimators are based on the assumption that the population of interest is generated by some process, a process that depends on a set of parameters to be estimated from the sample. The actual population to be surveyed is but one random realization from this process. We cannot observe the assumed process, but our sample allows us to estimate the parameters of the assumed model. Population estimates are obtained by combining the sample estimates with the model-based predictions for the nonsampled part of the population (Valliant et al. 2000). Thus, the issue of model bias looms large over these estimators and convincing support for the chosen model must come from previous surveys or from the sample data themselves.

Parameters of a population are designated by capital letters from the Latin and Greek alphabets. Lowercase letters are reserved for individual unit/element values.

3.2.2 Probability Sampling

The general principle of sampling (Fig. 3.2) is to select a subset of units (i.e., a sample) from a population, to measure this subset intensively, and to draw inference from the sample to the entire population.

There exist countless approaches to select a sample from a population. Intuitively it is obvious that the sample should represent the entire population. The term representative as used in everyday language suggests that the sample should be a tailgate miniature or a scaled-down replica of the population.

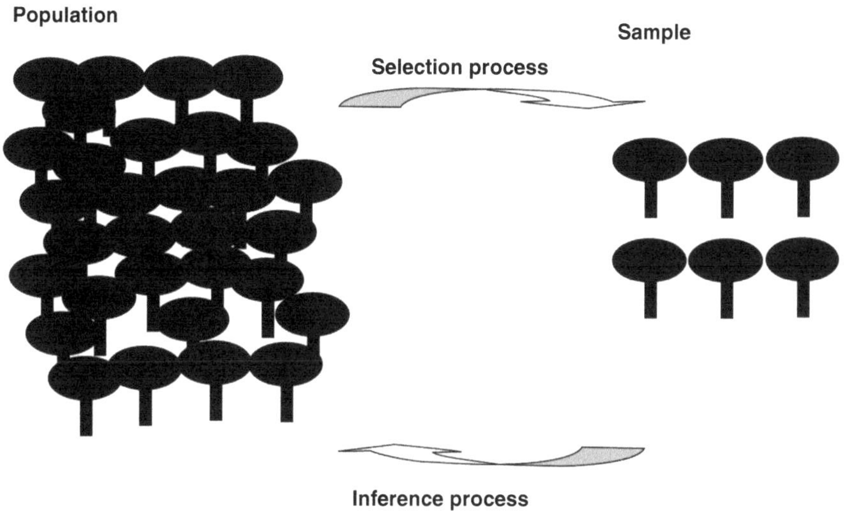

Fig. 3.2. The principle of sampling

Unless each unit in the population has an equal chance of being selected, this intuitive concept is inappropriate. Many widely used sampling methods assign varying selection probabilities to the individual units; the chance of being selected can be assigned with respect to a known attribute or quantitative measure of the units. A selection method complies with the conditions of probability sampling when a procedure is followed that ensures that each unit in the population has exactly the predetermined probability of being selected for the sample. The selection probabilities are used in the estimators of parameters of interest and in estimators of sampling variance (Thompson 1992). The choice of selection probabilities and estimators is called a sampling strategy (Särndal et al. 1992).

Given a specific population of N units the set of all possible distinct samples, $s_1, s_2, \ldots, s_v$ can be defined and the units making up each sample can be designated (two samples are distinct if their union minus their intersection is not empty). If n units out of N are to be selected without replacement (a unit can only be selected once) there are $\binom{N}{n} = \frac{N!}{n!\,(N-n)!}$ possible distinct samples (Levy and Lemeshow 1991). Note, $n! = n(n-1)(n-2)\ldots(1)$ and $0!=1$. For example, if a population contains 200 elements and we wish to take a sample of 25 elements, then the total number of possible samples is approximately 4.5×10^{31} (exact number is 45,217,131,606,152,448,808,778,187,283,008), quite an astronomical figure. For each possible sample, say s_i, a selection probability π_{si} can be specified. The sum of these selection probabilities over all possible samples is 1. A selection probability tells us how frequent a particular sample s_i will be selected. We shall see later how we use these selection probabilities to derive unbiased estimators of population attributes and parameters. Further details on the use and significance of selection probabilities are in, for example, Brewer and Hanif (1983). The term probability sampling refers to sampling with a known selection probability of all sample units making up the population. In practice it is hardly possible to list all possible samples s_i and their associated selection probabilities π_{si}. For sample estimators based on a probability sample it is sufficient to know how to assign selection probabilities to the sampled units. We denote the actual sample by the symbol *s*, where *s* is one of the possible distinct samples. Estimates of population totals and averages are normally obtained by an *expansion* of individual sample attributes/variables to an estimate(s) of the population total (Levy and Lemeshow 1991). Let y_i be the attribute/variable obtained from sample *i* with selection probability π_i. The expanded estimate is y_i/π_i. Since the probability of obtaining this expanded estimate is π_i the expected value of the *i*th expanded value is simply y_i. Thus, expansion estimators of totals and averages are unbiased (if sampling was exhaustive the total, or average, would be equal to the true total, or average).

Probability sampling methods employ a thorough selection process that ensures that each unit in the population to be sampled has exactly its designated probability of being selected. In practice that means that any unit being selected as part of the sample has to be accepted, irrespective of any problems or difficulties in assessing it. A common problem in forest surveys is the accessibility of terrain. Figure 3.3 shows a forest patch that is inaccessible by a normally equipped field crew. Where these areas are not excluded from the sampling frame (i.e., inventory results refer to accessible forests only) they need to be surveyed when selected under a probabilistic sampling scheme. Often it is cheaper, quicker, or more comfortable to omit those units. This leads to in the problem that there is no longer control over the probability with which the units comprising the population are selected. Some units have little or no chance of being selected or are selected with uncontrolled or subjective probabilities. Such samples are called nonprobability samples.

3.2.3 Definitions and Notations

In sample surveys, data on one or more variables/attributes are collected for each selected unit of the population. Values reflecting a variable of a unit/element forming a finite population are defined by $y_i (i = 1,\ldots,N)$ and N is the number of population units/elements. Through sampling, a sample s composed

(a)

Fig. 3.3. Inaccessible forest areas in **a** Switzerland and **b** Germany

of n units is selected from the population units/elements. Variable/attribute values associated with a sampled unit are denoted by $y_{i|i \ni s}$, where $i \ni s$ means that the ith population value is sampled. To simplify notation we will drop $i \ni s$ when warranted. The ratio n/N specifies the proportion of units selected from the N population units and is termed the sampling fraction; the symbol f is often used to denote this fraction. In infinite populations the sample fraction is nil by definition.

(b)

Fig. 3.3. (Continued)

Sampled units are used to estimate parameters for the population. The four most important population parameters are:

1. The mean $\overline{Y}$ (e.g., the mean standing reserve in the inventory area)
2. The total Y (e.g., the total standing reserve in the inventory area)
3. The ratio R between two means or totals (e.g., volume per hectare)
4. The proportion P of units with a specific attribute (e.g., proportion of units with a given tree species).

The sample provides us with estimates of population parameters. Estimates are distinguished from their true population values by adding a caret above the associated symbol. The relationship between population values and sample estimators is given in Table 3.1 for the most common population parameters.

3.2.4 Properties of Estimators

Whereas the term "estimate" signifies the value of a parameter, an "estimator" denotes a rule according to which a parameter is derived from the sample data. Estimators based on sampling surveys must display certain qualities.

Table 3.1 Population and sample estimators of common parameters

Parameter	Population value	Sample estimator
Mean	$\bar{Y}$	$\hat{\bar{Y}} = \frac{1}{n}\sum_{i \ni s} y_i$
Total[a]	Y	$\hat{Y} = N \times \hat{\bar{Y}}$
Ratio[b]	R	$\hat{R} = \frac{\bar{y}}{\bar{x}} = \frac{\hat{\bar{Y}}}{\hat{\bar{X}}} = \frac{\sum_{i \ni s} y_i}{\sum_{i \ni s} x_i}$
Proportion	P	$\hat{P} = \frac{1}{n} \times \sum_{i \ni s} \delta_i$, where $\delta_i = \begin{cases} 1 \textit{ if the ith unit has the attribute} \\ 0 \textit{ otherwise} \end{cases}$

[a] finite populations only
[b] of means viz. totals

An estimator is called a "consistent estimator" if the larger the sample size n, the closer the estimate, say $\hat{Y}$, is to the true population parameter value Y. When the expected value of the estimator $E(\hat{Y})$ equals the true parameter Y, the estimator is unbiased. Estimators not meeting this condition are termed "*biased estimators*." Bias is defined as the difference between the expected value of an estimator of a population parameter and the true value of this parameter. For the estimator $\hat{Y}$ the bias is given by $bias(\hat{Y}) = E(\hat{Y}) - Y$. An estimator that is unbiased for a given sample design (if correctly implemented) is said to be design-unbiased. Model-based estimators are said to be model-unbiased if the model is true and the expectations of model predictions equal the expectations of the population units for which predictions are made.

Unbiasedness is a desirable property of an estimator. Important is also the accuracy of an estimator. In repeated sampling of a single population using the same sampling design the estimates obtained from an estimator will vary between samples. Accuracy refers to the size of the deviations of the sample estimates from their true value (Cochran 1977). Normally, though, we would not know the true value. Different estimators of the same population parameter can have different accuracy. Normally, though, we do not know the true parameter value, which precludes a correct assessment of accuracy. We can, however, estimate the precision of an estimator. Precision is a measure of the deviations of individual sample estimates in repeat sampling from their mean (average). Precise estimators produce estimates that cluster tightly around their average. That means that we can have a high degree of confidence in the value of a sample-based estimate. If we were to repeat the sampling process we would likely obtain a result quite similar to the one we already have. Precision is commonly quantified as the inverse of the estimated variance of an estimate (Cochran 1977). To assess precision of an estimate we need an estimate of its sampling variance (var). Estimators of sampling variance have been developed for all practically relevant sampling designs and population parameters including model-based predictions.

Bias and precision of estimators are both important attributes to consider in planning a survey. Estimators for various design alternatives (viz., model alternatives) may produce different amounts of bias and vary in precision. Actually, competing estimators often display a trade-off between bias and precision (Congdon 2001). The usual criterion for comparing two estimators is the mean square error (MSE). The MSE of an estimator, say $\hat{Y}$, is defined as

$$\mathrm{MSE}\left(\hat{Y}\right) = \mathrm{var}\left(\hat{Y}\right) + \mathrm{bias}\left(\hat{Y}\right)^2$$

Note, the true variance of a sample-based estimate and the true bias will never be known in practical applications. Instead we use available estimators of variance and approximations to the bias (Särndal et al. 1992). A survey analyst normally prefers an estimator with the lowest expected MSE.

Robust estimators are also desirable (Staudte and Sheather 1990). Robust estimators are less sensitive to a few outlying sample values and to violations of assumptions than are nonrobust estimators. The ideal estimator is unbiased, highly precise, and robust. It is a challenge for the survey analyst to optimize the sampling strategy, i.e., the choice of sample design and estimators.

3.3 Survey Design and Sampling Design

In planning a forest inventory, a range of methodological issues have to be considered. What data are to be collected? For which units of the population should they be collected? Which system of nomenclature (including measurement rules or definitions for each attribute to be assessed) is to be applied? How should data be captured and processed in order to derive the requested information? Additionally operational, organizational, and administrative issues have to be resolved. It is the objective of the *survey design* to settle these issues with respect to the available budget and the information needs. Ideally this process could be formulated as an optimization problem. What is the best design under a fixed set of resources and precision target? Examples of how this problem is resolved in various settings are found in, for example, Mandallaz 2001, Brus et al. 2002, and Arner et al. 2004). Historic material in the form of data from previous or related surveys provides a good source of prior information of what the intended sampling may produce.

A second set of methodological issues deals with the question of how to select the sample from the population (i.e., the sample selection) and how to derive suitable estimates from the data collected (i.e., the estimation procedures). Based on sampling theory, a variety of techniques have been developed for sample selection and estimation. It is the objective of the *sampling design* to select the most appropriate sampling methods in light of a set of overarching objectives and constraints. The sampling design itself is part of the survey design.

Sampling designs can be divided into two main groups depending on whether data and information outside the variables of primary interest (auxiliary information) are used to shape the design and/or the estimators:

1. Sampling designs without auxiliary information
2. Sampling designs with auxiliary information

In sampling designs without auxiliary information, only the observations on the variables of interest are used to derive the parameters.

The populations we deal with in forest inventory can be described by a long list of attributes, some associated with the trees, others with the environment in which they grow or have grown. Information on several of these attributes may be available to the survey analyst at the time a survey is planned. Available information that is in some way associated with the attribute of interest for the survey can often be incorporated in the design for stratification or assignment of selection probabilities and in the estimators in the form of predictors. Common examples of auxiliary information in forest inventory include aerial photography, satellite imagery, and various thematic maps. As a rule, sampling designs/estimators that exploit auxiliary information optimally are more efficient than design/estimators that ignore this information. The most common sampling designs in the two groups are listed in Fig. 3.4.

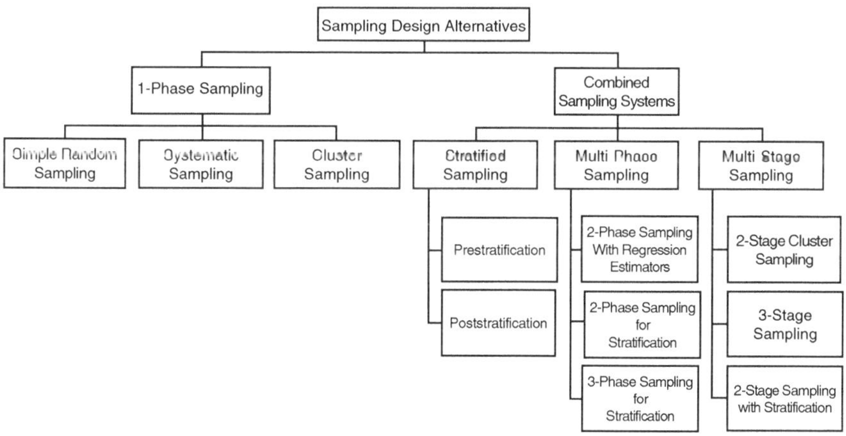

Fig. 3.4. Sampling design alternatives. The *boxes* on the *rightmost branch* list designs that incorporate auxiliary information in the design and/or the estimation phase. The *boxes* on the *leftmost branch* list designs that do not incorporate auxiliary information. (Courtesy of Pelz and Cunia 1985)

The following sections describe the major sampling designs currently used in forest inventories. The designs can apply to any population large or small, spatially contiguous, or spatially dispersed. Large forest enterprises may conduct several different inventories, each using a different design and estimators. Results from an inventory or several inventories may be used in postinventory analyses to provide estimators for specific subpopulations, updating, and forest modeling (van Deusen 1996; McRoberts et al. 2002; Tuominen et al. 2003). The utility of inventory results for these additional uses should be factored into the sampling strategy.

3.3.1
Simple Random Sampling

We begin the detailing of common inventory sampling designs with simple random sampling (SRS) not because it is particular widely used in its simplest form but because a presentation of SRS and its estimators will make it easier to comprehend and appreciate more complex designs and their estimators.

In finite-population SRS n units are selected at random from the N units comprising the entire population. The selections are done in such a way that all possible distinct samples of size n have the same selection probability. Since there are $\binom{N}{n} = \frac{N!}{n!(N-N)!}$ possible samples the selection probability of each (sample) becomes $1\Big/\binom{N}{n}$. The probability that a single unit/element is in the sample is n/N. The principle of equal sample inclusion probabilities extends naturally to infinite populations but we have, of course, no means of calculating these probabilities.

The simplest way of selecting units is to number all the elements of the population, to choose n numbers randomly, and to include the elements with the corresponding numbers in the sample. Here, however, it must be ensured that all the elements are listed – a circumstance that practically never occurs in forest inventories. In forest inventories using SRS, aerial photography, satellite imagery, or a map is needed to establish a frame from which the sample is to be taken. X/Y coordinates are randomly chosen and the survey is then conducted at the corresponding points. These coordinates may designate the centers of fixed-area plots, point samples, stem distance methods, or a fixed number of trees located nearest to the randomly selected coordinate whether it is sampled or not (Sect. 3.4).

There are two types of SRS: SRS with replacement and SRS without replacement. In SRS with replacement the same element may be drawn twice or more often and thus the elements are given the same selection probability at every draw, i.e., n/N for each draw in a finite population. In SRS without replacement a selected unit/element is removed from the sampling frame before the next unit/element is selected. Thus, for a distinct element remaining in the sampling frame after completion of k draws the probability of selection at the $(k+1)$th draw is $(n-k)/N$ and so on for $k = 0, \ldots, n-1$. Whenever a unit/element is selected more than once the sample will contain "copies" of the sample record associated with the unit. Copies of a sample record provide no new information about the population; hence, sampling with replacement is considered potentially wasteful and less efficient. The rationale for detailing with-replacement sampling is that some variance estimators can only be derived if we assume sampling with replacement (Brewer and Hanif 1983). These with-replacement estimators are then used as approximations to an estimator for sampling without replacement. When the sample sizes are small compared to the size of the population of interest the differences will often be trivial. In the following only SRS without replacement is considered.

3.3.1.1 Estimating the Population Mean

The population mean is given by

$$\bar{Y} = \frac{1}{N}\sum_{i=1}^{N} y_i.$$

Under SRS the sample mean $\hat{\bar{Y}}$ is an unbiased estimator of the population mean:

$$\hat{\bar{Y}} = \frac{1}{n}\sum_{i=1}^{n} y_i.$$

3.3.1.2 Sampling Error

In natural resource surveys the variable values associated with a sampling unit vary from unit to unit. The degree of variability depends on the variable and the population in question. Variability is thus an essential characteristic of survey sampling. The standard error and its square, the variance, are useful measures to quantify the variability or dispersion of values for individual population units about their mean. The variance of individual unit values of a variable, say y_i, is defined by

$$\operatorname{var}(y_i) = \frac{\sum_{i=1}^{N}\left(y_i - \bar{Y}\right)^2}{N-1}.$$

The standard deviation of the population attribute is the square root of the variance:

$$SD(y_i) = \sqrt{var(y_i)}.$$

Under SRS, a sample-based design-unbiased estimator of the population variance is

$$\hat{\text{var}}(y_i) = \frac{\sum_{i=1}^{n} (y_i - \bar{y})^2}{n-1}.$$

A sample-based estimator of the population standard deviation of y_i, $\hat{SD}(y_i)$, is obtained by taking the square root of $\hat{var}(y_i)$.

It is often convenient to remove the effect of the measurement scale from estimators of variability. Variances expressed in relative terms with respect to the mean of the variable to which they refer are scale-invariant. The coefficient of variation is a popular scale-invariant measure of variation:

$$\text{CV}(y_i) = \frac{SD(y_i)}{\bar{Y}}.$$

Sample-based estimators of the coefficients of variation are obtained by replacing the population quantities by their respective estimators.

The interunit variation means that the sample mean based on a sample of size n will also vary from one sample to the next if we repeat the sampling. Let us assume a population of size N, from which we take five samples each of size n. For each sample we calculate, say, the mean $\hat{\bar{Y}}_j$, $j=1, \dots ,5$. Obviously, the five means will vary. Consequently any sample estimate of a parameter is subject to an error due to the randomness of the sample. This error is termed the standard error of sampling, or simply the standard error.

The larger the variability of the units, the larger is the standard error in separate estimates. Luckily, it is not necessary to take several samples from the same population in order to determine the standard error. We can make use of the *central limit theorem* which says that the mean of n randomly selected population values of a variable is asymptotically $(n, N \rightarrow \infty)$ normally distributed with a variance that is the variance of the random variable divided by n (Casella and Berger 2002). With SRS an estimator of the sampling variance of an estimate, say, $\hat{\bar{Y}}$ is

$$\hat{\text{var}}\left(\hat{\bar{Y}}\right) = \left(\frac{N-n}{N}\right)\frac{\hat{\text{var}}(y_i)}{n}.$$

An unbiased sample-based estimator is obtained by replacing the population variance var (y_i) by a sample-based estimate of this variance. The square root of the sampling variance is the standard error of sampling:

$$\widehat{\text{SE}}\left(\hat{\bar{Y}}\right) = \sqrt{\hat{\text{var}}\left(\hat{\bar{Y}}\right)}.$$

The quantity $(N{-}n)/n$ accounts for the changes in selection probability for sampling without replacement and is termed the *finite-population correction factor*. If the sampling fraction n/N is small, the finite-population correction factor will be close to 1. We get $\hat{\text{var}}\left(\hat{\bar{Y}}\right) \simeq \frac{1}{n}\,\hat{\text{var}}\,(y_i)$ when $n/N \simeq 1$. Omitting the finite-population correction factor results in a slight overestimation of the true variance. For practical purposes, the finite-population correction factor needs not be considered if the sampling fraction is smaller than 5%. Note the previous formula extends naturally to population parameters other than the one chosen here.

3.3.1.3
Confidence Intervals for Sample Estimates

The concept of standard error is often not intuitively clear to many users of inventory data and they may find it difficult to assess the significance of a standard error and interpret it correctly. Estimates arising from a sample-based inventory ought to include a measure of uncertainty of the estimates. Confidence intervals for sample estimates provide an intuitive easily understood measure of the impact of a standard error.

A confidence interval for an estimate gives the range within which one can expect the true population parameter to be located. The bounds of the confidence interval are termed confidence limits. The interval should have the property that the probability of the true value being located within the confidence limits is known, say $1-\alpha$. The quantity $1-\alpha$ is called the confidence coefficient and the interval is called the $100(1-\alpha)\%$ standard interval. Typically the 95% confidence interval is presented (α=0.05). The 95% confidence interval covers for 95 out of 100 replicate samples of size n the true value of the population. Conversely, there is a 5% chance that the true value is outside this interval. A specific sample-based estimate of the confidence interval either includes the true value or does not.

The distribution of sample estimates under repeat sampling is usually assumed to be normal (invoking *the central limit theorem*) with a mean equal to the estimate obtained and a variance equal to the estimated variance divided by the sample size. Under this assumption and continuing with the previous example with a population mean as the parameter of interest, the lower $\left(\hat{\bar{Y}}_{\mathrm{L}}\right)$ and upper $\left(\hat{\bar{Y}}_{\mathrm{U}}\right)$ limits of the $100(1-\alpha)\%$ confidence interval for a sample-based estimate are

$$\hat{\bar{Y}}_{\mathrm{L}} = \hat{\bar{Y}} - t_{n-1,1-\alpha/2} \times \widehat{\mathrm{SE}}\left(\hat{\bar{Y}}\right)$$

and

$$\hat{\bar{Y}}_{\mathrm{U}} = \hat{\bar{Y}} + t_{n-1,1-\alpha/2} \times \widehat{\mathrm{SE}}\left(\hat{\bar{Y}}\right).$$

$t_{n-1,q}$ is the qth quantile of Student's t distribution (Casella and Berger 2002). Values of t for n=50 and some common values of α are given in Table 3.2. For large sample sizes, say larger than 50, the quantiles of the t distribution are very close to the corresponding quantiles of a standard normal distribution z_q.

Table 3.2. Values of t for some common values of α (n=50)

	Confidence coefficient (%)				
	50	80	90	95	99
α	0.50	0.20	0.10	0.05	0.01
$t_{49,\,1-\alpha/2}$	0.68	1.30	1.68	2.01	2.68
$t_{49,\,\alpha/2}$	−0.68	−1.30	−1.68	−2.01	−2.68

It is customary to use the standard normal quantile for large n. For large n and 95% confidence probability, t is approximately 2, and the confidence interval is called the 95% confidence interval. For a 68% confidence probability, t is approximately 1.

A word of caution is appropriate. Large sample sizes are usually needed to assure that a $100(1-\alpha)\%$ confidence interval has the desired properties. We say that the nominal coverage is asymptotically correct. Distributions of sample statistics obtained from small samples may be skewed and not well described be either Student's t distribution or the normal distribution. Confidence intervals obtained from such distributions may be liberal (they cover the true parameter below the nominal rate) or conservative (they cover the true parameters above the nominal rate). Various resampling schemes and improved approximations of the sampling distribution have been suggested to remedy problems of this nature (Davison and Hinkley 1988; Barndorff-Nielsen and Cox 1989; Shao 2003).

3.3.1.4 Estimating the Population Total

The population total, Y, is obtained by multiplying the population mean $\overline{Y}$ by the number of elements in the population:

$$Y = \sum_{i=1}^{N} y_i = N \times \overline{Y}.$$

An unbiased estimator of the population total $\hat{Y}$ is

$$\hat{Y} = \frac{N}{n} \sum_{i=1}^{n} y_i = N \times \overline{Y}.$$

As Y is N times the estimator $\bar{Y}$, an unbiased estimator of the variance of $\hat{Y}$, $\hat{\mathrm{var}}\left(\hat{Y}\right)$, is

$$\hat{\mathrm{var}}\left(\hat{Y}\right) = N^2 \times \hat{\mathrm{var}}\left(\hat{\overline{Y}}\right).$$

The standard error of $\hat{Y}$ and the upper and lower confidence limits are

$$\widehat{\mathrm{SE}}\left(\hat{Y}\right) = \sqrt{\hat{\mathrm{var}}\left(\hat{Y}\right)} = N \times \sqrt{\hat{\mathrm{var}}\left(\hat{\overline{Y}}\right)},$$

$$\hat{Y}_L = \hat{Y} - t_{n-1,1-\alpha/2} \times \widehat{SE}\left(\hat{Y}\right),$$

and

$$\hat{Y}_U = \hat{Y} + t_{n-1,1-\alpha/2} \times \widehat{SE}\left(\hat{Y}\right).$$

Note, in the above we assumed N is known without error. At times we do not know N but we may have an estimate of N, an estimate with a sampling error. To account for the added uncertainty, we must add an estimate of the error stemming from imperfect knowledge of N. The general technique for obtaining a variance estimator of an estimate that depends on several random variables is based on a Taylor series approximation (Kotz and Johnson 1988). The technique also goes under the name of the delta technique (Kendall et al. 1983). Specifically, let Y be a function f of a set of predictor variables, i.e., $Y = f(X_1, X_2,\ldots,X_v)$. A first-order Taylor series approximation of the variance of, say, a mean $\overline{Y}$ is

$$\operatorname{var}\left(\bar{Y}\right) \simeq \sum_{j=1}^{v} \left(\frac{df}{dX_j}\right)^2_{X_j = \overline{X}_j} \operatorname{var}\left(\overline{X}_j\right)$$

when the predictors $X_1,X_2,\ldots,X_v$ are independent (no covariance). Approximations to variances of a sum, a ratio, or a proportion are found by straightforward extensions. At times the predictors $X_1,X_2,\ldots,X_v$ will not be independent of each other. Under these circumstances the covariance of all possible pairs of predictors must be added to right-hand side of the previous equation.

3.3.1.5 Determining Sample Size

The sufficient SRS sample size is determined by the inherent (natural) variability of the attribute values of the population, the degree of precision required for the results, and the confidence coefficient we wish to apply to confidence intervals of sample estimates. In SRS the sample size needed to satisfy a desired precision (E) with a confidence coefficient of $100(1-\alpha)\%$ is calculated, for say a mean, according to

$$n = \frac{t^2_{n,1-\alpha/2} \times \operatorname{var}(y)}{E^2}.$$

In practice we would normally not know the variance of the variable of interest. It must be replaced by an estimate derived from historic information, related surveys, or simply from a qualified expert guess. Prudence dictates a conservative guess.

As the t value depends on the degrees of freedom, i.e., the sample size, calculations for small sample sizes must be done iteratively. During each iteration, the n value determined in a previous iteration is used to determine the appropriate t value. Iterations are stopped when the upwardly rounded value of n no longer changes. Prodan (1965) suggested an alternative estimator of the variance derived from knowledge about the range of values y_i in the population (Beyer 1968). Provided that an estimate of the maximum and minimum values of y_i can be obtained, an approximate estimator of the variance is

$$var(Y) \simeq \left(\frac{\max(y_i) - \min(y_i)}{4} \right)^2 .$$

A usually very conservative "guess" is obtained by assuming that y_i is uniformly distributed between the maximum and the minimum. If this is indeed the case, the variance of y_i is $\left[\max(y_i) - \min(y_i)\right]^2 / 12$ (Snedecor and Cochran 1971).

3.3.1.6
Sampling for Proportions and Percentages

Some forest inventory results may be presented in terms of counts, proportions, or percentages. Examples would be tree count, proportion of burned forest, and percentage of teak volume in a forest. Counts, proportions, and percentages usually involve elements/units belonging to a defined class or exhibiting a given characteristic. Additional examples might include ownership types, proportion of trees with stem damage, or the percentage of the forest area that is difficult to access. When there are more than two mutually exclusive classes for an attribute/variable, the term "multinomial variable" is used. The results obtained on the basis of multinomial variables are presented as classifications on a nominal (e.g., tree species, soil type, access) or on an ordinal (e.g., stand layer, timber quality, burned status) scale. Multinomial variables are frequently analyzed and presented on the basis of a sequence of proportions or counts.

The following discussion is limited to the relatively simple case of SRS where each sampled unit exhibits a binary class value. There is extensive literature on the more complex analysis of proportions and percentages in other sampling designs (Kish 1965; Cochran 1977; Sukhatme et al. 1984; Agresti 1992; Lloyd 1999).

Binary variables assume one of two values, typically the value y_i=1 when the element/unit belongs to a given class and y_i=0 otherwise. The number of population elements in the class assigned a value of 1 is given by

$$Y = \sum_{i=1}^{N} y_i = N \times P,$$

where P is the proportion of elements/units in the population with a class value of 1. The proportion of elements that do not have the class attribute with the value of 1 is of course 1–P.

The population variance of y_i is

$$\mathrm{var}(y_i) = \frac{1}{N-1} \times \sum_{i=1}^{N} (y_i - P)^2 = \frac{N}{N-1} \times P \times (1-P).$$

Sample-based estimators of the population proportion and its variance are obtained from the previous equations after substituting n for N and adding carets to distinguish them from the true population values. When n is very large a good approximation to the standard error of $\hat{P}$ is given by

$$\widehat{\mathrm{SE}}(\hat{P}) \simeq \sqrt{\frac{\hat{P} \times (1-\hat{P})}{n-1}}.$$

The confidence interval for $\hat{P}$ can be calculated from

$$\hat{P} \pm \left(\left| t_{n-1,\alpha/2} \right| \times \sqrt{\frac{\hat{P}(1-\hat{P})}{n-1}} + \frac{1}{2n} \right),$$

where the last term is a correction for continuity, which is necessary as P is not a continuous variable. According to Cochran (1977) the omission of the continuity correction leads to confidence intervals which are too small (liberal). A detailed discussion of alternatives for the construction of confidence intervals for P is given by, for example, in Dees (1988) and Burk (1991). The standard normal approximation for binary confidence intervals is sufficient in situations, where $\hat{P}$ and n are not too small. Cochran (1977, p. 58) gives the smallest values for $n \times \hat{P}$ to which the normal approximation can still be applied. For example, for P=0.2 the smallest n is 200, while at P= 0.1 n should be larger than 600.

If more than one observation is made on the binary trait in every sampled unit the estimation of proportions has to be modified since the number of observations per sample unit can vary. Fixed-area plots are a typical example. The number of, say, trees per plot varies naturally between plots, The estimators of P and the variance of P are now given by

$$\hat{P} = \frac{\sum_{i}^{n} \sum_{j=1}^{m_i} y_{ij}}{\sum_{i=1}^{n} m_i} = \frac{\sum_{i=1}^{n} a_i}{\sum_{i=1}^{n} m_i},$$

where subscript i refers to sample unit and subscript j to the population elements in the ith sample unit. There are m_i population elements in sample unit i of which a_i elements belong to the binary class given a value of 1. The corresponding estimator of variance becomes

$$\hat{\mathrm{var}}(\hat{P}) = \frac{1}{\overline{m}^2} \times \frac{\sum_i a_i^2 - 2\hat{P} \times \sum_i a_i m_i + \hat{P}^2 \sum_i m_i^2}{n-1},$$

where $\overline{m}$ is the mean number of population elements per sample unit.

Selection of an equal number of trees per plot as done in nearest-neighbor (NN) methods (Pielou 1970) can be treated as a special case where m_i is constant. In this case a_i is the area occupied by the selected trees (de Vries 1986).

The estimation of proportions when the attribute/parameter of interest has more than two classes can be treated as a special case of a binary class problem. In brief, one obtains simultaneously an estimate of the proportion for each class as if it was a binary class. Specifically, if the kth class out of K is considered, then elements/units in class k are given a value of 1 and all other elements/units a value of 0. The variance for each class is calculated as before for a binary class. In balanced samples the sum of the estimated proportions for the K classes will sum to 1, as they should. However, sample imbalances may cause a violation of the sum-to-1 constraint. Formulae for postestimation calibration of proportions to meet a sum-to-1 constraints and formulae for variance and covariance estimation for proportions are found in, for example, Wu (2003); Angers (1989); Agresti and Caffo (2000); Sison and Glaz (1995).

The estimation of confidence intervals for K classes has to take into account that estimates of precision are simultaneously given for K classes. Doing this requires us to distribute the global significance level α across the K interval estimates. Multinomial confidence intervals have been described by Gold (1963), Quesenberry and Hurst (1964), and Goodman (1964, 1965). The methods presented are based on the normal approximation (Fienberg and Holland 1973; Angers 1989; Sison and Glaz 1995) and they differ with respect to the approach to calculate simultaneous probability estimates. We limit ourselves to the presentation of the Bonferroni method (Miller 1981), which can be used for the adjustment of significance levels in multiple significance tests. The Bonferroni method distributes the global significance level equally across the individual classes; hence, the individual confidence coefficients become $1-\alpha/K$. For a simultaneous confidence coefficient of 0.95 ($\alpha=0.05$) the Bonferroni-adjusted significance level for each of four classes is $\alpha_4=0.05/4=0.0125$. The t value associated with this simultaneous Bonferroni-type confidence coefficient is obtained from tables or from one of the many statistical software programs available today.

Estimates of P are of course restricted to the interval from 0 to 1. At times an estimate of the lower bound of a confidence limit will be negative or the upper limit could exceed 1. A logistic transformation of P to $\log\left(\frac{P}{1-P}\right)$ and an estimation of the confidence limits on this transformed scale and a subsequent back transformation to the original scale resolves this type of problem since the back transform of a logistic variable is always between 0 and 1 (Lloyd 1999). For example, if we have an estimate of 0.1 for P and the standard error of this estimate is 0.0948 with nine degrees of freedom ($n=10$), the limits of our 95% standard interval would be −0.086 and 0.286. The logistic transform of 0.1 is 2.197 and the standard error on the logistic scale is found by application of the previously mentioned delta technique to be $1/\sqrt{n \times \hat{P}(1-\hat{P})}$ or 1.054. The standard interval on the logistic scale is therefore (−4.263, −0.131). After a back transformation to the original scale the interval is (0.0139, 0.467).

3.3.1.7
Ratio Estimators

Ratio estimators are widely used in forest inventories. Any attribute related to an area such as, for example, the number of trees per hectare or the volume per hectare, is defined as a ratio of two different attributes. The population ratio R is obtained by dividing the population total of the attribute (total volume, total number of stems) in the numerator of the ratio by the population total of the attribute in the denominator of the ratio. A sample-based estimator of R is the ratio of the two sample estimates of population totals. This is a *ratio of means estimator*, which has a bias of the order of $1/n$. There is no unbiased sample-based estimator of R.

$$\hat{R}_{yx} = \frac{\sum y_i}{\sum x_i} = \frac{\hat{Y}}{\hat{X}},$$

where summation is over the sampled values. For sample sizes n over 30 the bias if often negligible, but skewed population distributions of Y and especially X can introduce a serious bias in a sample estimate (Hess and Bay 1997; Rao 1988). The variance of a ratio of means estimate is

$$\hat{\text{var}}\left(\hat{R}\right) = \frac{\hat{\text{var}}\left(y_i\right) + \hat{R} \times \hat{\text{var}}\left(x_i\right) - 2\hat{R} \times \hat{\text{cov}}\left(y_i, x_i\right)}{n \times \hat{X}^2},$$

where $\text{cov}\left(y_i, x_i\right)$ is the covariance between the two attributes/variables y and x. The covariance is estimated as

$$\hat{\text{cov}}\left(y_i, x_i\right) = \frac{\sum y_i x_i - n^{-1} \sum y_i \sum x_i}{n-1},$$

where summation is over the sampled values.

It is often possible to estimate a plot-level ratios R_i for each plot in a sample. For example, the number of trees per hectare in each plot. However, the estimation of the population ratio should not be based on these individual ratios because the mean of these individual ratios as an estimate of R has more bias than the ratio of means even if n is large (Cochran 1977). The exception is when $y_i = R \times x_i$ for all elements in the population. In this case the mean of Y_i/X_i is obviously R everywhere. Furthermore, the per unit ratio is often unstable and exhibits a large variance and a very skewed sampling distribution.

3.3.1.8
Advantages and Disadvantages of SRS

Strict adherence to the principles of simple random selection guarantees unbiased and consistent estimates of population parameters and their standard errors. Yet there are often other sampling designs for which the expected sampling error for a given sample size is lower than the sampling error expected under SRS. The ratio of expected sampling variance under a design, say $\mathfrak{I}$ to that of SRS is called the efficiency or design effect of $\mathfrak{I}$ (Särndal et al. 1992). Relative to a more efficient design a SRS requires a greater sample size for a given expected standard error. This usually also means that the cost for a forest inventory based on SRS would be higher than the costs incurred under a more efficient sampling design. Note, however, that the expected efficiency of a design relies on theoretical expectations. The survey planner has to obtain estimates of the expected sampling error under different competing designs and their cost implications before a rational choice can be made.

A SRS design often requires a surprisingly large investment in organization, checking, and location of the samples, investments that can be more time-consuming and therefore more expensive than for other, more efficient procedures. Also, through random selection an irregular spatial distribution of sample locations may result, so the population as a whole is not uniformly represented (Fig. 3.5). Although these outcomes are fully expected under the SRS design it is clearly unsatisfactory and perhaps even unacceptable to proceed with a sample that one suspects will yield estimates that are far from the true population values. For these reasons, the SRS design is commonly applied only to smaller homogeneous subpopulations as part of a more complex design. In general, a stratified sampling design with many homogeneous strata and just two samples per stratum offers the most efficient design (Royall 1998). Finally, a SRS design offers few opportunities for a postsampling correction/calibration to mitigate the negative impact of a "poor" sample.

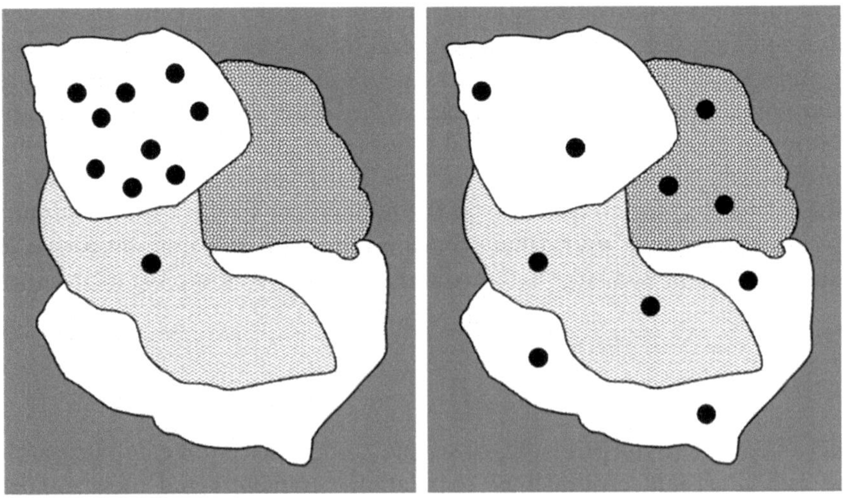

Fig. 3.5. Location of sample plots in two random samples

3.3.2 Systematic Sampling

As the term implies, the sample units are not randomly distributed across inventory area, but are drawn from the sample frame according to some systematic procedure. In systematic sampling the population is often subdivided into an exhaustive list of spatial units and n sample units are selected from this list by first choosing one unit at random, and then from this random starting position a systematic selection of the remaining n–1 units is made. Thus, the sample consists of one randomly selected unit. It is possible to derive unbiased estimators of totals and means from this sample but not of variance (Thompson 1992). The template for the spatial subdivision is often a regular grid of square cells or an equilateral triangular network. A major advantage of systematic sampling is that it is easy to locate the sample locations, the population is uniformly covered, and the efficiency is generally better than using SRS. As a rule, sample designs which are more "spatially balanced" will have a lower root-mean-square error when sampling from a population with patterned variation (Matérn 1980; Olsen et al. 1999; Stevens and Olsen 2004). As the joint selection probability of selecting two distinct units in the sample is either positive or zero, depending on the systematic sampling protocol, the selected elements are not independent of each other. This feature makes systematic sampling fundamentally different from SRS. Large-scale forest inventories, such as, for example, national inventories, often adopt a systematic design for the selection of sample location (Pelz and Cunia 1985; EC 1997)

In systematic sampling a sampling frame is constructed, which is a list of all sets of elements that are available for selection. When the basic sampling frame is in the form of a list (e.g., plant rows in a plantation) or consists of elements passing a certain point during a period of time (e.g., logs in a sawmill) the sample is generated by choosing elements from the frame that are separated by a constant interval L. If the population size N is a whole-number multiple of L then the sampling intensity n/N is equal to $1/L$ and all possible samples are size n. For example, the sampling frame for a plantation made up of 12 rows of plants could be the list [1, 2, 3, 4, 5, 6, 7, 8, 9, 10, 11, 12], where the number refers to the sequential position of a row of plants. For L=4 the sampling intensity is 1:4 and the sample will be composed of one of the following four possible subsets of rows [1, 5, 9], [2, 6, 10], [3, 7, 11], or [4, 8, 12]. In the sampling process one and only one of these four subsets would be selected as a sample.

When the basic sampling frame is in the form of a map, the universe is two-dimensional and we have several options for tesselating the population into a set of mutually exclusive and jointly exhaustive sets of units. Sometimes the tesselation is already provided to us in the form of suitable administrative units, like, for instance, counties or postal code districts. An example of a simple geometric tesselation is in Fig. 3.6. The population displayed is completely contained inside a regular square grid of ten rows and ten columns. There are two spacing intervals to be determined for the sample selection, one for the rows, S_r, and one for the columns, S_c. The intervals chosen may or may not be equal. A random starting point for selecting the systematic sample is chosen by choosing at random one integer between 1 and S_r, say rS_r and one integer between 1 and S_c, say rS_c. Once found, the sample consists of the sample unit with coordinates $rS_r + i \times S_r$ and $rS_c + i \times S_c$ for $i = 0, \ldots, n-1$. For a square or rectangular sampling frame the sampling design is a 1 in $(S_r{\times}S_c)$ systematic sample. For $S_r{=}S_c{=}5$ in a 10×10 grid (N=100), for example, this systematic sample design results in a sample size of 4. There are 25 possible samples of size 4. We only list five of them interleaving those samples that follow logically from a preceding sample. In the row–column notation given in Fig. 3.6 the possible samples of size 4 are [A1, A6, F1, F6], [A2, A7, F2, F7], . . ., [A5, A10, F5, F10], [B1, B6, G1, G6], . . ., and [E5, E10, K5, K10]. In this simple example all of the possible 25 subsets subset would have the same size n=4. An irregular spatial outline of the population or an irregular existing tesselation of the population may produce subsets of unequal size. When the sampling frame is not a square or a rectangle the discrepancy between the desired sampling intensity $1/(S_r \times S_c)$ and the actual sampling intensity could become large.

Systematic sampling can be implemented as a special form of either a simple cluster sampling or a two-stage cluster sampling. Implemented as a simple cluster sampling, each of the 25 possible subsets is considered as a single cluster and only one is selected (Fig. 3.7).

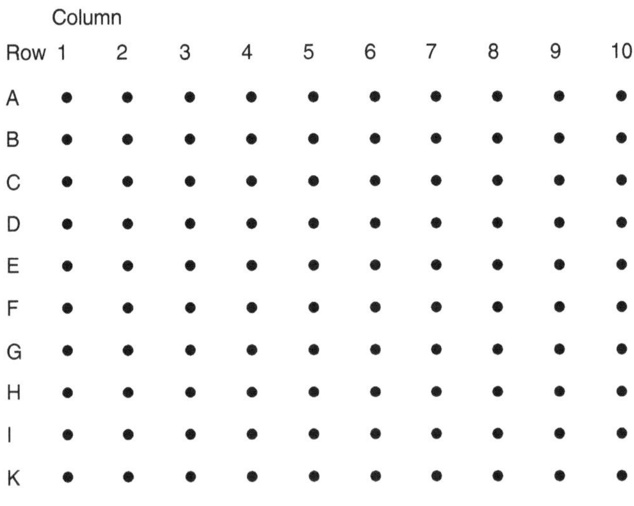

Fig. 3.6. A two-dimensional basic sampling frame

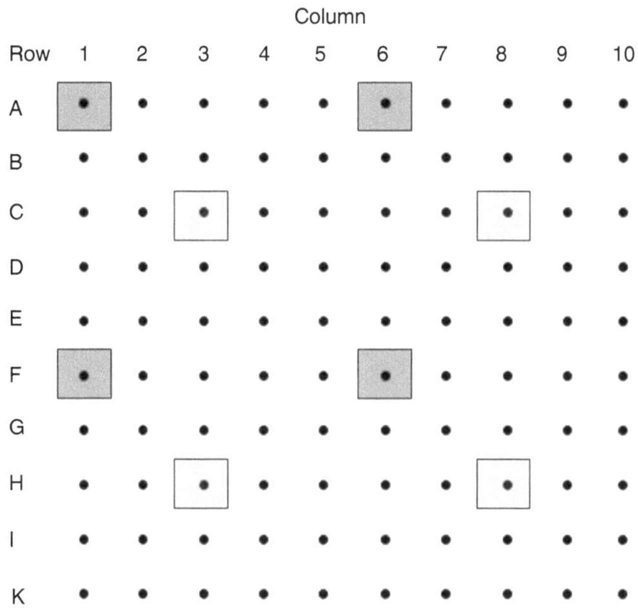

Fig. 3.7. Systematic sampling implemented as a cluster sampling. A cluster is composed of four shaded squares. The population contains 25 clusters, of which two are selected

When systematic sampling is implemented as a two-stage cluster sampling the sampling frame is divided into n compact clusters, each containing $S_r \times S_c$ elements. A subsample of size 1 is drawn from each of the n clusters. In the previous example n=4 compact clusters would be generated and one element selected from the 25 units inside each cluster (Fig. 3.8).

Estimators for the variance of the overall total and mean vary according to the way the systematic sampling is implemented (Cochran 1977). However, it is generally impossible to provide unbiased estimators of the variances when systematic sampling is used. Attempts have been undertaken to find estimators with little bias and low variance (Bellhouse 1985; Wolter 1985; Sherman 1996). The most commonly used approach is based on the assumption that a systematic sample is equivalent to a random sample; however, this assumption holds only when population attributes are randomly distributed over the population. With the assumption of SRS equivalency, means and totals are computed using the formulae applicable to SRS (Sect. 3.3.1). SRS estimators of standard errors applied to estimates from systematic sampling are usually conservative; they overestimate, on average, the actual error. An overestimation of about one third is not unusual, but more extreme results have been reported (Hartley 1966; Bellhouse 1988; Stehman 1992).

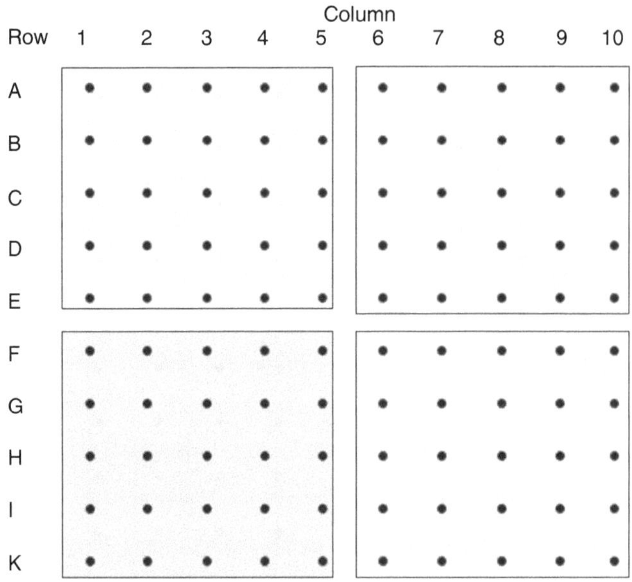

Fig. 3.8. Systematic sampling implemented as a two-stage cluster sampling

A series of variance estimators under systematic sampling have been proposed as more attractive than those flowing from the assumption of SRS. Jessen (1942), Yates (1949, 1981), and Cochran (1977) suggested a procedure that involves grouping of pairs of adjacent sample units (clusters). Each grouped pair of units is considered as a single stratum in a stratified sampling scheme, which allows the computation of a within-stratum variance, an estimate needed for the purpose of correcting the "inflated" SRS variance estimate. This procedure leads to corrected variance estimates that are biased, the bias being either positive or negative. Yates (1981) suggested for the situation of two-dimensional sampling to combine sample units into blocks of 2×2 units. Each block is then considered as a stratum in stratified random sampling. Instead of using squared residuals for computing a sampling variance, von Neumann et al. (1941) proposed using the sum of squared differences between adjacent sample units. This method is known as the method of squared differences (Ekström and Sjöstedt de Luna 2004; Stevens and Olsen 2004). The ideas of Neumann and Yates have been developed further in the form of NN estimators of local expectations and estimating the variance from the squared differences between the sample values and NN "predictions" (Ekström and Sjöstedt de Luna 2004; Stevens and Olsen 2004).

In systematic sampling the sample size is usually determined on the basis of optimizing a SRS design, which, for the reasons outlined earlier regarding the conservative nature of SRS estimators, eventually leads to somewhat more accurate inventory results than originally anticipated. In a systematic sampling from a square grid the sample size determines the scale of the grid to be used for implementing the systematic selection process. For a population occupying a square with an area F and a desired sample size of n, the appropriate grid spacing between sample locations should be $\sqrt{F/n}$. It gets a bit more complicated for population with an irregular outline, but in principle one finds the smallest rectangle that completely contains the population and then finds the grid spacing as before. Since some sample units may fall completely outside the population, the sample size is increased by trial and error until a satisfactory solution is obtained. For those who decide to use a triangular grid as the frame for a systematic sampling process, the distance between points becomes $\sqrt{2/3 \times F/n} \simeq 1.075 \times \sqrt{F/n}$ when the population occupies a square. In a triangular grid the distance between rows and columns is $0.557 \times \sqrt{F/n}$. To facilitate location of the sample units in the field, it is common practice to round off distances to the nearest 0.1 m. In establishing the grid, however, it should be borne in mind that some points may fall outside the target population. The density of the grid should be increased according to the proportions of sample units expected to fall outside the population of interest.

Semisystematic sampling designs have been suggested as a compromise between a systematic sampling design and SRS. In a semisystematic design the selection of spatially close sampling units is made less likely than by SRS or the number of selections from a single row, or column, is constrained to help achieve a more representative sample. Cox et al. (1997), Stein and Ettema (2003), and Stevens and Olsen (2004) provide examples of these semisystematic designs.

3.3.3 Cluster Sampling

Every sampling design is based on the division of the population into clearly defined units. The smallest units into which a population can be divided and which can be used for sample selection are the elements. Forest stands, lakes, road segments, and geometrically defined spatial polygons are but a few examples of units. Trees, snags, orchids, and deer are examples of elements. In cluster sampling two or more elements or two or more units are included in the sample at each sample location. The inclusion of two or more units/elements at each sample location intensifies the sampling effort at each sample location. The cost of including more than one unit/element at each single sample location is often modest compared with the cost associated with travel and measurement of one unit/element per sample location. One example for such clustering is the surveying of trees (elements) on sample plots (cluster/unit). Another example is the establishment of three or four sample plots in a fixed geometric configuration at each sample location instead of just a single plot. The grouping of elements into clusters lent the procedure its name.

From the viewpoint of expense, such grouping of sampled elements/units is justifiable. However, adding two or more "extra" units/elements at each sample location does not necessarily mean that the sample size can be reduced by a factor equal to the number of added elements/units at each sample location. The trade-off between the number of elements/units sampled per sample location and the number of sample locations to visit in order to achieve a target precision on estimates of population parameters depends on the distribution of the variance of attribute values across spatial and temporal scales. In general, the less the variance of the attribute value is within a cluster, the less is the efficiency of clustering sample observations relative to a SRS of individual units/elements. This is intuitively clear. If units/elements in a cluster are more alike than units/elements selected at random then we do not learn as much about the population from one cluster with m units/elements as we would learn from m independent units/elements. Thus, for a cluster sampling approach to be attractive in terms of precision for a given overall number of sampled units/elements, the variation within a cluster must be large relative to the among-cluster variance.

In forests the variation in tree attribute within sample plots of, say, 100 m^2 is often as large as the variance between such plots (Correll and Cellier 1987; Saborowski and Smelko 1998; Barnett and Stohlgren 2003; Gray 2003). We can exploit this large small-scale variation by adopting a cluster sampling design for our forest inventories. In most forest surveys, the use of clusters with several elements (more then 10) and three to four plots is often fully justified in terms of both cost and overall precision. Conversely, in homogenous forest areas a cluster composed of more than a few elements per sample location would be inefficient.

An efficient cluster sampling design offers an attractive balance between the cluster size (m) and the number of sample locations to visit (n). The balance is a function of the spatial distribution of attribute values. The survey designer must have, at least, a working knowledge of how this distribution will affect the efficiency of cluster sampling with different cluster sizes and different spatial configurations of elements/units in a cluster (Smith 1938; Kleinn 1996; Magnussen 2001).

Cluster sampling is common in forest inventory. Examples are the national forest inventories of Sweden, Finland, Austria, France, the USA, and Germany (Köhl 1990). Clusters are square, rectangular, or have more complex shapes. Sample plots are placed in a fixed geometric configuration within each cluster.

The term cluster is rarely used explicitly for the spatially grouped plots. Instead the word tract has become widely accepted as a quasi-synonym for a cluster (of sample plots). A classic example of cluster sampling is provided by the Camp Unit System(Fig. 3.9), introduced in Thailand for inventorying teak stands (Loetsch 1957). A camp located in the center of a cluster is surrounded by what is termed satellites, each satellite comprising several sample plots that can be surveyed by a field team within a single day.

The simplest form of cluster sampling is the survey with clusters of constant size. To facilitate the understanding of cluster sampling this version of cluster sampling is detailed. Note, however, that cluster size in forest surveys is rarely constant. Fixed-area sample plots are in effect clusters of trees. It is obvious that the cluster size, i.e., the number of trees per plot, changes from cluster to cluster. Even clusters (tracts) designed with a fixed number of plots in a fixed geometrical configuration usually have no sample data for units/elements that are outside or straddle the population boundary.

The sample-based estimator of the population mean for one-stage cluster sampling with n clusters of equal size (m) or equal numbers of sample units per cluster selected by SRS is

$$\hat{\bar{Y}}_{clust} = \frac{1}{n}\sum_{i=1}^{n}\frac{1}{m}\sum_{j=1}^{m} y_{ij} = \frac{1}{n}\sum_{i=1}^{n}\hat{\bar{Y}}_i,$$

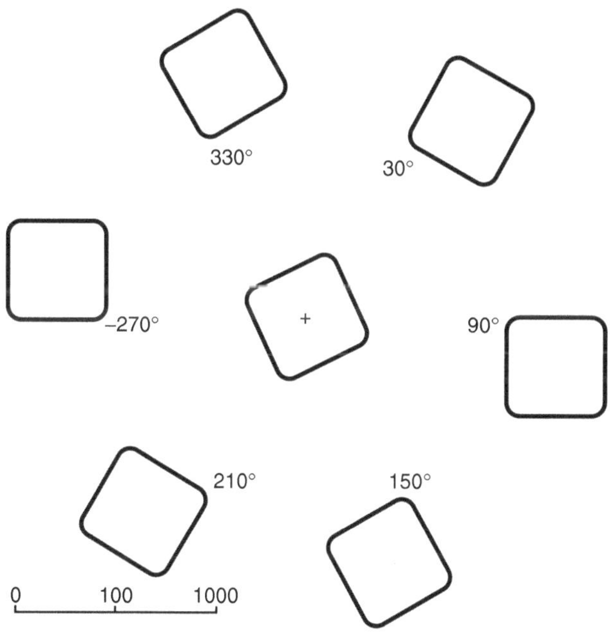

Fig.3.9. Camp Unit System

which is simply the average of cluster averages. The corresponding estimator, for sampling with clusters of unequal size, is a weighted average of cluster means with weights proportional to cluster size (Cochran 1977). The estimator of the sampling variance of the estimated population mean in a finite population composed of N clusters is

$$\hat{\mathrm{var}}\left(\hat{\bar{Y}}_{\mathrm{clust}}\right) = \frac{\left(1 - \frac{n}{N}\right)}{n-1} \sum_{i=1}^{n} \left(\hat{\bar{Y}}_i - \hat{\bar{Y}}_{\mathrm{clust}}\right)^2$$

or simply the among-cluster variance of cluster means. Estimators for population totals are obtained by dropping the averaging of within-cluster observations.

When sample clusters are located at random locations or when the population is not tessellated uniquely by the clusters, the finite-population correction factor is dropped from the variance estimator. Many cluster shapes (tracts) used in forest inventory do not tessellate the population completely. The tessellation would produce overlaps of clusters or leave gaps between clusters. The selection probability of each cluster would no longer be equal and joint selection probabilities would have to be calculated for every pair of possible clusters. Unequal cluster sizes would change the variance estimator to account for unequal weighting of the squared deviations from each cluster. Again, the weight would be proportional to cluster size.

Estimators for clusters of unequal size are described in Cochran (1977) and Sukhatme et al. (1984). The optimization of forest inventory cluster sampling designs with clustering of fixed-area and variable radius sample plots is discussed by Scott (1981) and Köhl and Scott (Scott 1994).

3.3.3.1 Two-Stage Cluster Sampling

In two-stage cluster sampling, or simply two-stage sampling, the entire population is divided into N clusters. A sample of n clusters is selected. The ith cluster is assumed to be subdivided into M_i equally large smaller units called elements or simply second-stage units. A sample of m_i second-stage units is taken from the ith cluster. Thus, the sample is taken in two steps: first n clusters are selected, then a sample of secondary units is taken from each cluster. In a forest sampling context the clusters could be, for example, 1-km^2 Advanced Very High Resolution Radiometer (AVHRR) pixels and the secondary unit could be a 25×25 m^2 Landsat Enhanced Thematic Mapper Plus (ETM+) pixel, and if selected the attributes of trees in this pixel would be measured by some field procedure. A forest stand can also be viewed as a cluster and the inventory plots placed within this stand acting as secondary units. Two-stage cluster sampling is a very flexible design and applies well to a variety of applications. At each stage a different selection scheme can be applied; this includes stratified selection of clusters and selection of secondary units chosen with equal probability, and selection of the secondary units with probability proportional to size (PPS)/prediction.

Two-stage procedures are generally preferable where localized systematic trends are expected. Systematic trends may occur in mountainous regions, along rivers and water bodies, and where natural history or anthropogenic effects have shaped vegetation and land use into distinct mosaics. The first-stage sampling then primarily serves to isolate the systematic variation to the first-stage units.

Two-stage sampling designs are frequently employed in forest inventory (Fig. 3.10). In its simplest incarnation every cluster contains the same number of secondary units, and both clusters and secondary units within clusters are randomly selected at each stage. Two-stage sampling is particularly attractive

Fig.3.10. Two-stage cluster sampling

when access to individual sample locations is time-consuming and costly; a topical situation in many tropical forest regions.

Two-stage sampling is distinctly different from two-phase sampling. In the latter, an auxiliary variable is sampled and measured in the first phase and a target attribute/variable is measured in the second phase on a small subset of the first-phase sample units. The association between the auxiliary and target attribute is exploited, usually via linear regression, to allow a prediction of the target attribute for all those units/elements sampled in the first phase for which only the auxiliary attribute/variable was recorded. In two-stage sampling the stratification of the population into clusters and second-stage units serves only as a conduit for determining selection probabilities.

Aerial photography, for example, can be used in both design types. In a two-phase sampling design the photographic images would be used to estimate, by some form of interpretation, the value of an auxiliary attribute for a series of sample units (stands, plots). The attribute of interest would be measured by standard field procedures on a small subset of these units. In a two-stage design, however, the photographic images would only serve to stratify the population into clusters (stands, tracts) and then a number of clusters would be selected at random, after which a number of secondary units would be selected from each cluster for recording of the attribute of interest. If a full coverage of the inventory area by aerial photographs is not possible, then a modification of the two-stage estimators given next is required (Saborowski 1990).

With SRS sampling of n clusters out of N population clusters and sampling of m_i secondary units within the ith cluster composed of M_i secondary units the unbiased two-stage estimator of the population total is

$$\hat{Y}_{\text{clust2}} = \frac{N}{n}\sum_{i=1}^{n}\frac{M_i}{m_i}\sum_{j=1}^{m_i} y_{ij}$$

and the estimator for the population average per first-stage unit is obtained by division by N. We see that each second-stage sample is scaled to an unbiased estimate of the total in the sampled first-stage unit. For clusters of equal size and equal second-stage sample sizes the estimator of the total is greatly simplified. We leave the simplification as an exercise for our readers. The sample-based estimator of variance of the estimated total is

$$\hat{\text{v}}\text{ar}\left(\hat{Y}_{\text{clust2}}\right) = \frac{N^2\left(1-\frac{n}{N}\right)}{n(n-1)}\sum_{i=1}^{n}\left(\hat{Y}_i - \hat{Y}_{\text{clust2}}\right)^2 + \frac{N}{n}\sum_{i=1}^{n}\frac{M_i^2}{m_i}\left(1-\frac{m_i}{M_i}\right)\hat{v}ar\left(y_{ij}\right).$$

The two-stage estimator of variance is simply the variance of the first-stage totals plus the average of the second-stage variances scaled to the first-stage expectations.

Cochran (1977) provides a mnemonic way to construct variance estimators for multi-stage sampling. Simply put, the expected value of a population parameter θ in a *k*-stage sampling is $E_1\left(E_2\left\{\left[\ldots E_k(\theta)\right]\ldots\right\}\right)$, where E_l, $l = 1,\ldots,k$ is the expectation of θ over all possible samples at the *l*th stage of sampling with all units at higher levels fixed. The expected variance of this expectation is $\text{var}_1\left(E_2\left\{\ldots\left[E_k(\theta)\right]\ldots\right\}\right) + E_1\left\{\text{var}_2\left[E_3(\theta)\right]\right\} + E_1\left(\left\{\cdots E_{k-1}\left[var_k(\theta)\right]\ldots\right\}\right)$, where $\text{var}_l(\theta)$ is the *l*th-stage variance of θ with all higher stages fixed.

When all secondary units in all clusters are sampled ($m_i = M_i$ for all *i*) we revert to estimators appropriate for single-stage cluster sampling. For $n=N$, that is all clusters are sampled, but $m_i < M_i$ for at least some *i*, the two-stage estimator is identical to the estimator for stratified random sampling with clusters acting as strata.

At times we may have interest in estimating the attribute mean per first-stage cluster (unit). We can get the estimator for this average by dividing the estimator of the total by *N*. Similarly, the estimator for the variance of this mean is the previous variance estimator of the total divided by N^2. For large *N* the first-stage sample fraction n/N is negligible and can be set to zero without incurring more than a trivial bias in the resulting variance estimator. With small first-stage sample fractions, and equal secondary sample sizes in each cluster, the two-stage variance estimator for the mean simplifies to the variance of the first-stage (cluster) mean values divided by *n*. In this specific situation, the variance can be estimated from knowledge of first-stage cluster means only, a useful result for two-stage designs with systematic subsampling of second-stage units since we would not have an unbiased estimator of the second-stage variance.

Two-stage sample estimators of totals (mean) and variance for designs with unequal selection probabilities of first- and second-stage units have been developed (Mahalanobis 1946; Bowden 1979; Cochran 1977; Nusser et al. 1998).

The previous two-stage estimators assumed that the population was divided into a unique set of *N* clusters that, in turn, were subdivided into a fixed number of secondary units. *N* would be known in this situation. When first-stage clusters are merely a fixed-area-sampling device located at random in the population *N* is unknown and must be estimated by dividing the area of the population by the area of a first-stage unit. Also, when and a small number of inventory plots are placed at random or in some geometric configuration inside a first-stage unit we do not a priori know *M* but we can estimate *M* by dividing the area of the first-stage unit (cluster) by the area of an inventory plot. Possible errors in estimates of *N* and *M* must accounted for by extending the previous variance estimator to include these potential sources of variation.

3.3.3.2
Two-Stage Cluster Sampling for the Estimation of Proportions

When the sampling objective is to estimate the proportion P of secondary units with a specific attribute class value we are dealing with sampling for the estimation of a proportion. To facilitate the derivation of estimators of means and variance we define the attribute value (y_{ij}) of the jth secondary unit in the ith primary unit as 1 when the unit has the class value of interest, and 0 otherwise. Under SRS of n primary units and m_i secondary units out of a total of M_i units in the ith primary unit, the estimators of P and the sampling variance are

$$\hat{P}_{2st|SRS} = \frac{1}{n} \times \sum_{i=1}^{n} \sum_{j=1}^{m_i} \frac{y_{ij}}{m_i}$$

and

$$\hat{\text{var}}\left(\hat{P}_{2st|SRS}\right) = \frac{1 - \frac{n}{N}}{n(n-1)} \text{var}\left(\hat{P}_i\right) + \frac{1}{n \times \sum_{i=1}^{n} m_i^2} \sum_{i=1}^{n} m_i^2 \hat{P}_i\left(1 - \hat{P}_i\right)\left(1 - \frac{m_i}{M_i}\right),$$

respectively, where $\hat{P}_i$ is the estimate of the proportion P in the ith primary unit. These formulae extend to cases with more than two classes by simply replacing an estimate of P by an estimate of the vector of class-specific population proportions.

3.3.3.3
Two-Stage Cluster Sampling with Stratification of the Primary Units

When the first stage is a sample of units in an aerial photograph or in a remotely sensed image we can often associate each first-stage unit with a specific land use or vegetation cover-type class. Within each first-stage unit we may sample one or more second-stage units. Second-stage units could be, for example, conventional forest inventory ground plots. Two-stage designs of this type are important for forest resource inventory. The first-stage units of a given stratum, say, h, are usually assumed to have a constant size and shape, but size and shape may vary from stratum to stratum. We shall consider a population with $N = \sum_h N_h$ first-stage units and M_h second-stage units in each first-stage unit of stratum h.

The estimator of the population mean per second-stage unit under SRS of n_h first-stage units and m_h second-stage units within each first-stage unit in stratum h is

$$\hat{\bar{Y}}_{2st|STR} = \frac{\sum_h \frac{N_h \times M_h}{n_h} \sum_{i=1}^{n_h} \frac{1}{m_h} \sum_{j=1}^{m_h} y_{hij}}{\sum_h N_h \times M_h} = \sum_h W_h \times \hat{\bar{Y}}_h,$$

where w_h is the relative size of stratum h in terms of second-stage units. For unknown N_h and M_h, w_h is replaced by n_h/n (Cochran 1977, p. 328). With SRS in both stages an unbiased estimator of the variance of $\hat{\bar{Y}}_{2st|STR}$ is

$$\hat{\text{var}}\left(\hat{\bar{Y}}_{2st|STR}\right) = \sum_h W_h^2 \times \left[\text{var}\left(\hat{\bar{Y}}_h\right)\left(1 - \frac{n_h}{N_h}\right) + \frac{\left(1 - \frac{n_h}{N_h}\right)\left(1 - \frac{m_h}{M_h}\right)}{n_h} \hat{\text{var}}\left(\hat{\bar{Y}}_{hi}\right)\right],$$

where the term $\hat{\text{var}}\left(\hat{\bar{Y}}_{hi}\right)$ is the average weighted within first stage unit (cluster) variance with weights proportional to the number of second-stage units selected within a first-stage unit. Estimators for the population total and its sampling variance are obtained from the previous estimators by scaling to the number of second-stage units in the population.

Note, if the secondary units are drawn systematically from within the primary units, the design is not a true two-stage cluster sampling. In effect, the appropriate estimators to use in this case would be those given for single-stage cluster sampling.

3.3.4 Stratified Sampling

In stratified sampling we use auxiliary information to stratify the entire population into, say, H strata. Stratification aims at forming groups of elements (units) with more or less similar attribute values. The ideal stratification eliminates the within-stratum variation, hardly possible in practice. In the ideal case a single sample from each stratum would suffice to gain complete knowledge about the population parameter of interest since there is no within-stratum variance, all elements (units) would have the same attribute value. As we move away from the ideal case more samples are needed to precisely estimate the mean (total) of a stratum. In other words, stratification aims at dividing a population into a number of parts which are as homogeneous as possible.

Besides improving the variance efficiency of estimators other reasons to choose a stratified sampling design are (1) estimates for homogeneous subpopulations (strata) may be required, (2) the desired precision is not the same for all subpopulations, (3) assessment cost and/or attributes of interest are not the same for all subpopulations, and (4) different sampling protocols apply to different subpopulations.

A diverse spectrum of criteria can be used to stratify a population. Some examples are major timber type, vegetation type, stand structure, species mixtures, site quality, protective status, habitat, ecological sensitivity, wetland status, recreational use, nontimber resource values, and political and administrative units. Where satellite imagery provides the auxiliary information the stratification is often done on the basis of the value of various indicators of vegetation types, such as, for example, the normalized-difference vegetation index (Sims and Gamon 2002; Wulder et al. 1996; Carlson and Ripley 1997; Gholz et al. 1997; Ricotta et al. 1999) or tessellated cap indices (Gustavson and Parker 1992; Bettinger et al. 1996).

Cochran (1977) and Dalenius and Gurney (1951) give generally valid rules for an optimum stratification of a single population parameter of interest when the distribution of an attribute value (y) or some auxiliary variable (x) related to y via a linear model is used directly for stratification. Given assumptions about the distribution of y and the relationship between x and y they develop an optimal solution for the number of strata and the selection of strata boundaries. The solution is optimal in terms of a minimum variance of the estimator of a population total (mean). For many typical distribution models for y, Gaussian, rectangular, triangular, and exponential, some five to ten strata appear to give substantial reductions in variance.

Once the criterion upon which to base the stratification has been decided, the inventory designer needs to consider the allocation of sample sizes to the strata. Again, both Dalenius and Cochran provide solutions for fixed total sample sizes or conversely for a fixed total cost of the inventory. When both stratification and sample size allocation are considered simultaneously the optimum design is often to maximize the number of strata and then take two samples per stratum. When the attribute values within a population are clustered, spatially or temporally, it is often good practice to define strata for each cluster.

Elements within a stratum are selected independently from the selections of elements in other strata. This accommodates stratum-specific sample sizes, selection criteria, and survey methods. When SRS is applied in all strata, the procedure is termed stratified random sampling.

In assessment, the strata are first evaluated separately and the results are then compiled to give overall estimates. The fact that stratified sampling renders it possible to compute estimations for subpopulations together with their precision is a distinct advantage.

In most practical situations auxiliary information suitable for a stratification of the population is readily available when planning for an inventory begins. By using this auxiliary information the efficiency of the inventory estimates of population parameters can be greatly improved. On the basis of the auxiliary information the population is divided into H distinct nonoverlapping strata or conversely the population is divided into N distinct nonoverlapping units and each unit is assigned to one and one only stratum. The strata cover the whole population without overlap, i.e., $N = \sum_h N_h$.

3.3.4.1
Sample Allocation

Deciding the number of samples to take from each stratum is perhaps the most important decision the inventory designer has to make once the decision to adopt a stratified sampling design has been made. Allocation of samples to strata may be done in various ways. The decision is often one of allocating a fixed total sample of size n to individual strata. The expected precision and cost of the resulting design can then be approximated from subject knowledge, experience, or qualified guesses. Design alternatives for different n can then be compared and the one judged most attractive against a set of global objectives is then favored.

One simple solution to the allocation problem prescribes an equal number of samples to be taken in each stratum. The sample size in stratum h is thus $n_h = \lceil n/H \rceil$, where H is the number of strata and $\lceil n/H \rceil$ denotes the smallest integer larger than n/H. Equal strata sample sizes, however, are seldom effective, as small strata are sampled with an disproportionately higher intensity than a large stratum.

A popular allocation scheme is the allocation of samples in proportion to the size of the strata. The size of stratum h (N_h) is measured in the number of elements (units). With this approach the sample size in stratum h becomes

$$n_h = N_h \times n \times N^{-1} \text{ with } N = \sum_h N_h .$$

At times the inventory designer will have some ideas or estimates of the expected stratum-specific variance of the attribute of interest. When both the within-stratum variance and the stratum size are considered together in the allocation problem and the objective is to minimize the expected variance of an estimate of a population total (mean), the solution is termed optimal allocation. Cost constraints may of course necessitate a shift away from this optimum towards an affordable design. The optimal allocation or Neymann allocation (Cochran 1977, p. 99) becomes

$$n_h = \left\lceil \frac{N_h \times \ddot{\sigma}_h}{\sum_h N_h \times \ddot{\sigma}_h} \right\rceil ,$$

where $\ddot{\sigma}_h$ denotes an a priori estimate of the standard deviation of the attribute of interest in stratum h.

Alternatives to the criteria of these allocation schemes include consideration of differences in costs for different strata and various survey methods. The overall importance of a stratum may further modify the allocation. Imposing limits on the minimum and maximum sample size in each stratum is also a popular scheme akin to a "minimax" strategy (minimize the risk of an extreme low precision; Amrhein 1995).

3.3.4.2
Estimation of Population Means and Totals Under Stratified Sampling

The estimators of the mean and the variance for stratum, say h, are as follows:

$$\hat{\bar{Y}}_h = \frac{1}{n_h}\sum_{i=1}^{n_h} y_{hi}$$

and

$$\hat{\text{var}}(y_{hi}) = \frac{1}{n_n - 1}\sum_{i=1}^{n_h}\left(y_{hi} - \hat{\bar{Y}}_h\right)^2.$$

The estimator for the population means under stratified random sampling is

$$\hat{\bar{Y}}_{\text{STR}} = \sum_h \frac{N_h}{N} \times \hat{\bar{Y}}_h = \sum_h W_h \times \hat{\bar{Y}}_h.$$

Often we know neither the total population size N nor the size of individual strata (N_h, h=1, ..., H). If we then replace the stratum weights W_h by the sample-based weights $W_h = n_h/n$ we obtain a biased estimate. The bias remains constant as the sample size increases. When the attribute of interest is expressed in units per unit area, the area of the population (A) and the area of individual strata (A_h) is used instead of N, or $N_{h.}$ Area-based strata weights then replace the weights based on size in units, or elements. For a stratum area A_h and total area $A = \sum_h A_h$ the area weight for stratum h becomes $wa_h = A_h \times A^{-1}$. When strata areas are known to within a negligible error, a situation that is common when the strata information comes from a classified remotely sensed image, the bias arising from using estimated weights $\widehat{wa_h}$ in place of the true area weights wa_h can safely be ignored. If proportional allocation is used and W_h is replaced by the area proportion of stratum h most of the potential gain of stratification compared with SRS is nevertheless retained.

The estimator for the variance of $\hat{\bar{Y}}_{\text{STR}}$ is

$$\hat{\text{var}}\left(\hat{\bar{Y}}_{\text{STR}}\right) = \sum_h \frac{W_h^2 \times \hat{var}(y_h)}{n_h}\left(1 - \frac{n_h}{N_h}\right).$$

The variance estimator is simplified if the sample fraction in each stratum is negligible and if the sample allocation is proportional to stratum size (area). We leave the simplifications as an exercise for the interested reader.

Estimates of population totals and variance of population totals are obtained from the previous estimators of the population mean and variance by multiplying the former by N and the latter by N^2.

3.3.4.3
Estimation of Proportions Under Stratified Random Sampling

For attributes on a nominal or an ordinal scale it is often desired to give their proportion within a stratum. Let $P_{h.c}$ be the proportion of sample units in stratum

h with attribute class value c in the sample from the hth stratum. An estimator of $P_{h.c}$ is

$$P_{h.c} = \frac{1}{n_h} \sum_{i=1}^{n_h} \delta_{hi.c},$$

where $\delta_{hi.c}$ is an indicator variable taking the value 1 if the ith sample in the hth stratum has attribute class value c and zero otherwise. The estimate of the proportion in the population is

$$\hat{\bar{P}}_{\text{STR.c}} = \sum_h \frac{n_h}{n} \times \hat{\bar{P}}_{h.c}.$$

With proportional allocation, and assuming that the stratum-specific final population correction factors can be ignored, the variance estimator for the estimated population proportion is

$$\hat{\text{v}}\text{ar}\left(\hat{P}_{\text{STR.c}}\right) = \sum_h w_h^2 \frac{\hat{P}_{\text{STR.c}}\left(1 - \hat{P}_{\text{STR.c}}\right)}{n_h - 1}.$$

Again, area weights (known or estimated) could also be used in place of sample size weights.

3.3.4.4
Design Effect

In many situations the surveyor will have a choice of sample design. We have already mentioned how variance efficiency, costs, and other practical considerations play a role in the final choice. A survey planner will often lack one or more critical pieces of information needed to make a truly optimal choice. Instead of optimizing a design it may be informative to know what variance to expect under a given design and how this variance compares with the variance of a "benchmark" design. The benchmark design is commonly the SRS design. The ratio of expected variance under the candidate design and the expected variance under the benchmark design is called the design effect, or DEFF for short (Kish 1965; Cochran 1977; Särndal et al. 1992). The candidate design is favored when the design effect is less than 1.

The design effect of stratified sampling with proportional allocation and SRS as the benchmark is

$$\widehat{\text{DEFF}}_{\text{STR: SRS}} = 1 - \frac{(n-1)\sum_h W_h\left(\hat{\bar{Y}} - \hat{\bar{Y}}_{\text{STR}}\right)^2}{\sum_h \sum_{i=1}^{n_h} \left(y_{hi} - \bar{Y}_{\text{STR}}\right)^2}.$$

We gather from this expression that the design effect is below 1 when the among-strata variance is made relatively large compared with the total variance. As the second term in the equation for the design effect is positive and less than 1, the design effect for a proportionate stratified design will always be less than 1, i.e., stratified random sampling with proportional allocation of samples to strata is always more efficient than SRS. To the extent that the stratum means differ from each other, the second term will increase with a corresponding decrease in the design effect. Conversely as the among-strata variance increases the within-stratum homogeneity is on the increase. In conclusion then, stratified random sampling with proportional allocation of samples to strata may produce a significant decrease in the sampling variance relative to the variance expected for a SRS design with the same total sample size.

3.3.4.5 Poststratification

The term poststratification applies to a procedure for which SRS samples are stratified to a set of known strata after completion of the sampling. In other words, the auxiliary strata information was not used during the sampling process. Poststratification may apply to a field survey completed before a remote-sensing-based stratification becomes available. Poststratification facilitates forest surveys, as field sampling and analysis or interpretation of remote-sensing data can be done independently.

In its simplest form poststratification applies to data from a SRS. Using the previous notation for stratified random sampling but with the addition of ".ps" to distinguish a poststratum from an a priori stratum. The poststratification estimator of the population mean is

$$\hat{Y}_{\text{STR.ps}} = \sum\nolimits_{h.\text{ps}} W_{h.\text{ps}} \times \hat{Y}_{h.\text{ps}}$$

and, assuming we can ignore the finite-population correction factor, the estimator of the sampling variance of the poststratified estimate of the population mean is

$$\hat{\text{v}}\text{ar}\left(\hat{Y}_{\text{STR.ps}}\right) =$$

$$\frac{1}{n}\left[\sum\nolimits_{h.\text{ps}} W_{h.\text{ps}} \times \hat{\text{v}}\text{ar}\left(y_{h.\text{ps}}\right) + \sum\nolimits_{h.\text{ps}} W_{h.\text{ps}}\left(1 - W_{h.\text{ps}}\right) \times \frac{1}{n_{h.\text{ps}}}\,\hat{\text{v}}\text{ar}\left(y_{h.\text{ps}}\right)\right].$$

The first term in the variance estimator is identical to the variance under a stratified sample with proportional allocation of sample sizes and within-stratum SRS. The second term reflects an increase in the variance due to the random nature of the strata weights. The strata weights in poststratification are the expected value of a random binary variable taking the value of 1 if a sample is in stratum h and 0 otherwise. The term $W_{h.ps}\left(1 - W_{h.ps}\right)$ is the well-known variance of a binomial random variable (Snedecor and Cochran 1971). As before, we can replace the strata weights that are based on sizes of strata in terms of population elements (units) with area-based weights.

The above poststratification estimators are not changed if the initial sample is not obtained under SRS but from a systematic sample. The implicit proportional allocation is more likely to be satisfied in this case than when the initial sample is obtained under a SRS scheme. According to Cochran (1977, p. 134) poststratified sampling is almost as precise as proportional stratified sampling, provided that the poststratified sample is reasonably large in each stratum ($n_{h.ps}$>20), and the effects of possible errors in the weights $W_{h.ps}$ can be safely ignored. As the increase in variance will be small if the average poststratification sample size across strata is sufficiently large, the application of the equations presented here without further adjustments for poststratification is defensible.

3.3.4.6
Pros and Cons of Stratified Sampling

Among the advantages of stratified sampling is the fact that estimates for subpopulation means or totals and their sampling variances are readily available. As the survey procedures for separate strata must be independent (otherwise there may be a covariance between results from different strata) sampling designs and sample sizes can be chosen freely to fit separate strata. In that regard the stratified sampling design is indeed flexible. In almost all cases, a gain in precision of population estimates of means, totals, ratios, and proportions is possible from either a prior stratification or a poststratification of the population.

On the other hand, stratified sampling can also convey certain disadvantages. The effect of inaccurate determination of the sizes of the strata has been mentioned. This problem is manifest when aerial photographs are used as the basis to define strata. Strata boundaries are transferred to the aerial -photographs and the sizes of the strata are determined either planimetrically or through some form of point grid counting (Loetsch and Haller 1964; de Vries 1986). Either way, the procedure is costly and time-consuming and impractical for large-scale surveys. Small spatially scattered strata (leopard pattern) increase the likelihood for errors in the determination of the sizes of the strata. Also, dated aerial photographs make estimates of the sizes of the strata imprecise. In all cases, the effect of the error in the strata area estimation must be carefully considered as it can greatly diminish the gains otherwise expected from stratified random sampling.

While stratification for the estimation of a single population attribute is generally advantageous, the situation is less clear for a multipurpose inventory with its many attributes of interest. For each variable of interest, a different optimum stratification rule is more likely to emerge than not. Consequently, the final stratification becomes a compromise, a compromise flavored by certain threshold and limits of precision not to be imperiled in any attribute or subpopulation.

The utility of stratified sampling is to be particularly critically examined when the inventory is to be repeated on successive occasions. When different sampling designs have been used for the individual strata or the samples have been allocated to strata using a procedure other than proportional allocation, the planning and execution of successive inventories can become very difficult or in extreme cases outright jeopardized. If sample units change from one stratum to another and the inventory design has also changed over time, new strata must be defined to accommodate a possible strata × design effect in the estimates. A resultant dramatic increase in the number of strata becomes a distinct possibility. To the extent it materializes, it affects adversely the design effect of stratified random sampling.

Furthermore, if the population attributes of interest have changed between two inventories the stratification used in the older inventory may be inopportune. An example of such a shift in focus is the current emphasis towards nonproductive functions of the forest at the expense of a narrower focus on timber values.

Most problems that arise from shifting strata and attributes can be mitigated effectively by proportional allocation of sample sizes and a systematic sampling within strata. Inventory designs with these characteristics may be suboptimal for determining an actual condition but they offer the advantage – not to be underestimated – of flexibility and permanence.

3.3.5 Two-Phase Sampling

Two-phase sampling or double sampling is a sampling procedure where two samples are taken from the population. The idea is to exploit an association between the attribute values in the two samples. In the first sample, a large number of easy-to-assess or low-cost sampling units are taken in order to measure one or more auxiliary variables. From this sample a second, smaller, sample is taken for the purpose of assessing the attribute/variable of interest. The statistical link between the auxiliary variable(s) and the variable(s) of interest can be established either by a linear regression (two-phase sampling with regression estimators) or by using the auxiliary variable to estimate the size of the strata (two-phase sampling for stratification). The two-phase design extends naturally to three and more phases (Magnussen 2003).

3.3.5.1 Two-Phase Sampling with Regression Estimators

Two-phase sampling with regression estimators is similar to single-phase sampling with regression estimators, with the difference being that the auxiliary variable, say x, is not measured on all N population elements but on a subsample of

N. In the first phase a large sample of size n_1 is selected; in the second phase a random subsample of size $n_2 \leq n_1$ is selected where the auxiliary variable, *x*, and the variable of interest, say *y*, are measured. The two-phase sample with regression estimator of the population mean for a single variable of interest and a single auxiliary variable is

$$\hat{\bar{Y}}_{2prgr} = \hat{\bar{Y}}_2 + \hat{\beta} \times \left(\hat{\bar{X}}_1 - \hat{\bar{X}}_2\right),$$

where $\hat{\bar{X}}_1$ and $\hat{\bar{X}}_2$ are the sample-based estimates of the mean of *x* in the first-phase sample and in the second-phase sample, respectively, $\hat{\bar{Y}}_2$ is the estimate of the population mean of *y* obtained from the second-phase sample, and $\hat{\beta}$ is the least-squares regression coefficient of *y* on *x* computed from the second-phase sample. Note, $\hat{\beta}$ can be improved by recognizing that the first-phase sample of *x* gives a better estimate of the variance of *x* than does the smaller, second-phase sample (Särndal et al. 1992). The second term in the two-phase sampling with regression estimator of the population mean is a term that corrects the SRS estimate by an amount that is proportional to the difference between the first-phase and second-phase estimates of the population mean of the auxiliary variable and the average effect of a one unit change in the auxiliary variable on the expected value of the variable of interest. Unequal probability sampling in the second phase must also be taken into account when the regression coefficient is computed. This is done by weighting the second-phase sample pairs of *x* and *y* by the inverse of their inclusion probability.

An estimator of the variance of $\hat{\bar{Y}}_{2prgr}$ is

$$\hat{\text{var}}\left(\hat{\bar{Y}}_{2prgr}\right) = \hat{\text{var}}\left(y|x\right) \times \left(\frac{1}{n_2} + \frac{\left(\hat{\bar{X}}_2 - \hat{\bar{X}}_1\right)^2}{\sum_{j=1}^{n_2}\left(x_j - \hat{\bar{X}}_2\right)^2}\right) + \frac{\hat{\text{var}}\left(y\right) - \hat{\text{var}}\left(y|x\right)}{n_1} - \frac{\hat{\text{var}}\left(y\right)}{N}$$

where $\hat{\text{var}}(y|x)$ is the conditional variance of *y* given *x*, or simply the variance of the expected values of *y* once *x* is known. For a linear relationship $y = \alpha + \beta x + \text{error}$, the conditional variance of *y* is the variance of the linear prediction $\alpha + \beta x$. The first term on the right-hand side of this variance expression contains the variance of the linear predictions of *y*, while the second term adds the variance of the prediction errors. The third term is a correction factor for finite-population predictions. In infinite populations the last term drops out.

An alternative approximation of the variance of $\hat{\bar{Y}}_{2prgr}$ is

$$\hat{\text{var}}\left(\hat{\bar{Y}}_{2prgr}\right) \simeq \frac{\hat{\text{var}}\left(y\right)\left(1 - \hat{\rho}_{xy}^{\,2}\right)}{n_2} + \frac{\hat{\rho}_{xy}^{\,2} \times \hat{\text{var}}(y)}{n_1} - \frac{\hat{\text{var}}(y)}{N},$$

where $\hat{\rho}_{xy}$ is the second-phase sample-based estimate of the product moment correlation coefficient between the second-phase sample values of y_i and x_i. Again the computational details of $\hat{\rho}_{xy}$ depend on the second-phase sampling design.

Two-phase sampling with regression is typically used in forest inventories where remote-sensing data and field assessments are to be combined While two-phase sampling with regression estimators is often more efficient than two-phase sampling for stratification, it has specific problems when used in practical settings. The cost relationship between the assessment in the first and the second phase is one factor to consider carefully. The other is the strength of the relationship between the variable of interest (y) and the auxiliary variable (x). This strength is often measured in the fraction of the variance in y that is explained by x. The coefficient of determination (ρ^2_{xy}) quantifies this strength; ρ^2_{xy} is a real number between 0 and 1. A higher ρ^2_{xy} means a stronger relationship and conversely a lower variance of the two-phase sampling with regression estimator. In large areas or in forests with a large spatial variability, ρ^2_{xy} values around 0.4 are not uncommon. Thus, only 40% of the variation in y can be explained through variation in x. In homogenous or small-scale forest areas higher R^2 values are commonplace. To expect ρ^2_{xy} values larger than 0.9, however, is unrealistic given the natural variation of the variables and the lack of perfect relationships between common inventory attributes. Furthermore, such high values are probably questionable and could be the result of transformations of x, y, or both or the result of forcing the regression through the origin when an intercept term is significant. The interpretation of estimates of ρ^2_{xy} should always be prudent. A few nontypical observations in the secondary sample or simply a nonrepresentative sample can grossly inflate the sample-based estimates of the population value of the correlation (Royall 2001). A cursory glance at the equation for the approximation of the variance may give the misguided impression that any correlation can be exploited advantageously in two-phase sampling with regression estimators.

De Vries (1986, pp. 117–120) provides an illustrative numerical example of two-phase sampling with regression estimators. In phase 1 the volume in 90 randomly selected photographed plots is determined by a trained interpreter. A subsample of 30 plots is randomly selected and their volume is determined from field measurements of height, diameter, and local volume tables. The correlation between the two volume estimates was strong (0.94) and the 90 photographic interpretations of voume brought about a 60% reduction in the estimated variance relative to a SRS with 30 ground plots.

Attributes measurable in remotely sensed images like tree height, crown diameter, or the number of trees within a defined area are often used as independent variables in a regression function to estimate the growing stock of a forest or forest stand. These types of applied regression estimators, however, have the distinct disadvantage that the independent variables can only be determined correctly from the aerial photograph under a set of specific conditions and appropriate resolution, conditions that are difficult to meet. Schade (1980), for instance, believes that a scale of 1:10,000 is too small to determine the crown size. In dense multilayered forests, typical in many tropical regions, the assessment of the number of trees or the crown diameter is difficult and in fully stocked stands the direct measurement of tree heights is impossible. Consequently, a double-sampling design for estimation of volume or biomass derived from volume functions, conversion factors, and presumed auxiliary measurements in aerial photographs remains, for many practical applications, a risky proposition.

We have so far regarded the application of two-phase sampling with regression estimators to a single attribute/variable of interest. Forest resource inventories are usually planned and implemented for the estimation of a large number of population parameters, all of interest to owners, managers, and stakeholders. In addition, estimates are often desired for the both the forest as a population and for one or more subpopulations. Detailed representation of estimated parameters might include, for example, a total broken down by tree development stage, by tree species, and by ecological zonation. For each estimate, a new regression relationship has to be derived from the sample data. The independent estimation of a large number of regression models from a single data set invariably entails nonadditive results. Even the probability of obtaining nonsensical results is not trivial. Users of inventory expect and should expect additivity of results. The nonadditivity problem is akin to that encountered in sampling with partial replacement (SPR). The need to derive a multitude of regression relationships and consequently the need to adjust the results to ensure additivity means that the analysis of inventory results based on double sampling with regression estimators quickly becomes very complex if not awkward. Furthermore, the correctness of the regression analysis depends on satisfying a set of rather strong assumptions regarding the residuals and the variables x and y. Also, not all target variables can be estimated with a single estimator. A number of variables on nominal or ordinal scale require a transformation, if possible, to meet the assumptions of the regression model. A nonlinear relationship between x and y would exclude the use of two-phase sampling with regression estimators.

An implicit requirement for the application of regression analysis is that the assessment of the variable of interest and the auxiliary variable is done on the same element. This can only be safeguarded if the sample plots in the two phases coincide exactly. Studies of the positional accuracy in the Swiss National Forest Inventory (NFI) found that the centers of the aerially photographed plots and the terrestrial sample plots were, on average, 5 m apart (Keller 2001). Since great care and much expenditure were applied to obtain an accurate location of the terrestrial plots in the first Swiss NFI, it is reasonable to assume that their accuracy is the best possible under current practical conditions. Future generations of Global Positioning System (GPS) sensors may raise the bar. In inaccessible forests, with difficult terrain and possibly with crews not fully trained in surveying techniques the accuracy will worsen. In tropical forests sample plot centers rarely coincide with centers marked on a photograph or a satellite image. A close relationship between the correct auxiliary variable and the variable of interest should therefore not be expected. The fact that a relationship between two attributes/variables often changes across locations (Gertner 1984; Walters et al. 1991) further strains the notion of a single linear relationship across the entire sample. Finally, measurement errors in the auxiliary variable attenuate the slope (Fuller 1987), and our estimate of the slope is biased.

In conclusion, in forest inventory a double sampling with regression estimators is only a realistic option for the analysis of a few major high-priority attributes.

3.3.5.2
Two-Phase Sampling for Stratification

Two-phase sampling for stratification is similar to stratified sampling except for the fact that the sizes of the strata are not known without error. Strata sizes are estimated by the larger, first-phase sample and the variable/attribute of interest is assessed in the second phase. In two-phase sampling for stratification an auxiliary categorical variable that can take one of H distinct values is sampled in the first phase for the purpose of estimating the proportion of the population elements/units in each category (stratum). The second-phase sample can be a subsample of units sampled from the first phase (dependent) or an independent sample. We shall limit ourselves to the dependent sample with SRS in each phase. De Vries (1986) treats the rare case of independent sampling in the first and second phases. The national forest inventory in the USA, for example, obtains sample-based estimates of the proportion of the land base that is in the forest stratum, possibly forest stratum (i.e., status is uncertain), and nonforest stratum through an intensive sampling and classification of plots located on aerial photographs (Spencer and Czaplewski 1998). A less intensive ground sampling provides the attributes of interest. The stratum of each ground plot is known at the time of ground sampling. Population parameters are then estimated through a combination of estimates obtained for each of the three strata. The uncertainty surrounding estimates of stratum size has to be included in the estimators of sample variance. In two-phase sampling for stratification with H strata and sample sizes n_1 in the first phase and n_2 in the second phase the estimator for the population mean, for example, is as follows (Cochran 1977):

$$\hat{\bar{Y}}_{2pstr} = \sum_{h=1}^{H} \frac{n_{1h}}{n_1} \times \hat{\bar{Y}}_h = \sum_{h=1}^{H} \hat{W}_h \times \hat{\bar{Y}}_h,$$

where n_{1h} is the first-phase sample size in stratum h (h=1, . . . , H), $n_i = n_{i1} + n_{i2} + \ldots + n_{iH}, i = 1,2$, and $\hat{\bar{Y}}_h$ is the second-phase estimate of the mean of the hth stratum. We see that the population mean is estimated as a weighted sum of the means of strata with weights (w_h) equal to the first-phase estimates of the relative frequencies of units (elements) in each stratum. A sample-based estimator of the variance of the estimated mean is

$$\hat{\mathrm{var}}\left(\hat{\bar{Y}}_{2pstr}\right) = \frac{N-1}{N} \sum_{h-1}^{H} \left(\frac{n_{1h}}{n_1 - 1} - \frac{n_{2h} - 1}{N-1}\right) \frac{W_h \, \hat{\mathrm{var}}\left(y_{2h}\right)}{n_{2h}}$$
$$+ \frac{N - n_1}{N\left(n_1 - 1\right)} \sum_{h=1}^{H} W_h \left(\bar{Y}_{2h} - \hat{\bar{Y}}_{2pstr}\right)^2,$$

where $\hat{\mathrm{var}}\left(y_{2h}\right)$ is a second-phase sample-based estimate of the variance of variable y in the hth stratum. As expected, the variance is the weighted sum of within-stratum and among-strata variances corrected for sample fractions and a finite population size.

For large n_1 and N the previous variance estimator approximates the variance in stratified sampling, i.e.,

$$\hat{\mathrm{var}}\left(\hat{\bar{Y}}_{2pstr}\right) \times \sum_{h=1}^{H} \frac{W_h \times \hat{\mathrm{var}}\left(y_{2h}\right)}{n_{2h}} \approx \hat{\mathrm{var}}\left(\hat{\bar{Y}}_{str}\right)$$

This approximation can often be used with impunity instead of the more complex estimator when the first-phase sample consists of a very large number of classified pixels (say over 1,000) taken from a very large remotely sensed image with, say, over 100,000 population pixels (N). This result ought to be intuitive. With a large sample size in the first phase, the variances of estimates of relative sizes of strata become small and can be neglected and the difference between $\hat{\mathrm{var}}\left(\hat{\bar{Y}}_{2pstr}\right)$ and $\hat{\mathrm{var}}\left(\hat{\bar{Y}}_{str}\right)$ has no practical importance.

The advantage of double sampling for stratification over stratified sampling is that a laborious assessment of the sizes of the strata is replaced by a quicker and less costly sampling procedure (Sutter 1990). Strata may be defined exclusively for the purpose of estimation and they may not otherwise form any meaningful subdivision of the population. The within-stratum variance, however, must be smaller than the variance in a nonstratified population before there can be a pay-off from the first-phase stratification in the form of a lower sampling variance. In comparison with double sampling with regression estimators, the derivation of regression functions has been eliminated. In many practical situations there is no suitable auxiliary variable that is uniformly and strongly correlated with the variable of interest across the entire population. Two-phase sampling for stratification would be a logical alternative in these circumstances. This argument extends to the case where several variables of interest are to be determined and where field samples cannot be located with any great precision or linked with a desired accuracy to observations made in the first phase.

3.3.6 Multiphase Sampling

Two-phase inventories can be extended to multiphase sampling procedures through the inclusion of additional assessment levels (Köhl and Kushwaha 1994). As the number of phases increases the number of possible pairs of two phases that each could be used as an estimation of a desired parameter increases exponentially. The optimum use of the sampled information requires that we combine all possible estimators optimally, i.e., weighted with respect to their sampling variances. While the extension of two-phase estimators to multiphase sampling is theoretically straightforward, the fact remains that the thinning of sample sizes down through the hierarchy of phases quickly erodes any tangible gains in precision by going from two to three or even more phases (Magnussen 2003).

Three-phase sampling with regression estimators is, as expected, a complex design with few practical applications (Pfeffermann et al. 1998). It can hardly be recommended for other than special purpose surveys owing to considerable estimation problems. Therefore, only three-phase sampling for stratification is detailed. As before, we limit details to random subsampling in the second and third phases Examples of practical applications of three-phase sampling for stratification are given in Cherrill and Fuller (1994), Kirkman (1996), Williams (1996), Lunetta et al. (1998), Vogelmann and Howard (1998), Cannell et al. (1999), Lunetta and Elvidge (1999), Brown et al. (2000), Cruickshank et al. (2000), Flores and Martínez (2000), Moran et al. (2000), Chong et al. (2001), Franklin (2001), and Magnussen (2003). As general references Bowden (1979) and Johnston (1982) can be recommended.

Estimators for three-phase sampling for stratification use an extension of the notation of two-phase estimators. The first two phases provide information for estimation of the relative frequencies of the strata in the population. This includes estimation of the proportions of second-phase strata in each of the first-phase strata. Given N, these estimators of strata proportions can be used to estimate the number of population units in each stratum and the combination of first-phase and second-phase strata. These estimates of the sizes of the strata are needed to scale estimates from subsamples in the second and third phases to estimates for first-phase sampling. In summary, separate estimators of the population parameter of interest are produced for each phase and then combined to give one final estimate. The unbiased estimator of the population mean for L first-phase strata and H second-phase strata and sample sizes n_1, n_2, and n_3 in the three phases is

$$\hat{\bar{Y}}_{3pstr} = \sum_{l=1}^{L} \hat{W}_l \sum_{h=1}^{H} \hat{W}_{lh} \times \hat{\bar{Y}}_{3lh} = \sum_{l=1}^{L} \hat{W}_l \times \hat{\bar{Y}}_l,$$

where $\hat{w}_{lh}$ is the proportion of second-phase sampleøs in first-phase stratum l and second-phase stratum h, and $\bar{Y}_{3lh}$ is the third-phase mean of sample values of y in first-phase stratum l and second-phase stratum h. Again, we see that the mean is simply a doubly weighted average of third-phase estimates with weights equal to the relative stratum frequencies in the first and second phases. First-phase sample sizes in first-phase strata l and second-phase strata h are n_{2lh} with $l = 1, \ldots, L$ and $h = 1, \ldots, H$. The total number of second-phase samples in first-phase strata l is $n_{2l} = \sum_{h=1}^{h} n_{2lh}$. Under SRS in each phase, simple ratios of sample sizes are used to estimate the number of population units in each stratum. For example, if we have n_{2lh} second-phase samples out of a total of n_{2l} second-phase samples in first-phase strata l and second-phase strata h, and we have n_{1l} first-phase samples out of a total of n_1 first-phase samples in first-phase strata l, and we would estimate the number of population units in the $l \times h$ strata to be $N \times (n_{1l}/n_1) \times (n_{2lh}/n_{2l})$. After these preliminaries we can write the approximate variance of the estimate of the population mean as

$$\hat{\text{var}}\left(\hat{\bar{Y}}_{3pstr}\right) \simeq \hat{\text{var}}_{\text{among first}-\text{phase strata means}} + \hat{\text{var}}_{\text{among second}-\text{phase strata means}} + \hat{\text{var}}_{\text{within first}-\text{phase} \times \text{second}-\text{phase}}$$

where

$$\hat{\text{var}}_{\text{among first}-\text{phase strata means}} = \frac{N(N-1)(n_1-1)}{n_1^2(N-1)^2} \sum_{l=1}^{L} \hat{W}_l\left(\hat{\bar{Y}}_l - \hat{\bar{Y}}_{3pstr}\right)^2,$$

$$\hat{\text{var}}_{\text{among second}-\text{phase strata means}} = \frac{N(N-1)(n_1-1)}{n_1^3(N-1)^2} \sum_{l=1}^{L} (n_{1l}-1)\tilde{V}_l,$$

and

$$\hat{\text{var}}_{\text{within first}-\text{phase} \times \text{second phase strata}} = \frac{N-n_1}{n_1(N-1)} \sum_{l=1}^{L} \hat{W}_h \left(\hat{W}_h - \frac{N-n_1}{n_1(N-1)}\right) \times \ddot{V}_l,$$

with

$$\tilde{V}_l = \left[\ddot{V}_l - \frac{1}{n_{2l}} \sum_{h=1}^{H} \hat{W}_{lh} \times \left(\frac{n_{2l}}{n_{2lh}} - 1\right) \Big/ \left(\frac{1}{n_{2lh}} - \frac{1}{n_{1l}}\right) \times \hat{\text{var}}(y_{hl})\right],$$

and (Cochran 1977)

$$\tilde{V}_l = \frac{n_{1l}-1}{n_1} \sum_{h=1}^{H} \left[\frac{N \times \frac{n_{1l}}{n_1} \times \frac{n_{2lh}}{n_{2l}} - 1}{N \times \frac{n_{1l}}{n_1} - 1} - \frac{n_{2lh}-1}{n_{2l}-1}\right] \frac{\hat{W}_{lh}}{n_{2lh}} \hat{var}(y_{lh}) + \dddot{V}_l,$$

and

$$\ddot{V}_l = \frac{N \times \frac{n_{1l}}{n_1} - n_{1l}}{n_{1l}\left(N \times \frac{n_{1l}}{n_1} - 1\right)} \sum_{h=1}^{H} \hat{W}_{lh}\left(\hat{\bar{Y}} - \hat{\bar{Y}}_l\right)^2.$$

3.4 Errors in Forest Surveys

In sampling surveys, two types of errors are distinguished: sampling errors and nonsampling errors. Sampling errors result from the facts that only part of the population is surveyed and population parameters are estimated from the sample. The estimated parameters may deviate from the true population values. One way in which sampling error may be expressed is through the standard error of the mean. This ought to be given for all estimators, as it is essential for the correct interpretation of inventory data. Nonsampling errors, on the other hand, are not connected with the problem of dealing with only part of the population but arise from inaccurate measurements, less-than-perfect measuring devices, mistakes in the execution of the sampling plan or protocol, and sampling the wrong units/elements. Nonsampling errors of this nature are likely to inflate the apparent sampling variance and to introduce a bias in the estimates. When the sample observations are derived from model-based predictions and not a direct measurement per se our data are subject to model error. Since models only predict the expected value, the apparent sampling variance of model-based predictions will be too low; the residual model variance has been left out.

From a statistical point of view the reliability of results can be quantified by giving their precision, accuracy, Mean Square Error (MSE), or bias. We shall give definitions as there continues to be considerable confusion about the correct use of these terms:

Precision: Precision refers to the size of deviations in the estimate of a population parameter in repeat application of a sampling procedure. The standard error or confidence interval quantifies precision. Increasing the number of observations increases the precision of a statistical estimate.

Accuracy:Accuracy refers to the size of deviations between an observed value and the true value. Thus, if we know the true value of a population parameter then we can also define the accuracy of a survey estimate as the deviation between the estimate and the true value.

Bias: Bias is directly related to the accuracy of an estimate. Bias is the difference between the estimated value and true value of a parameter. We cannot estimate bias unless we know the true value of a parameter. In practice we do not have this knowledge.

The effect of precision and bias can be seen in Fig. 3.11.

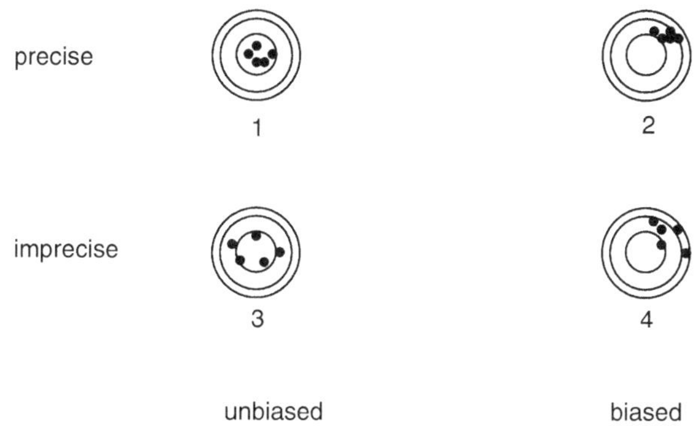

Fig. 3.11. Example for the concept of accuracy and precision of an estimator. If we assume that the target center is the true but unknown population value and that the position of each shot represents the estimate obtained by a random sample, it follows that (*1*) the estimator is precise and accurate, (*2*) the estimator is precise but biased, (**3**) the estimator is imprecise but unbiased, (*4*) the estimator is imprecise and biased. (After Vanclay 1994)

Mean Square: MSE is a useful measure of reliability. It combines the precision of an estimate with the square of the bias. The MSE is a useful criterion for the comparison of two or more competing estimators, possibly with different amounts of bias. A direct comparison of estimators in terms of precision only may skew the choice towards estimators that generate highly precise but biased estimates. According to Cochran (1977, p. 15) the MSE of an estimate of, say, a population total is

$$\mathrm{MSE}\left(\hat{Y}\right) = \hat{v}ar\left(\hat{Y}\right) + \left(\hat{Y} - Y\right)^2 .$$

For unbiased estimators MSE and precision are asymptotically $\left(\{n, N\} \to \infty\right)$ identical. In the following chapters mostly unbiased estimators for population parameters will be presented. The MSE and the precision of the estimates derived from unbiased estimators should, therefore, be asymptotically equivalent. Alternative estimators such as, for example, Bayesian and Stein estimators, seek an attractive balance between bias and precision, often pursuing a minimum variance at the expense of some small amount of bias (Box and Tiao 1973; Congdon 2001).

The standard procedure for calculating sampling error does not allow for the influence of bias. Nevertheless, bias may multiply the sampling error by several magnitudes (Gertner 1984). Increasing the sample size may certainly decrease the sampling error but it can also increase the relative influence of bias. Consequently, possible bias should be guarded against from the earliest stage of planning a survey. It is often possible to somehow assess various sources of inventory errors (Gernter und Köhl 1992) and then gauge the effect of potential bias arising from different sources during the inventory or perhaps owing to the choice of estimators. Simulation studies may be needed to quantify bias introduced by an estimator.

3.4.1 Non-Sampling inventory errors

Estimators of population attributes and their sampling variances have so far been presented as if the observations were not only complete but also the best possible. Best possible means that the most accurate technique for obtaining data is applied everywhere. Practical surveys can rarely live up to this ideal; forest inventories are no exception (Goelz and Burk 1996; Chen 1998; Lesser and Kalsbeek 1999; Chen and Cowling 2001).

3.4.1.1 Nonobservation

The sample can be incomplete owing to nonobservation of sample units or errors in the population frame from which sample units are selected (Särndal et al. 1992). Nonobservations of sample units can occur when some units are not visited because (1) access is denied, (2) access poses a danger to the survey crew, (3) the sample unit could not be located, and (4) sampling was terminated owing to time and budgetary constraints. Errors in the population frame usually result in an undercoverage, certain population units have a zero probability of inclusion in the sample, or conversely an inflated inclusion probability if the unit appears more than once in the frame. Regardless of cause, an incomplete sample will, as a rule, result in a biased estimate of both the population attribute value and its sampling variance. We can easily appreciate this result if we subdivide a population of N units into two parts, one with attribute values y_{1i}, $i = 1,\ldots,N_1$ for which observations can or will be made, and the second with attribute values $y_{2i}, i = 1,\ldots,N_2$ for which observations will be missing regardless $(N_1 + N_2 = N)$. Let us assume $w_2 = N_2 \times N^{-1}$, our average sample-based estimator of, say, a population mean is biased since $E_s\left(\hat{\bar{Y}}\right) - \bar{Y} = w_2(\bar{Y}_1 - \bar{Y}_2)$, where the expectation is with respect to the sampling distribution of sample estimates. Only in the rare case when observations are missing completely at random (MCAR; Little and Rubin 1987) can we expect no bias since under MCAR $\Pr(Y \mid \text{missing}) = \Pr(Y) \times \Pr(\text{missing})$ and $E(\bar{Y}_1) = E(\bar{Y}_2)$. To mitigate a potential bias from missing observations we must either make assumptions about either y_2 or the mechanism leading to a missing observations, or perhaps both. We can impute the missing values by suitable substitutes by specifying a data model for y_2 to complete the sample and then obtain the usual estimates as if the sample was complete. There are many ways to do the imputation and to obtain the statistical estimators from samples with imputations, each relying on a specific set of assumptions. Readers are referred to Rubin (1987), Efron (1994), Schafer and Scheinker (2000), and McRoberts (2001) for details. Alternatively, if data are missing at random and there exists a quantifiable relationship between data values and the probability of a missing value we may be able to adjust our sample estimates by obtaining a new set of sample inclusion weights from estimates of the probability of a missing value $\Pr(Y_1, X_1, X_2)$, where x is an auxiliary variable known for all population elements. The adjustment option is only available if the sample contains at least some elements from across the entire range of x. For example, if the probability of obtaining a sample is a monotone function g of the slope ϕ of the terrain at the sample location and $0 < g(\phi) < 1$ then the inverse $1/g(\phi)$ can be used in conjunction with the original design-based inclusion probabilities to reduce the bias due to missing observations.

Estimators of sampling variance obtained from surveys with missing data do not account for a potential bias. The MSE (MSE = var + bias^2) would be a preferred estimate of precision in the presence of a potential bias; however, we will rarely know the magnitude of the bias. A rule of thumb (Cochran 1977) says that unless the absolute bias is less than 0.1 times the sample-based estimator of the standard error of an estimate, the reported standard error and any conventional confidence interval for the estimator could be seriously misleading.

In sampling for proportions we can at least impose limits on the missing data and use these limits to construct a conservative confidence interval for the estimated population proportion. Cochran (1977) gives an example with $n_2 = 200$ missing binary observations out of $n = 1{,}000$ target observations (i.e., $n_1 = 800$) and 80 positive responses ($y = 1$). By assuming that the missing data at one extreme would be all 0 and all 1 at the other extreme, the following conservative 95% confidence interval of the sample estimate of the population proportion becomes

$$CI\left(\hat{P} = 0.08\right) = \left\{\frac{80+0}{1000} - 1.96 \times \sqrt{\frac{0.08 \times 0.092}{1000}}, \frac{80+200}{1000} + 1.96 \times \sqrt{\frac{0.08 \times 0.092}{1000}}\right\}$$
$$\cong \{0.063, 0.308\}$$

We can apply the same rationale towards estimating the sample size needed to achieve a given target precision when it is known that some binary observations will be missing. Since bias does not decrease with sample size, even a modest rate of nonresponse (less than 10%) can have a serious impact on the quality of survey estimates and every possible effort should be made to obtain a complete sample.

3.4.1.2
Measurement Errors

Directly observed or compiled attribute values of a sampled population element (unit) are rarely, if ever, completely free of errors or bias. An observation deviating from the best possible observation is said to be in error and possibly biased. Conceptually we can write an observed, viz., a compiled, attribute value

y'_i for the *i*th population element as a linear sum of the best possible value y_i plus a series of error and bias terms arising from various sources (k). For a univariate attribute we can write our observations (compilations) as

$$y'_i = y_i + \sum_k e_{ik} + \sum_k b_{ik} = y_i + e_{i\cdot} + b_{i\cdot},\ E(e_{ik}) = 0,\ b_{ik} \neq 0,$$

where $e_{i\cdot}$ *and* $b_{i\cdot}$ denote the sum of error and bias in the observation (compilation) y'_i. The errors e_{ik} will depend on the attribute and on the entire process, including the design that generates the observed, viz., compiled, values y'_i. Extension to multivariate attributes is straightforward. Only the univariate case will be detailed.

For surveys with errors and possibly bias in observed/compiled attribute values the sample-based estimate of, say, a population mean $\bar{Y}$ is $\hat{\bar{Y}}' = \hat{\bar{Y}} + \bar{e} + \bar{b}$, i.e., the estimate we would have obtained if there were no errors or no bias $\left(\hat{\bar{Y}}\right)$ plus the sum of the average error and the average bias. We would normally not know the average error nor the average bias, but it is often assumed that the errors are independent and cancel across the sample, which would leave us with an estimate with a bias of $\bar{b}$. With these assumptions the expected MSE of a sample estimate is

$$E_s\left[\text{MSE}\left(\hat{\bar{Y}}'\right)\right] = \text{var}\left(\hat{\bar{Y}}\right) + \text{var}(\bar{e}) + \frac{1}{n}\sum_{i=1}^{n}\left(b_{i\cdot} - \bar{b}\right)^2 + 2 \times \text{cov}\left(\bar{e}, \hat{\bar{Y}}\right),$$

where expectation is over the sampling distribution of $\hat{\bar{Y}}'$. The expected MSE is thus the sampling variance in the absence of errors and bias $\left[\hat{\text{var}}\left(\hat{\bar{Y}}\right)\right]$ plus the sum of the variance of the average error, the average of the squared bias differential of individual samples, and twice the covariance between the average error and $\hat{\bar{Y}}$. The covariance term accounts for a possible correlation of errors and attribute values. For example, a given surveyor may introduce a constant observer bias to all samples assigned to this individual. Likewise, an instrument may introduce an error that is a function of the read-out value.

The practical consequence of the expected MSE equation is that only the first two terms decline at a rate of $1/n$; the bias contribution is usually independent of n but may actually increase with n if an increase in sample size somehow compromises the quality of measurements or compilations. The covariance term may decline with increasing n but only under a set of rather specific and restrictive circumstances. Consequently, even a modest bias or error covariance can greatly inflate the relative standard error of a survey. Gertner and Köhl (1992) gave a clear demonstration of this in their analysis of the errors and possible bias in the Swiss NFI. A modest 1% bias in the volume compilations, for example, would double the relative standard error of the estimate of the total volume in a strata dominated by Norway spruce. When the covariance term $\text{cov}\left(\bar{e}, \hat{\bar{Y}}\right)$ is negative, the observed variance can be less than the actual sampling variance of the best possible estimate. When the observable values are constrained or categorical, the covariance term is generally negative.

Unless we somehow produce estimates of the measurement error variance, the bias, and the error covariance, we will not know by how much our sample-based estimates of population attributes and estimates of sampling error have been biased. If we are willing to assume a constant bias and no error covariance the variance of $\hat{\bar{Y}}'$ is

$$\operatorname{var}\left(\hat{\bar{Y}}'\right) = \frac{1}{n}\,\hat{\operatorname{var}}\left(\hat{Y}\right)\left(1-\frac{n}{N}\right)+\frac{1}{n}\,\hat{\operatorname{var}}\left(e\right).$$

Thus, the expected sampling variance of a real-valued population attribute is inflated by the variance of measurement errors; the bias term vanishes since it was assumed to be constant. On the other hand, the expected sample-based estimate of the sampling variance is

$$E\left[\hat{Var}\left(\hat{\bar{Y}}'\right)\right] = \left[\frac{1}{n}\hat{Var}\left(\hat{\bar{Y}}\right)+\frac{1}{n}\hat{Var}\left(e\right)\right]\left(1-\frac{n}{N}\right)$$

Hence, a sample-based estimate of the sampling variance of a population attribute observed/compiled with an error and possibly a constant bias is only approximately unbiased if the sample fraction *n/N* is negligible.

Measurement errors may, however, not be independent across samples. In presence of a correlation of some or all measurement errors (and constant bias) the expected sample-based estimate of sampling variance becomes

$$E_s\left[\hat{\operatorname{var}}\left(\hat{\bar{Y}}'\right)\right] = \left[\frac{1}{n}\,\hat{\operatorname{var}}\left(\hat{\bar{Y}}\right)+\frac{1}{n}\,\hat{\operatorname{var}}\left(e\right)\left(1-\hat{\rho}_e\right)\right]\left(1-\frac{n}{N}\right),$$

where the expectation is over the sampling distribution and $\hat{\rho}_e$ is an estimate of the intrasample correlation between errors of measurement with

$$\hat{\rho}_e = E_s\left[\left(e_i \times e_j \middle| \{i,j\} \in s\right)\right] \times \hat{\operatorname{var}}\left(e\right)^{-1}.$$

Since the correlation is mostly positive a sample-based estimate of sampling variance appears more precise than it is. Positively correlated errors are not unusual in attributes derived from remotely sensed data (Since we have ignored the covariance of errors in our calculations of variance) (Congalton 1988; Dobbertin and Biging 1996).

3.4.1.3 Estimating Nonsampling Errors and Bias

Bias and measurement errors can seriously compromise the quality and precision of a survey. Diligence and high standards are required in all aspects and all phases of a forest inventory to keep measurement errors and bias within acceptable bounds. Quality standards and quality checks are integral parts of a forest inventory. Still, measurement errors and bias are virtually impossible to stamp out of a forest inventory. It is therefore good practice to investigate the impact of measurement errors and bias on survey results. An error budget discloses all possible sources and the expected impact of error in the entire process that begins with a visit to a sample unit and ends with a set of survey estimates of population attributes and is ideally suited for this purpose (Gertner and Köhl 1992; Kangas 1996). The error budget will ideally not only disclose the possible bias in estimators but also suggest where and how better standards can mitigate the impact of bias and measurement errors. In well-designed inventories with high measurement and compilation standards the contribution of natural intrinsic variation in attribute values to the overall sampling variance is usually orders of magnitude larger than the contribution of measurement errors and bias (Gertner and Köhl 1992; Kangas 1996).

Survey observations (viz., compilations) include measurement errors and bias. Estimates of the measurement errors and possible bias can be obtained either by repeated measurements and compilations performed on a subset of sample units or by model-based Monte Carlo simulations of the entire process that produced the desired estimates.

The repeated measurement option is simple but costly, and also potentially flawed unless great care is taken to ensure that the repeated measurements allow an unbiased estimation of the error structure and possible bias. Several pairs of two independent repeat observations y'_{i1} and y'_{i2} of the (best possible) sample attribute value y_i in the ith sample unit allow, under the assumption of equal bias and equal error variance, the estimation of the measurement error variance and the covariance of errors associated with a single sample. Estimates are obtained by solving for var(e) and cov (e_1, e_2) the following equations:

$$\frac{\sum_{i=1}^{m}\left(y'_{i1}-y'_{i2}\right)^2}{m} = 2\times\left[\text{var}\left(e\right)-\text{cov}\left(e_1,e_2\right)\right]$$

and

$$\frac{\sum_{i=1}^{m}\left(y'_{i1}+y'_{i2}\right)^2}{m} = 2\times\left[\text{var}\left(e\right)+\text{cov}\left(e_1,e_2\right)\right],$$

where m is the number of representative sample units selected for repeat observation. Under SRS, n large, and $m = 0.5 \times n$, the average of two replicate estimates of the sampling variance provides an approximate estimate of the sampling variance of $\hat{\bar{Y}}$ via

$$\hat{\text{var}}\left(\hat{\bar{Y}}\right) \cong 0.5\left[\hat{\text{var}}\left(\hat{\bar{Y}}'_1\right)+\hat{\text{var}}\left(\hat{\bar{Y}}'_2\right)\right]-\hat{\text{var}}\left(e\right)\left(1-\hat{\rho}_e\right).$$

Fewer than 30 repeat samples will not provide reliable estimates (Cochran 1977). The potential of a bias in the estimate of the measurement error variance arises from the fact that if the same person is asked to do a task twice a recall from the first execution is likely to influence the second execution in some way or other. If a different person is assigned to the second observation the estimate of the measurement error variance will be confounded by interpersonal bias. A direct sample-based estimate of bias can only be obtained from multiple repeat observations (more than six) of an attribute value using the adopted survey methodology and one final observation of the best possible value. The large number of repeat measurements ensures that the average measurement error will be close to zero and therefore $\hat{b}_i \cong \overline{Y'_i} - y_i$.

In forest inventory the repeat measurement approach to estimate measurement error variance and bias is often either impractical or too costly. Instead the surveyor attempts to produce model-based estimates of measurement errors based on Monte Carlo simulations. The measurement error budget starts with the estimator of the (best possible) attribute value of interest. The estimator is then expanded into a model that includes variables for all the basic attribute values observed, viz., measured, on a sample unit. The model also includes all functions and their parameters needed to transform basic sample attribute values into the desired attribute value. Once the model has been established, a measurement error distribution and possible bias are specified for each variable and parameter in the model based on either results from specialized studies or subject knowledge. The actual survey data are then often assumed to be the best possible (no measurement error and no bias). Errors and bias are then added to all sample observations and model parameters according to the specified models in order to simulate a new data set from which the desired estimates are obtained. By repeating this process a large number of times (more than 500) one can estimate the relative contribution of bias and measurement errors to the overall estimate of sampling error since we "know" both the best possible value and the simulated observed value of all inputs to an estimator. An example of a measurement error budget for volume estimation follows since it will contain most of the commonly encountered features and problems.

Let us assume that we are interested in estimating the total volume of stem wood in a population area (PA) of 300-ha (strata) from a simple random sample of size $n = 40$ with fixed-area circular sample plots with a nominal area of $A = 100$ m^2 for all plots. In each plot we measure the diameter at breast height (DBH) of the n_{ti} trees in the ith plot for which DBH≥5 cm with a tape. The height of a maximum of 12 trees representing the range of DBH in the plot is measured on each plot. A function that predicts tree height (HT) from DBH is estimated from pairwise observations of HT and DBH. A predicted height $\widehat{HT}$ is obtained for all trees with no measured height. A volume equation transforms paired values of DBH and HT, viz., DBH and $\widehat{HT}$, into a stem-volume prediction of a single-plot tree. Stem volumes of single trees are summed on a per plot basis and expanded to plot-specific estimates of the population total. The desired estimate is the average plot-specific estimate of the total and the sampling variance of the estimated total is the variance of the plot-specific estimates of the total divided by n and multiplied by the finite-population correction factor. With no measurement errors and no bias the ideal (theoretical) sample estimate of total stem volume with DBH≥5 cm would be

$$\widehat{\text{vol}} = \frac{\text{PA} \times 10^4}{A} \sum_{i=1}^{n} \sum_{j}^{n_{ti}} \hat{\alpha}_{oij} + \hat{\alpha}_{1ij}\,\text{DBH}_{ij}^2 + \hat{\alpha}_{2ij}\,\text{HT}_{ij} + \hat{\alpha}_{3ij}\,\text{DBH}_{ij}^2\,\text{HT}_{ij},$$

where vol is the total volume of stem wood and α_{kij}, $k = 0,\ldots,3$ are the best possible parameters of a volume equation that generates the best possible predictions of stem volume for the jth tree in the ith plot. All design variables (PA, A, n), and all variables and parameters used in the volume compilation $\{n_{ti}$, DBH, HT, $b_{0ij}, \ldots b_{3ij}\}$ are subject to both measurement errors and bias. The actual observed volume estimate is based on

$$\widehat{\text{vol}'} = \text{PA}' \times 10^4 \sum_{i=1}^{n} \frac{1}{A'_i} \sum_{j}^{n_{ti}} \hat{\alpha}'_0 + \hat{\alpha}'_1\,\text{DBH}'^2_{ij} + \hat{\alpha}'_2\,\text{HT}'_{ij} + \hat{\alpha}'_3\,\text{DBH}'^2_{ij}\,\text{HT}'_{ij},$$

where, as befor e, X' is an attribute value with measurement error and possible bias and X is the best possible counterpart with no error and no bias. Note that the PA may also be in error and that only one volume equation is used to predict stem volume from DBH and HT. We could have used plot-specific volume equations or volume equations specific to various subsets of plots (Lappi 1991) but to keep this example relatively easy we opted for just a single equation. The error-free and bias-free volume estimate and the actual observed estimate are linked through the following set of measurement equations (carets have been suppressed to avoid cluttering):

$$\text{PA}' = \text{PA} + e_{\text{PA}},$$

$$A'_i = A + e_i^2(r) \times \pi + 2e_i(r)\sqrt{\pi},$$

$$n'_{ti} = n_{ti} + e_i(n_{ti}),$$

$$\alpha'_0 = \alpha_0 + e_i(\alpha_0) + e_{ij}(\alpha_0) + b_i(\alpha_0) + b_{ij}(\alpha_0),$$

$$\alpha'_1 = \alpha_1 + e_i(\alpha_1) + e_{ij}(\alpha_1) + b_i(\alpha_1) + b_{ij}(\alpha_1),$$

$$\alpha'_2 = \alpha_2 + e_i(\alpha_2) + e_{ij}(\alpha_2) + b_i(\alpha_2) + b_{ij}(\alpha_2),$$

$$\alpha'_3 = \alpha_3 + e_i(\alpha_3) + e_{ij}(\alpha_3) + b_i(\alpha_3) + b_{ij}(\alpha_3),$$

$$\text{DBH}'_{ij} = \text{DBH}'_{ij} + e_i(\text{DBH}) + e_{ij}(\text{DBH}) + b_i(\text{DBH}) + b_{ij}(\text{DBH}),$$

$$\text{HT}'_{ij} = \begin{Bmatrix} \text{HT}_{ij} + e_i(\text{HT}) + e_{ij}(\text{HT}) + b_i(\text{HT}) + b_{ij}(\text{HT}) \text{ if measured} \\ \text{HT}_{ij} + e_i(\widehat{\text{HT}}) + e_{ij}(\widehat{\text{HT}}) + b_i(\widehat{\text{HT}}) + b_{ij}(\widehat{\text{HT}}) \text{ if predicted} \end{Bmatrix},$$

where $e(X)$ denotes a measurement error of attribute or parameter X with an expected value of zero and $b(X)$ the bias of attribute X. Subscripts refer to plot level (i) and tree level (j). No covariance between any of the errors or error processes was specified in the equations. They were left out intentionally. One should, however, expect covariance between some errors and between some error processes but they would be inventory-specific and we wished to keep the example simple. A model-based transformation of basic observations invariably introduces covariance between errors and processes. Kangas (1996) illustrates how one can use Taylor series approximations to obtain model-consistent correlated multivariate errors for the Monte Carlo simulation. Another approach to obtain correlated errors is to specify only a set of basic errors (as in n_{ti}, DBH, and HT) and then propagate these errors through the estimation process using, again, Taylor series approximations to the best possible estimate or conversely the observed estimate. Dedicated textbooks give the necessary details (Fuller 1987; Goodchild and Gopal 1989; Carroll et al. 1995).

To arrive at an error budget we need to specify the distribution of all measurement errors at both the plot level and the tree level and all bias terms and then conduct Monte Carlo simulations of repeat sampling of actual+error+bias attribute values followed by an estimation of total volume for each repeat sample. It is customary to perform a sequence of Monte Carlo simulations, each with a specific set of errors and biases set to zero. This allows a separate assessment of various sources of errors and bias.

Plot-level errors and bias are determined by plot-specific characteristics ranging from topographical attributes to stand/forest conditions, in general, and plot-specific aspects of the measurement process, in general. For example, the plot area may be in error owing to an error in the slope correction to a horizontal reference area (Gertner and Köhl 1992). The number of plot trees may be in error because inappropriate corrections were done to adjust for boundary effects (Gregoire and Scott 2003) or mistakes were made when it was decided whether a tree located at the plot boundary was inside or outside of the plot. Change in surveyors, surveyor diligence, weather conditions, and time of day may also introduce plot-specific errors and/or bias.

3.5 Selection of Trees on Sampling Units

The sampling units of forest inventories are usually not individual trees but rather a group of trees satisfying some criterion. Sample trees may be selected by, for example, satisfying the criterion of location inside a sample plot, exceeding a distance-weighted size threshold in point sampling, proximity to a survey location or a survey line, or satisfying a rank proximity criterion at a sample location.

The rank proximity criterion includes a fixed number of trees closest to a sample location. For example, the six trees closest to a random sample location may be selected (Prodan 1965). Estimators based on this type of tree selection will often be biased, especially if the spatial distribution of trees is aggregated or in some other way displays distinct spatial patterns. Further, in dense or trackless forests it is time-consuming and expensive to determine the ranking of tree distances to a random point. Consequently, this procedure is not to be recommended for tropical forests and is not discussed here further. Those who would like to know more about it are referred to the literature (Prodan 1965; Payandeh and Ek 1986; Pollard 1971).

3.5.1 Tree Selection with Fixed-Area Sampling Units

Fixed-area sampling units are the simplest intuitive basis for selecting trees to be assessed in forest inventories. The term plot is applied to small circular, rectangular, square, or triangular areas. A strip is a rectangular sample area, whose length is a multiple of its width. Unbiased estimates can be computed for all sample areas, no matter what their shape. In planning an inventory, survey costs must be weighed against desired precision to determine the optimum size and shape of the sample plots.

The shape of a sample plot is mainly determined by cost and other practical considerations. In temperate latitudes, circular plots are usually employed as having the smallest periphery in relation to area and consequently the lowest number of borderline trees. In tropical forests, where the undergrowth hinders both access and visibility and large areas must be surveyed, it is usual to take rectangles or squares because such plots are the easiest to establish. Very often strips of up to 30-m wide and several hundred meters long are recommended.

For a fixed total sample area, choosing a larger plot size means that the sample size goes down since the product of sample size and plot area is constant. A large plot is likely to produce a lower among-plot variance than a smaller plot since large plots in general display more within-plot variation than do smaller plots. Yet the lower number of larger plots afforded under a fixed total sample area may actually produce a higher standard error than sampling with a smaller plot (Correll and Cellier 1987; Magnussen 1998; Gray 2003). The optimum plot size in terms of minimum sampling variance for a fixed total sample area is determined by the spatial distribution and the variability of the forest to be surveyed. Small plots in homogeneous forests may furnish results with higher precision, as the number of independent observations for a given sampling intensity is higher. On the other hand, in heterogeneous forests, the coefficient of variation between small plots may increase so greatly that it would be better to use a larger plot. Consequently, not only the costs but also the variability of the inventory area must be taken into consideration. A key statistic to gauge the efficiency of different plot sizes is the intraplot correlation coefficient (Cochran 1977). The coefficient measures the similarity of observations within the sample plot. Basically, the more similar the observations are, the more efficient is the small plot, and vice versa (Correll and Cellier 1987; Saborowski and Smelko 1998).

Zeide (1980), as well as Mesavage und Grosenbaugh (1956), Tardif (1965), and O'Regan and Arvantis (1966), examined various methods for optimizing plot size. Zeide weighed the time needed to locate a plot against the specified precision and stated that the optimum plot design is the design with the lowest expenditure for the specified precision. The optimum plot size a_{opt} was computed from

$$a_{opt} = a_1 \left(\frac{t}{m}\right)^2,$$

where a_1 is the size of the plot used in a pilot survey, t is the average travel time between two neighboring plots, and m is the average measuring time on a plot of size a_1. Zeide concluded from this that the greater the distance between plots, the larger the plots should be.

To compare two plot designs with plot types 1 and 2, the relative efficiency of type 1 for the estimation of, say, a population total is

$$\text{eff}_{\text{type 1:type 2}} = \frac{\hat{\text{var}}\left(\hat{Y} \middle| \text{plot type 1}\right) \times \text{time of inventory with type 1}}{\hat{\text{var}}\left(\hat{Y} \middle| \text{plot type 2}\right) \times \text{time of inventory with type 2}}.$$

An efficiency ratio less than 1 means that plot type 1 is more efficient, and vice versa. Note that the efficiency depends on the population parameter of interest. It is possible that one plot type is more efficient for one parameter but less efficient for a different parameter.

As a rule of thumb, a plot should be large enough to contain enough trees per plot for the survey results to be representative of the population at large while at the same time keeping the time to complete a plot to a minimum. It follows that small plots should be employed for dense stands with small trees, and large ones for open stands and large trees. Very often, a distinction is made between unproductive relocation time between plots and productive survey times. When travel time is significant, as in a tropical forest, the size of inventory plots tends to be large, often in the 0.4–0.5-ha range.

The horizontal plane is the reference base for all inventory data, and sample plots in sloping terrain must be adapted accordingly. Three distinct procedures for this adaptation are given next; the first is general, while the second and third are for circular plots only:

1. All plots: Demarcate the plot on the incline and then expand the plot proportionally to the degree of inclination in such a way that an orthogonal projection of the expanded plot onto the horizontal plane matches exactly the nominal plot in area.
2. Circular plots: Demarcate an ellipse on a slope with the short axis and the long axis determined in such a way that when the ellipse is projected orthogonally onto the horizontal plane it coincides with the circular outline of the nominal inventory plot.
3. Circular plots: Measure or compute the horizontal distance of candidate plot trees to the plot center. Only trees within a distance of the nominal plot radius are included and measured.

In order to simplify the field survey and to facilitate an audit and a quality control of the inventory the enlargement of the plot in proportion to the degree of inclination appears most suitable. Köhl and Brassel (2001) showed for the Swiss NFI that there is no significant difference between the second and third expansion approach. In mountainous regions, an error in the corrected plot can induce a nontrivial error in the survey results. For this reason, corrections for slope inclination must be made by fully qualified personnel only, and all pertinent data about the expansion/correction process should be captured to allow control and possibly correction of the errors.

Sample plots in areas with a high stem density may contain a large number of trees. The aforementioned general principle about small plots being more efficient in areas with a high tree density and larger plots being more efficient in areas with a low tree density has led to a design with multiple concentric plots. Two or more plots of differing size are demarcated around a given sample point. In the smallest area, all trees with a diameter greater than a given minimum fixed by design (e.g., 12 cm) are surveyed. In the larger plots, the minimum diameter threshold is higher. This design often allows a considerable reduction of survey time with barely noticeable decreases in efficiency. Figure 3.12 shows the concentric plot design employed in the Swiss NFI. On the smaller, 200-m^2 plot, all trees with a DBH over 12 cm are measured, while on the larger, 500-m^2 plot only trees with a DBH of 35 cm or above are measured.

It is also important to consider the life span of an inventory plot when deciding on a plot size. Permanent fixed-area sample plots intended for multiple surveys are difficult to optimize. The number of trees and their size will naturally change over time. To ensure that the plot size is sufficient throughout the life of a plot, a permanent fixed-area plot tends to be relatively large. For continuous forest inventory and monitoring, fixed-area plots are to be particularly recommended, as they allow easy determination of growth components such as ingrowth, mortality, and cuts (Scott 1998). Also, fixed-area plots are usually

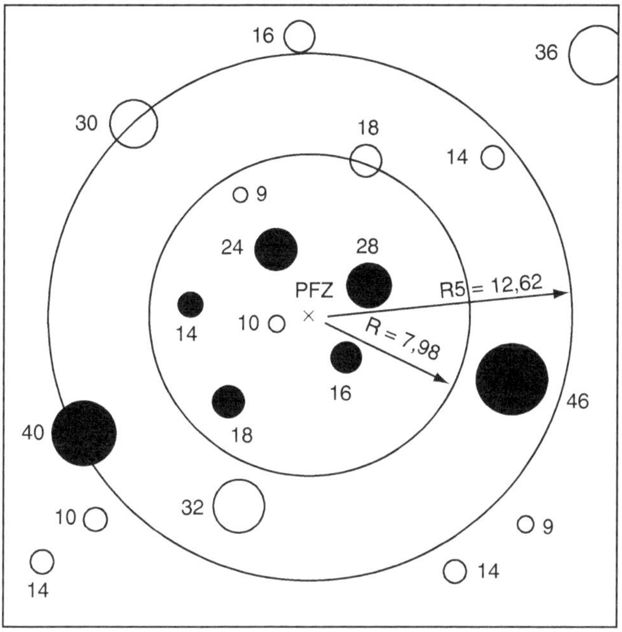

Fig. 3.12. Field plot of the Swiss National Forest Inventory (*dots* tallied trees, *circles* trees not tallied)

simple to survey, maintain, and analyze. For these reasons they are preferred over variable-radius plots.

3.5.2
Scaling of Individual Tree Data into Sample Plot Values

The statistical approach to sampling designs generally assumes that sample plots represent the smallest (natural) sample unit; however, actual sampling may not be done with this unit. Indeed plots of different size may be used or trees may be selected based on a criterion of inclusion. Thus, individual tree values sampled during a survey have to be scaled to this natural unit. It is common to take an area of 1 ha as the natural unit. The scaling is accomplished by area weighting of the attribute, say *Y*, of the *j*th tree on the *i*th sample location. Let a_{ij} denote the area of the sample plot used to sample the *i*th tree at the *j*th sample location. The meaning of sample plot area for trees selected by a criterion of inclusion will become clear as we later examine various sampling methods.

The area weight given to the attribute value Y_{ij} is $w_{ij} = a_{ij}^{-1}$, which becomes the attribute value per hectare. This area weighting is flexible as it extends naturally to sampling with unequal probability of inclusion for individual trees. A simple example ought to clarify the concept. Let us assume that we are sampling with a set of fixed-area concentric sample plots. Trees below a certain size threshold are measured on the smaller plot(s) and trees above a certain threshold are measured on the bigger plot(s). Thus, the selection probability of trees is not constant but depends on tree size. The effect of different plot sizes on selection probabilities has to be corrected through scaling to a common (natural) unit via area weights. For each concentric plot a separate scaling factor applies. If, for example, two concentric plots of sizes 0.02 and 0.05 ha are used, the scaling factors are calculated as

$$w = \frac{1}{0.05} = 20 \text{ for trees on the } 0.05 - \text{ha sample plot}$$

and

$$w = \frac{1}{0.02} = 50 \text{ for trees on the } 0.02 - \text{ha sample plot.}$$

In the previous example it was the size of the trees that determined whether they were measured on one plot or the other. Often DBH is used as the size criterion owing to ease of measurement. In that case the area weights (scale factors) become functions of DBH. If in the previous example trees with a DBH between 12 and 35 cm were tallied in the smaller, 0.02-ha plot and trees with a larger DBH were tallied in the 0.05-ha plot we can express the weights as

$$w_{ij} = \begin{cases} 50 \text{ if } 12\text{cm} \leq \text{DBH}_{ij} \leq 35\text{cm} \\ 20 \text{ if } \text{DBH}_{ij} > 35\text{cm} \end{cases} .$$

Note, the scaling of inventory estimates to a unit area (here 1 ha) allows us to assess the effect of plot size, plot shape, and selection criterion on statistical estimates of interest. Recall, we do not consider a scaled attribute value as a ratio of two random variables since we assume throughout that the scale factor is known without error. Measurement errors in a_{ij} are not considered.

After scaling individual attribute values Y_{ij} to a unit area (1 ha), we usually sum them to a single value Y_{i+} for the ith sample location:

$$Y_{i+} = \sum_j w_{ij} Y_{ij} .$$

3.5.3 Point Sampling

Compared with fixed-area plots, point sampling is a relatively new procedure. It was first presented by Bitterlich (1947, 1997) in 1947 and was further refined and theoretically substantiated by Keen (1950) and then Grosenbaugh (1952). Alternative names for this method are angle count, variable plot cruising, and plotless cruising, names that reflect one of the most important features of the method: the area surveyed varies.

The first step in point sampling is the same as that in surveying with fixed-area plots: the selection of sample locations (points). Then, attributes of interest are measured on all trees meeting a certain criterion of selection. Typically the criterion is DBH and the decision on whether to include or exclude the tree from selection is based on a measurement with an angle gauge instrument. The simplest form of an angle gauge has a cross-arm attached horizontally to a vertical stick held at a known distance from the eye. With this instrument, an angle in the horizontal plane and 1.3 m aboveground is aligned with the trunk of each tree visible from the sample location. All trees with a DBH forming an angle greater than the angle subtended by the crossbar are selected. Trees with a smaller DBH are ignored (Fig. 3.13). Assessments and measurements are then carried out on the selected trees. Refinements of this method include electronic verification of inclusions based on optical/electronic measurements of angles and distances at a fixed reference height (Bitterlich 1997).

The basal area per hectare at the sample location is determined through multiplication of the number of "in" trees by a constant factor derived from the given angle subtended by the horizontal crossbar; no extra measurements are needed. Thus, each tree assessed, independent of its diameter, represents the same basal area per hectare; a proof is given next.

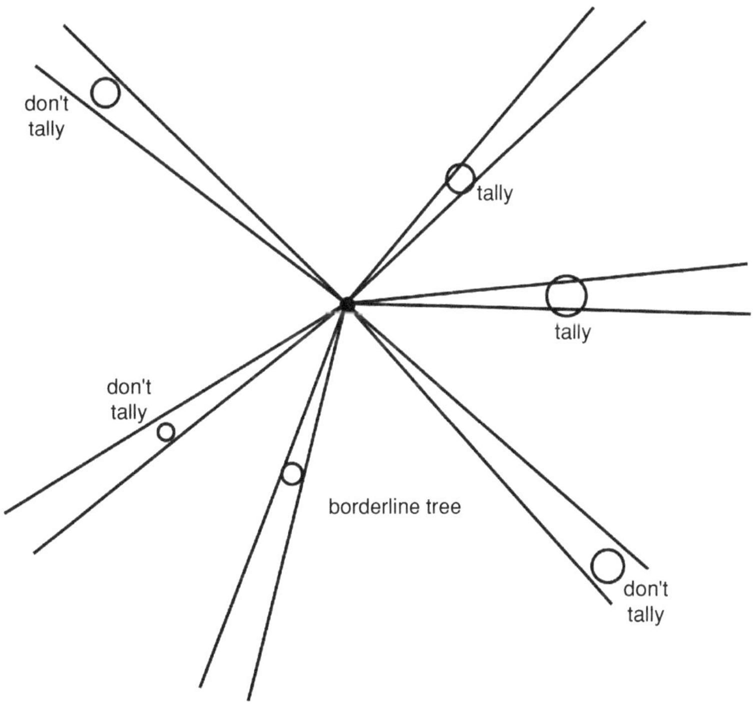

Fig.3.13. Point sampling

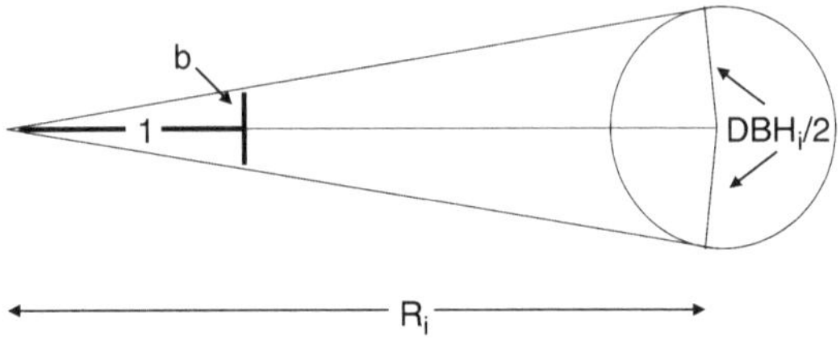

Fig. 3.14. Geometrical principle of point sampling

The geometrical principle of point sampling is illustrated in Fig. 3.14 with a cross-arm 2-cm wide (b=2 cm) attached to a stick 1-m long (l=1 m). For a tree at distance R_i with a DBH equal to DBH_i and subtending an angle equal to the angle subtended by the angle gauge cross-arm we get

$$\frac{b}{l} = \frac{2}{100} = \frac{\text{DHB}_i}{R_i} \Leftrightarrow R_i = 50 \times \text{DBH}_i.$$

Any sample location within a distance of $50 \times \text{DBH}_i$ from this tree would result in the tree being included in the sample. In other words, the sample area for this tree is $\pi \times (50 \times \text{DBH}_i)^2$. The attribute value of the tree is therefore given an area weight equal to the inverse of this area. If the attribute value is basal area $\left(\pi/4 \times \text{DBH}_i^2\right)$ the area-weighted attribute value is simply

$$\frac{\pi/4 \times \text{DBH}_i^2}{\pi \times (50 \times \text{DBH}_i)^2} = \frac{1}{10000} \frac{\text{m}^2}{\text{m}^2} = 1 \frac{\text{m}^2}{\text{ha}}.$$

This means that every selected tree represents a basal area of 1 m^2/ha. The basal area per hectare is estimated by simply counting the number of "in" trees.

The simple derivation just shown is only valid for b/l=2/100. If a gauge with a different subtended angle (α) is used a more general equation must be employed (see Fig. 3.14 for details and definitions):

$$R_i = \frac{d_i}{2 \times \sin(\alpha/2)}.$$

For any given gauge angle α and count N_{count} of "in" trees the basal area G per hectare is estimated as

$$\hat{G} = N_{\text{count}} \times 10^4 \times \sin^2(\alpha/2) = N_{\text{count}} \times \text{BAF},$$

where BAF is the basal area factor. The basal area factor is indicated on commercially available angle gauges. After a 360° sweep and deciding on "in"/"out" for every visible tree, one obtains the basal area per hectare by multiplying the number of "in" trees (N_{count}) by the basal area factor. The chosen angle and thus the factor determine the number of trees selected. The wider the angle, the higher the factor and the lower the number of trees selected. In tropical forests, factors between 8 and 10 are popular; they ensure reasonable counts (between 10 and 40).

As already illustrated, point sampling with an angle gauge is essentially sampling with PPS (basal area) (Fig. 3.15). In fixed-area sampling all trees have the same probability of selection, a probability that only depends on plot size. For any attribute related to size, a selection with probability of selection proportional to size will result in a more efficient sampling (Brewer and Hanif 1983; Särndal 1996). The estimated sampling error for a given number of selected trees will be lower than for sampling with equal selection probability.

It can happen that the angle subtended by a tree's DBH appears to be exactly equal to the gauge angle. Such trees are termed "borderline trees"; their diameter and distance from the point center must be measured accurately, and the decision as to whether to include them or not is based on the equation. Alternatively one could toss a coin and decide on the basis of the outcome of the coin toss.

Trees not visible from the sample location are obviously a potentially serious source of bias in point sampling with an angle gauge. Great care must therefore be taken to ensure that no tree has been missed.

Commercially available instruments for point sampling are based on one of two different principles. One uses the previously outlined principle of two divergent lines starting at the viewpoint and extending to a fixed reference distance and beyond until intercepted by an obstacle (Fig. 3.16a). The practical problem

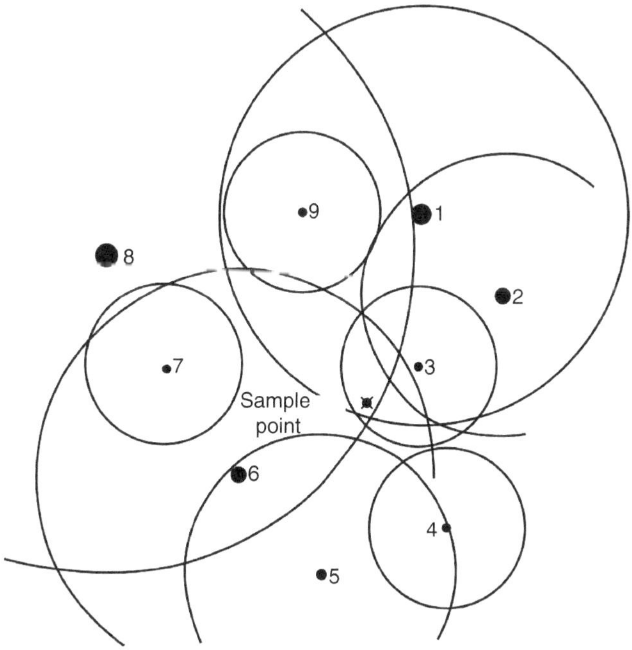

Fig. 3.15. Selection of trees in point sampling (Circles indicate the area within which a sample points needs to be located in order to select the corresponding tree)

that arises with this type of instrument is that a close object (reference distance) and a distant object (the tree) have to be focused and two lines (right and left side of the tree) observed simultaneously, often by a human observer. This renders decisions about whether or not to include borderline trees difficult. Angle gauges sold under the name of Relascope include a feature for automatic correction for inclinations from the horizontal. A wide-scale Relascope was developed for application in tropical forests (Loetsch et al. 1973).

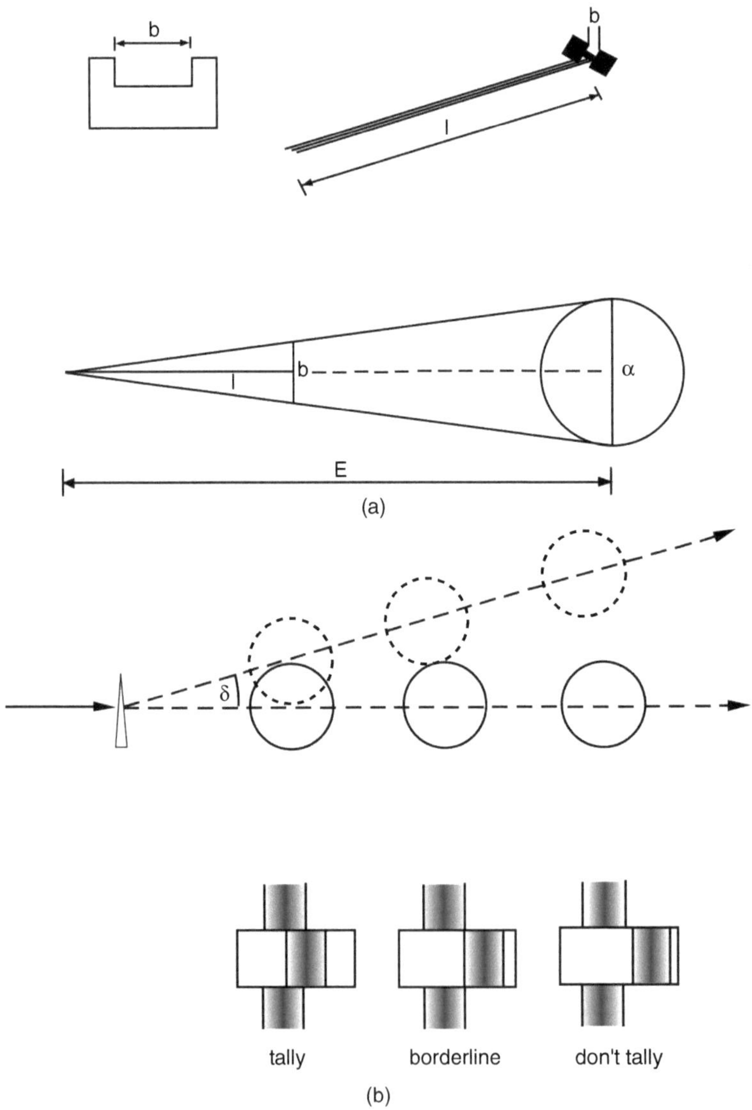

Fig. 3.16. a Stick-type angle gauges. **b** Wedge prisms for point sampling

The second principle is based on the diffraction of light rays, in our case diffraction of light rays from the tree as they go through a wedge prism in front of the observer (Fig. 3.16b). The observer will see two superimposed images of the tree stem: an actual nondiffracted image superimposed on a diffracted image. The tree is selected when its actual image overlaps with the diffracted image. Trees with a diffracted image displaced laterally relative to the actual image are not selected. It is much easier to decide about borderline trees with this type of instrument than with a Relascope or a stick-type angle gauge. Ease of use made them popular, especially in Canada and the USA.

With angle point sampling, any measured attribute (say population mean) of the trees counted as "in" should be expanded to a common reference area of 1 ha in order to remove the effect of differential inclusion probabilities. The expansion takes the form

$$\hat{Y}/\text{ha} = \frac{\text{BAF}}{n} \times \sum_{i=1}^{n} \sum_{j=1}^{N_{\text{count}}(i)} \frac{y_{ij}}{\text{BA}_{ij}},$$

where BA_{ij} is the basal area of the *j*th "in" tree at the *i*th sample location. Consequently the basal area of all "in" trees must be determined or, conversely, estimated from a measurement of DBH and the assumption of a circular outline of the stem cross section. Per-hectare estimates of stems and basal area deserve special attention. For stem count, the attribute value of each "in" tree is 1, so the estimator for the stem count per hectare becomes

$$\hat{N}/\text{ha} = \frac{\text{BAF}}{n} \times \sum_{i=1}^{n} \sum_{j=1}^{N_{\text{count}}(i)} \frac{1}{\text{BA}_{ij}}$$

and the estimator for the basal area per hectare is

$$\hat{\text{BA}}/\text{ha} = \frac{\text{BAF}}{n} \times \sum_{i=1}^{n} \sum_{j=1}^{N_{\text{count}}(i)} \frac{\text{BA}_{ij}}{\text{BA}_{ij}} = \frac{\text{BAF}}{n} \times \sum_{i=1}^{n} N_{\text{count}}(i).$$

There are many variations in horizontal point sampling methodology, some have already been described. Vertical point sampling and vertical line sampling (Strand 1958), the critical height method for volume assessment and angle counting by Wenk (Loetsch et al. 1973; Hush et al. 1982) have also become popular. Uebelhör (1988) describes the use of point sampling with the wide-scale relascope in the Philippines NFI and recommends point sampling for measurements in tropical rain forests to the reduce the cost of field surveys. Other applications of point sampling in tropical forests have been presented by Boon (1970), Puffenberger (1976), da Silva (1982), and Banyard (1987). Sampling for coarse woody debris has spurred new refinements of the Relascope idea for special-purpose sampling (Ståhl 1997, 1998; Ringvall and Ståhl 1999).

3.5.4 Point Sampling Versus Fixed-Area Plots

Forest resource sampling is a challenge owing to the frequently encountered problem of uniquely defining population elements/units and consequently the problem of defining a sampling frame. The point paradigm, by which a population is defined as all possible spatial locations of a sample unit, is adopted out of necessity. When there is no natural sampling unit the survey designer has to decide on how observations are gathered at each sample location. The decision as to whether to employ point sampling or sampling with fixed-area plots depends on the individual aims and needs of the inventory. In a study on an area of 60 ha in Surinam, Schreuder et al. (1987) compared the efficiency of fixed-area plots, point sampling, and horizontal line sampling. Fixed-area plots gave the best results in terms of root-mean-square error for tree number, horizontal line sampling for basal area, the sum of tree diameters, and the average tree diameter. Point sampling was superior for estimating the number of trees in the upper-diameter classes, while fixed-area plots fared better for the smaller-diameter classes. These findings apply in general.

With point sampling in stands with a high stem count, with clusters of big and small trees, or with dense undergrowth, there is a nontrivial risk that trees may be hidden, and consequently that a negative bias may be incurred. The time to implement repeated checks for hidden trees and their inclusion in the local sample quickly erodes any practical advantage of point sampling. In such cases, it is preferable to use fixed-area plots.

In the consideration of fixed-area plots versus point sampling the survey analyst must also take into consideration the life span of a sample location. Will the sample locations be used in future inventories or will they be part of an ongoing monitoring? If plots are to be used again and again over time for the purpose of estimating change and trends in population parameters, the fixed-area plot has some distinct advantages. In point sampling, the inclusion probability of a tree depends on the attribute value of the tree (commonly basal area) at the time of sampling. Thus, the inclusion probability does not remain constant over time for attributes/variables that change over time. In the context of estimating the population parameter "tree growth" the change in inclusion probabilities generates some unique estimation problems. At the time of remeasurement you can have trees included in the point sample that were not included at the previous time of sampling. Estimation of growth of individual trees becomes cumbersome when their inclusion probability has changed between the times of measurement (Flewelling and Thomas 1984; Scott 1998).

There are two distinct events that would allow a tree to enter the later measurement but not the earlier. First, the DBH of the tree exceeded the minimum threshold diameter on the first occassion but it was located beyond the critical inclusion radius. The growth estimate for this tree would equal its size. The common terminology for this type of growth is ongrowth. The second event that allows a tree to enter the sample between sampling times happens when its size exceeds the inclusion threshold on the second but not the first measurement occassion. The growth calculated for this tree is called ingrowth. Kuusela (1979) describes various estimators of growth based on point sampling with repeated measurements at a fixed set of sample locations. The complex nature of these estimators suggests that they should only apply in exceptional cases. Procedures for estimation of increment and growth components in fixed-area plot sampling are simple in comparison.

3.5.5 Sampling at the Forest Edge

Sample locations at the forest edge present a special estimation problem in forest inventories. Since the population of interest is restricted to areas classified as forest it can happen that the effective sample area for locations along the edge is less than the nominal area associated with sample locations in the interior of the forest. It would be wrong to simply disregard such sample locations as this would mean that trees growing in areas along the forest edge would have a different probability of being selected compared with trees growing further away from this edge (Williams 1996; Gregoire and Scott 2003). Since growing conditions and tree species along the forest edge are often distinctly different from the those in the interior forest, disregarding edge plots could lead to a serious bias in the inventory estimates.

The surveyor has several options for correctly dealing with the boundary effects. The problem and solutions are best understood if we adopt a tree-centered view of sample areas. For the fixed-area plot the sample area of a tree is simply the area covered by the sample plot when the center of the sample plot and the tree location coincide. For point sampling the sample area is the area covered by the circle centered at the tree location and with a radius equal to the critical distance of selection. Any sample location falling inside the sample area triggers the selection of the tree located in the center of the sample area. Trees located along the edge of the forest will have part of their sample area outside the population of interest. They are therefore less likely to be selected than a tree further away from the boundary. The solutions presented next are for sampling with fixed-area plots but they apply equally to point sampling by simply replacing the term sample plot by sample area. One recommended option involves finding the exact intersection of the forest edge with the inventory plot and then computing the area of the plot that is inside the population. The attribute values observed on this partial plot are scaled according to the area of the plot inside the population. The weighting scheme can also be applied to

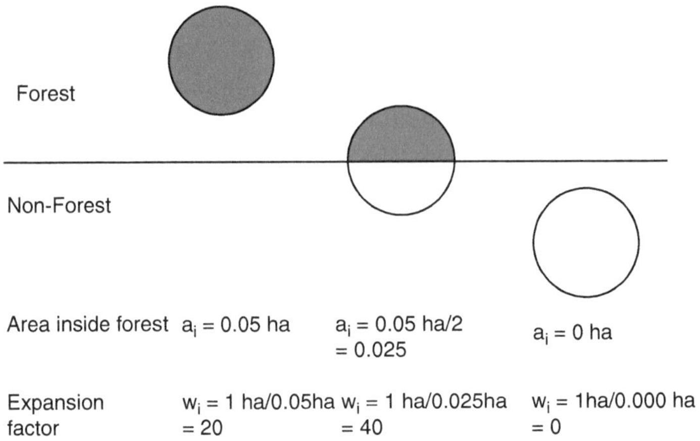

Fig.3.17. Sampling at the forest edge

trees in individual plots. For trees with a distance to the forest edge less than the radius of the appropriate sample area (e.g., 12.62 m for a circular plot of 0.05 ha) the part of the sample area inside the forest is determined for each tree and converted to an area weight w_{ij} (Fig. 3.17).

Another rather ingenious solution is the "fold-back" or the mirror reflection method (Schmid 1969). In a mirror reflection of a straddler plot, the part of the plot that falls outside the forest is projected orthogonally back into the forest with the forest edge serving as the axis of projection. The surveyor records all attributes for the part of the plot that is fully inside the forest and then all attributes on the mirror reflection of the part that was outside. In other words, a part of the plot is measured twice and occasionally three or four times if the boundary is on a corner and reflected portions of the plot overlap. Correctly applied, this method produces unbiased results, but it assumes that the forest boundary can be located accurately. In practice it is often quite difficult to decide on the exact location of a boundary. Any nonlinear edge will also generate practical problems with the mirroring. When the forest edge cannot be defined in precise and generally valid terms the method becomes problematic. Two easy, but resulting in biased estimates, solutions are to (1) relocate the straddling plot further away from the boundary to avoid overlap and (2) expand the part of the plot inside the forest to compensate for the area outside the forest. Gregoire and Scott (1990) compared four unbiased and three biased methods for dealing with sample plots at the forest edge in a mixed hardwood and mixed softwood stand in Maine, USA. They concluded that no single method was uniformly superior; the performance depends on the nature and magnitude of the "edge effect." Some biased methods of plot relocation performed, at times, better than the unbiased methods.

In practice usually only plots with a plot center inside the forest are accepted and tallied. The existence of sample locations outside the forest boundary, however, raises questions about the integrity of the sample frame and the multipliers to use to when you scale per hectare estimates to population totals. If a sample location is judged outside the forest but the outside location is actually a part of the forest estate (in an administrative or legal context) it can be argued that the "outside" location should have been included in the sample. In areas with illegal forest clearing, for example, the discarding of "outside" sample locations could lead to a serious overestimation of per hectare valued attributes.

3.6 Sampling on Successive Occasions

Sampling on successive occasions is done for the following main objectives:

- To determine the status of the forest resource at the time of the first inventory
- To determine the status of the forest resource at the time of the second inventory
- To determine changes in the forest resource between two successive inventories

The idea of quantifying change in a forest resource as the difference between two successive inventories was first applied to individual forest stands. Repeated measurements of a selected number of representative stands offered a way to verify the sustainability in terms of the yield of stands that were under a fixed forest management regime. This fundamental idea to quantify forest yield was born in the last century in Europe. In Germany, the first permanent plots were established in 1860 (Graves 1906). Foresters in France (Gurnaud 1878) developed a set of rules for how to estimate increment from successive measurement. In French-speaking countries the rules were given the name *la méthode du contrôle*. Biolley (1921) was the first to apply the rules. The forest of Couvet in the Swiss Jura, where the rules originated, was measured ten times between 1890 and 1946 in intervals every 6–7 years. The rule set has since been widely adopted. It is known as "the control method" in the English literature.

In the USA, the idea of obtaining quantitative estimates about the change in standing wood volumes through repeated measurements of the same set of plots gained support and acceptance during the years between 1929 and 1950 (Stott and Semmes 1962). The economic recession of the 1930s accentuated the need for reliable estimates of wood volume. A general increase in interest in primary production factors was instrumental in the pioneering application of sampling methods for estimating change. A direct adaptation of the European yield control methods, which were based on recording all trees within stands up to several hectares in size, would clearly not be feasible, the intensive control method of Gurnaud (1878) and Biolley (1921) even less so. As well, there was no representative network of "benchmark" stands to which a set of standard management regimes applies. The great expanse of the North American forests, mostly without any established stand structure, dictates that only a small fraction of the forest of interest could be surveyed. The favored approach was for the application of objective and scientifically sound sampling methods, a rare approach at that time.

3.6.1 Continuous Forest Inventory

In the 1930s, a sampling method, known as continuous forest inventory (CFI), was developed in the USA. CFI is based on repeated measurements of a set of sample plots (Stott and Ryan 1939). Stott and Semmens (Stott and Semmens 1962) give a historic overview of the CFI application. In the Midwest, between 1937 and 1938, a few hundred permanent sample plots in forests operated by the wood processing industry were established. In the Great Lakes and Central Plains states starting in 1939, approximately 3,700 permanent circular sample plots were set up in private, industrial, and public forest enterprises. In 1948, the inventory of forests in Ohio and Wisconsin took place with about 1,000 permanent sample plots. In 1952, the American Pulpwood Association (APA) became aware of the CFI and introduced it to its members. During the following years, a cooperation between the APA and the USDA Forest Service led to an extensive application of the CFI extending east of the Mississippi River. In 1962, approximately 50 enterprises associated with the wood processing industry managed 25 million acres according to the CFI method. Most CFI plots were established in what was termed "typical" timber-producing stands; as such they are not representative of the entire forest resource.

The pioneers of CFI in Germany were Krutzsch and Loetsch (1938). In 1936 they set up a series of permanent sample plots for yield control. In Sweden, CFI was pioneered by Patterson (1950) and early on applied to forest yield research at the Swedish forest experimental station. In Switzerland, CFI was introduced by Schmid (1967) and it was applied to forest management planning, in effect an extension of the classic control method to CFI. His intensive effort towards an applied survey method for permanent sample plots (Schmid-Haas et al. 1993) resulted in wide acceptance of the method in Swiss forestry.

With the CFI method, all sample plots, which are measured on the first occasion, are measured again in successive inventories. The estimators of population parameters under CFI are time-specific. We indicate the time dependency of

CFI estimators by a suffix t for time. The suffix takes on values 1,2,3, . . . for the estimates of population parameters at the first, second, third, and so on inventory. We are usually interested in the estimation of a change between two successive inventories. When the context is clear we simply refer to estimators and estimates of the "first" occasion and of the "second" occasion inventory, respectively. In continuation of our example for the estimation of a population mean under SRS at the first and the second occasion inventory, the CFI estimator is

$$\hat{\bar{Y}}_t = \frac{1}{n_t}\sum_{i=1}^{n_t} Y_{t,i}, t = 1,2.$$

Changes in a population parameter between two inventories can be derived as the difference between the estimates of the population parameter at two successive inventories. For the previous example we have

$$\Delta\hat{\bar{Y}}_{2,1} = \hat{\bar{Y}}_2 - \hat{\bar{Y}}_1.$$

When the same set of plots are remeasured on both occassions, the estimator of the variance of the change of the mean becomes

$$\hat{\mathrm{v}}\mathrm{ar}\left(\Delta\hat{\bar{Y}}_{2,1}\right) = \hat{\mathrm{v}}\mathrm{ar}\left(\Delta\hat{\bar{Y}}_2\right) + \hat{\mathrm{v}}\mathrm{ar}\left(\Delta\hat{\bar{Y}}_1\right) - 2\times\hat{\rho}\left(Y_2, Y_1\right)\times\sqrt{\hat{\mathrm{v}}\mathrm{ar}\left(\hat{\bar{Y}}_2\right)\hat{\mathrm{v}}\mathrm{ar}\left(\hat{\bar{Y}}_1\right)},$$

where $\hat{\rho}\left(Y_2, Y_1\right)$ is an estimate of the correlation coefficient between the observations on the second and the first occasion; $\hat{\rho}\left(Y_2, Y_1\right)$ is restricted to values between −1 and +1.

The higher the correlation is between paired observations from the first and second inventory, the smaller is the variance of their difference. For a large number of size-related attributes the temporal correlation between plot variables measured on two occasions will be positive. Autocorrelation is the single most significant contributor to this correlation (the attribute on the second occasion is equal to the attribute on the first occasion plus change). When the correlation is positive the variance of the change will be less than the sum of the variances on the first and second occasions. However, as time separates the two inventories the correlation tends to dissipate. The rate of decrease will depend on how well change is correlated with the attribute value on the first occasion. For trees growing in the absence of disturbances, in a homogenous environment free of competition, and with a nonrestrictive supply of nutrients, the correlation between change and initial attribute value may be quite strong over long periods of time. In heterogeneous environments with frequent disturbances and physiological stress the correlation may be weak, zero, or even negative. When the correlation is zero or perhaps even negative the variance of the difference will be equal to or larger than the sum of the respective variances. For example, if large trees are more prone to hurricane damage than small trees, and one or more hurricanes have gone over the forest since the previous inventory, the correlation between, say, plot volume at the two occasions could be negative. At the other extreme, when the correlation is 1, as is the case when the attribute value on the second occasion is equal to the attribute value on the first occasion times a constant plus a constant, the variance of the change is 0. A correlation coefficient of 1 is extremely rare. If the plots measured on the first and the second occasion are not the same and each set is selected independently of the other then the correlation of attribute values is by definition zero and the variance of the change estimate is simply the sum of the respective variances.

For a positive correlation coefficient a CFI estimate of change will have a smaller variance than an estimate change derived from two sets of independent observations. The advantage of using the CFI method rests with the reduction of the variance of estimated change.

The CFI method, despite its obvious advantage, encounters practical and inferential problems. Over time the locations of sample plots may become known beyond the surveyors and, as a result, they may evolve differently from the surrounding forest. This nontrivial risk is especially acute for visibly marked sample plots. The potential of an inferential problem is latent because, as paraphrased by Schmid-Haas (1983), "there is no guarantee that sample plots, visible or not, will remain representative of the target population." Schmid-Haas also believes that even the most experienced forester cannot be sure that he or she would not be influenced by the knowledge that certain parts of the forest are subject to the intensive scrutiny of repeated measurements. Consciously or unconsciously, it is possible that the sample locations are being treated differently in some way, shape, or form. A sample plot inventory, which cannot reliably eliminate this risk, may become biased and will quickly lose credibility and invested goodwill.

If only the net change has to be estimated, for example, volume growth, permanent sample plots would be more cost efficient than two independent surveys, which means that for the same cost they lead to a smaller sample error. This seems obvious, since the difference between two independent observations is not only caused by change alone, but also by the variation within the two populations. If only the current state is to be considered, temporary sample plots are often more cost effective than permanent plots, since the expenditures for marking the sample plot centers and the registration of sample tree locations do not exist.

The application of the CFI method can lead to inferential problems. All CFI inventory systems rely on the assumption that the permanent plots are representative. But are they? With time the plots may "drift" at a rate different from that of the population they are supposed to represent. This risk is especially acute in managed forests or in places with frequent land-use changes. As well, changes in the inventory objectives are difficult to accommodate in CFI with its system of plots established in the past and tailored to past objectives.

Practical survey objectives are often a blend of target precision on estimates of state and change. In this case a design with a mixture of permanent and temporary plots appears attractive. The idea of a survey design with both plot types arose from these considerations.

3.6.2 Sampling with Partial Replacement of Sample Plots

In SPR a fraction of the sample plots measured during a survey are replaced by new sample plots at the subsequent survey. This pattern of partial replacement is repeated over time. SPR was introduced into forest inventory around 1960. Kish (1964), Cochran (1977), and Sukhatme et al. (1984) also discussed the theory of SPR of sample plots. Bickford (1959) was the first to introduce the theory of SPR to forest inventory. The first to apply SPR was the USDA Forest Service in the Allegheny National Forest where it was combined with aerial photographs and modified accordingly (Bickford 1959).

We shall describe SPR estimators based on only two SPR occasions for the estimation of a population mean. SPR estimators for subsequent surveys are more complex. After the second SPR there are three types of plots available for the estimators:

1. n_{12} sample plots measured on both occasions (matched permanent plots)
2. n_1 sample plots measured only on the first occasion (unmatched first occasion plots)
3. n_2 sample plots measured only on the second occasion (unmatched second occassion plots)

The most precise unbiased linear estimator of the state on the first occasion $\hat{\bar{Y}}_1$, on the second occasion $\hat{\bar{Y}}_{2|1}$, and of change between the two occasions $\Delta\hat{\bar{Y}}_{21}$ is

$$\hat{\bar{Y}}_1 = \frac{\sum_{i=1}^{n_1} Y_{i1} + \sum_{j=1}^{n_{12}} Y_{j1|2}}{n_1 + n_{12}} = \frac{n_1}{n_1 + n_{12}}\hat{\bar{Y}}_1 + \frac{n_1}{n_1 + n_{12}}\hat{\bar{Y}}_{1|2},$$

where $Y_{j1|2}$ is used to denote the attribute value on the first occasion on the n_{12} matched plots. A corresponding estimator for the second occasion $\hat{\bar{Y}}_2$ is obtained by a simple switch of occasion subscripts. The best estimator of the status on the second occasion exploits the relationship between the attribute values on the first and the second occasion:

$$\hat{\bar{Y}}_{2|1} = \hat{c} \times \left[\hat{\bar{Y}}_{2|1} + \hat{\beta}_{2.1}\left(\hat{\bar{Y}}_1 - \hat{\bar{Y}}_{1|2}\right)\right] + (1 - \hat{c})\hat{\bar{Y}}_2,$$

where $\hat{\beta}_{2.1}$ is the ordinary least-squares regression coefficient obtained by regression of $\hat{\bar{Y}}_{2|1}$ on $\hat{\bar{Y}}_{j1|2}$ and $\hat{c}$ is an estimate of the optimal weight to be assigned to the first term, which is essentially the estimator used for double sampling with regression estimator. The optimal weight is

$$\hat{c} = \frac{n_{12}(n_1 + n_{12})}{(n_1 + n_{12})(n_2 + n_{12}) - \hat{\rho}_{2.1}^2 \times n_1 \times n_2},$$

where $\hat{\rho}_{2.1}$ is an estimate of the correlation coefficient between the second and the first occasion attribute values. The best unbiased linear estimator of change is

$$\Delta\hat{\bar{Y}}_{21} = \hat{\bar{Y}}_{2|1} - \hat{\bar{Y}}_{1|2},$$

where $\hat{\bar{Y}}_{1|2}$ is the double sampling with regression estimator of status on the first occasion computed as

$$\hat{\bar{Y}}_{1|2} = \hat{e}\left[\hat{\bar{Y}}_{1|2} + \hat{\beta}_{1.2} \times \frac{n_2}{n_{12} + n_2}\left(\hat{\bar{Y}}_2 - \hat{\bar{Y}}_{2|1}\right)\right] + (1 - \hat{e}) \times \hat{\bar{Y}}_1,$$

with

$$\hat{e} = \frac{n_2 + n_{12}}{n_1 + n_{12}} \times \hat{c}.$$

The variance estimator for $\hat{\bar{Y}}_1$ is obtained as for SRS, while that of $\hat{\bar{Y}}_{2|1}$ is

$$\hat{\text{var}}\left(\hat{\bar{Y}}_{2|1}\right) = \frac{\left(n_1 + n_{12} - \hat{\rho}_{2.1}^2 \times n_1\right) \times \left[\sum_{i=1}^{n_2}\left(Y_{i2} - \hat{\bar{Y}}_2\right)^2 + \sum_{j=1}^{n_{12}}\left(Y_{j2|1} - \hat{\bar{Y}}_2\right)^2\right]}{\left[(n_1 + n_{12})(n_2 + n_{12}) - \hat{\rho}_{2.1}^2 \times n_1 \times n_2\right](n_2 + n_{12} - 1)}.$$

Finally, an unbiased (but not minimum-variance) SPR estimator of change is

$$\hat{\text{var}}\left(\Delta\hat{\bar{Y}}_{21}\right) = \left(1 - \hat{k}_2\right)\frac{\hat{\text{var}}(Y_2)}{n_2} + \left(1 - l^2\right)\frac{\hat{\text{var}}(Y_1)}{n_1} + \frac{\hat{k}^2\,\hat{\text{var}}(Y_{2|1}) + l^2\,\hat{\text{var}}(Y_{1|2}) - 2\hat{k} \times l\text{cov}(Y_{1|2}, Y_{2|1})}{n_{12}}.$$

where

$$\hat{k} = \frac{n_{12}(n_1 + (n_2 - n_{12})\hat{\beta}_{2.1})}{n_1 n_2 - \hat{\rho}_{2.1}^2(n_1 - n_{12}(n_2 - n_{12})} \text{ and}$$

$$\hat{l} = \frac{n_{12}(n_2 + (n_1 - n_{12})\hat{\beta}_{2.1})}{n_1 n_2 - \hat{\rho}_{2.1}^2(n_1 - n_{12}(n_2 - n_{12})}$$

Ware (1960) examined inventory data from repeated measurements in the northeastern region of the USA and found that in six out of eight cases, the variance was not the same on both inventory occasions. It is therefore important not to simplify the change estimator by assuming equal variances if they are not. Violations of this assumption results in a biased estimator.

Ware and Cunia (1962) championed for a wider use of SPR in forest inventories. SPR at that time was mainly of theoretical interest and practical applications were few. Optimality of SPR for change estimation requires either the equality of population variance or the same sample size in successive inventories, or both. An optimal rate of replacement of sample units is only solvable for a single attribute. The problem became intractable for multivariate attributes. Different survey costs of new and repeatedly measured plots further increase the complexity of an already complex optimization problem.

Scott (1981, 1984) provided a similar set of estimators derived from the work by Meier (1953). Scott and Köhl (1994) applied the SPR concept to two-phase sampling for stratification on two and three occasions.

After more than two inventory occasions, the best SPR estimator becomes very complex and unwieldy (Scott 1986; Scott and Köhl 1994) thanks to a myriad of plot types and pairwise associations between plot measurement values exploited in the estimators. With only two inventory occasions, we have three different types of plots. With three inventory occasions, there would be a six types of plots with sample sizes n_1, n_2, n_3, n_{12}, n_{123}, and n_{23}. With four inventory occasions, we would have ten plot types. Therefore, the complexity increases rapidly with the number of observations over time. As design imbalance is also bound to creep in over time; even the most ambitious SPR design can barely stand the test of time.

The problems encountered in practical implementation with SPR are clear detractors. In some survey regions of the USA, SPR has recently been replaced by more flexible and less complex designs (Hahn and Scott 2003, personal communication) such as, for example, a semisystematic sampling design where plot location is random within a regular tesselation of the population into equal-sized hexagons.

3.6.3
Estimates for Subpopulations

Inventory results are not only needed for the entire population, but frequently also for thematic subunits, such as, for example, the forest area structured by ownership categories, by site quality, or by forest cover type. A tabular representation of subpopulation estimates arranged in one-way, two-way, or multi-way tables accommodates this need. The margins of these tables provide row, or column, totals of one or more thematic subunits. When cell and marginal estimates are obtained independently of each other, the additivity of estimates is no longer assured. It will depend on the sampling design and the estimators used to obtain estimates for individual cells. Only SRS and CFI estimators result in additive tables (Table 3.3). Two-phase sampling and SPR designs are

Table 3.3. Example of an additive table. Forest area by type of ownership and site quality in 1,000 ha

	Poor/moderate	Good/very good	Total
Public forest	404.1	408.0	812.1
Private forest	114.5	259.7	374.2
Total	518.6	667.7	1,186.3

Source EAFV (1988, p. 81)

Table 3.4. Example of a nonadditive table. Forest area by type of ownership and site quality in 1,000 ha

	Poor/moderate	Good/very good	Total
Public forest	409.1	407.0	824.9
Private forest	119.4	256.8	370.3
Total	503.1	671.9	1,186.3

notorious for creating nonadditive summary tables. For all other designs and estimators the cell values may not add to the row, or column, sums. An example is given in Table 3.4. Nonadditive tables are not a problem per se for a statistician. Nevertheless, we can hardly expect users of inventory results to accept nonadditive tabulated results. Consequently, a need to remove the nonadditivity comes around frequently to the inventory analyst. The most popular method is based on variants of iterative proportional fitting in which the row, or the column, discrepancies are distributed across cells in proportion to their row, or column, sums (Bishop et al. 1975; Li and Schreuder 1985; Zhang and Chambers 2004). Another popular approach computes EB posterior predictive estimates based on a model for the entire table (Green et al. 1992, Laird 1978). These model-based estimates are, by definition, additive.

3.7 Sampling for Rare and Elusive Populations

Sampling for estimation of population totals, density, or the total or mean attribute value when the population is rare (elusive) will require a large sample size to get the sampling error down to an acceptable level. Efficient sampling becomes paramount in order to control the cost and time needed to reach a target precision. Exploiting auxiliary information associated with the population of interest becomes especially attractive in this situation. Knowledge about the spatial distribution of the population can also improve the efficiency of sampling by choosing a design that is specifically tailored to do well under the assumed distribution (Kalton and Anderson 1986; Sudman et al. 1988; Christman 2000; Venette et al. 2002).

One of the real enigmas in sampling a rare/elusive population is the risk of an empty sample. To state that the estimated population total is zero with a sampling variance of zero (all sample values are zero) is a very strong statement that is bound to attract a lot of attention. Consider the interest in rare population and our concern about the disappearance of species and it is easy to fathom the questions that may flow from an estimate of zero. The surveyor can guard against a zero sample if one has a prior estimate p_0 of the probability that a sample unit will contain a population element. The chance of not having a zero sample if n units are sampled independently is then $(1 - p_0)^n$. Let us say that p_0 is estimated at 0.06 and that the surveyor wishes to keep the probability of an empty sample to 0.01 or less. A sample size equal to or greater than 765 would be needed.

Often sample sizes of this magnitude cannot be afforded and the surveyor may end up with an empty sample. The notion of a zero sampling variance is formally correct but clearly untenable in practice unless one is positive that the population has disappeared, but then why have a survey? Instead of a zero variance we suggest deriving a sampling error based on the rule-of-three (Jovanovic and Levy 1997). The rule states that $3/n$ is the upper 95% confidence bound for a binomial probability when in n independent trials no event has occurred. Since a $(1-\alpha)100\%$ upper bound (p) for a binomial probability can be found by solving $(1-p)^n \geq \alpha$ for p (n is fixed by design) with the approximate solution $p_{1-\alpha} \approx -\log\alpha \times n^{-1}$. For an upper 95% bound we get $-\log 0.05 = 2.996 \approx 3$, which is the essence of the rule-of-three. Assuming a half-normal distribution of sampling errors the approximate standard error of the zero estimate according to the rule-of-three would be $0.9227 \times n^{-1.5}$. To reach this result we first found the scale parameter of a half-normal distribution with a 95% quantile equal to $3/n$ and then obtained the standard deviation in a half-normal distribution with this scale parameter.

We shall now turn to potentially suitable design options for the sampling of rare/elusive populations.

3.7.1 Adaptive Cluster Sampling

Many rare and elusive populations are naturally clustered in space. The density of population elements can be quite high in a few scattered locations with favorable conditions. In surveys of rare/elusive populations the costs associated with traveling from one plot to another are often several orders of magnitude higher than the cost of measuring a plot. It would therefore seem to make sense to intensify sampling in areas with positive finds of the rare elements and reduce the time spent at plots with zero finds. Thompson (1990) has devised an adaptive design that achieves this goal. In adaptive designs the sampling effort is intensified at sample locations with a positive find. Strict adherence to a set of rules governing how and when sampling is intensified allows design-unbiased estimates of population attributes and their sample variances (Thompson 1990). There are many innovative adaptive designs for stratified (Thompson 1991), two-stage (Thompson 1991; Salehi and Seber 1997; Muttlak and Khan 2002; Smith et al. 2003), systematic (Acharya et al. 2000), line (Morgan 1997), restricted (Lo et al. 1997; Brown and Manly 1998; Christman 2001; Muttlak and Khan 2002), and point sampling (Roesch 1993).

Most of these designs can be integrated into one of the more conventional survey designs presented throughout this book. We shall take a closer look at adaptive cluster sampling as it is probably the one with the greatest potential for application in forest inventories.

In adaptive cluster sampling an initial simple random sample is taken with n fixed-area plots composed of one or more basic units of a fixed size and a geometry that completely tessellates the population. The population is viewed as composed of a regular array of N basic units. When a sample plot contains one or more of the desired population elements, the area surrounding the plot is searched for additional occurrence of elements. Around each plot with a positive find one imagines a regular grid of basic units with the actual plot located at the center. All surrounding units with a positive attribute value and connected directly or indirectly to the plot with a positive attribute value are included in the sample. All empty units along the outside edge of included nonempty units are also included, but only nominally because they have to be searched in order to determine whether they should be included or not. Two plots are directly connected if they share a common side and indirectly if they are connected through an unbroken chain of connected plots. In other words, an entire cluster of plots containing one or more population elements is included in the sample whenever the initial sample intersects a cluster. With this protocol more than one spatial cluster of elements may be included in the sample at a single sample location if the clusters are connected at the scale of the basic unit. Conversely, different sample plots may intercept the same cluster of connected units. The size and geometry of the basic unit will thus influence the connectivity of clusters and ultimately the efficiency of adaptive cluster sampling. The example given next will illustrate and clarify the sampling protocol.

To appreciate the design-unbiased sample estimators for this type of adaptive sampling it is helpful to view the population as composed of networks (Ψ) of basic units. There are two types of networks. One is made up of all empty units. Networks of this type are all of size $x = 1$ and have an attribute value $y = 0$. The other type of network consists of a set of connected (directly or indirectly) basic units each with one or more population elements. There are a finite number of networks in the population, say K, with sizes x_i, $i = 1,\ldots,K$ in basic units and attribute values of y_i, $i = 1,\ldots,K$. Thus, adaptive cluster sampling can be viewed as a SRS of networks and their associated attribute values (x and y). Our initial sample serves to intercept the networks. For plots composed of several basic units the adaptive sampling protocol is applied to each unit in the plot. The effect of multiunit sampling is in the number and mixture of intercepted networks. Alternative definitions of a network and connectivity are possible (Christman 2000), but the one just given is usually preferred and is the simplest to implement in the field.

A modified design-unbiased Hansen–Hurwitz (HH) estimator of the population mean under adaptive cluster sampling is (Thompson 1990)

$$\hat{\bar{Y}} = \frac{1}{n}\sum_{i=1}^{n}\frac{1}{X_{\psi_i}}\sum_{j\in\psi_i} Y_j,$$

where Y_j is the attribute total in the *j*th sample unit in the network(s) intercepted by the *i*th sample plot and X_{ψ_i} is the size in basic units of the network(s) intercepted by the *i*th sample plot. Empty networks of size 1 will dominate the sum when sampling rare/elusive populations. Note also that the empty edge units included in the sample do not appear in the estimate of the mean. In modified HH estimators the edge units are basically irrelevant. They do affect effective sample size since all units on the outside edge of an intercepted or connected nonempty unit have to be checked for inclusion or not.

The fact that only the average per unit attribute value of intercepted network(s) enters in the modified HH estimator of the population mean illustrates that the within-network variance of Y_j does not contribute to the sampling variance of the estimated mean. All sample plots intercepting a network(s) will record the attribute for that network(s). In SRS, recordings are strictly on a per plot basis. When the initial random sampling is without replacement the variance estimator of the population mean is

$$\hat{var}\left(\hat{\bar{Y}}\right) = \frac{(N-n)}{Nn(n-1)} \times \sum_{i=1}^{n}\left(\hat{\bar{Y}}_i - \hat{\bar{Y}}\right)^2,$$

where N is population size in basic units and $\hat{\bar{Y}}_i$ is the mean attribute value per unit in the network(s) intercepted by the *i*th sample plot. The modified HH estimators of the population mean and sampling variance are basically identical to the corresponding estimators under SRS. Again, if one considers adaptive cluster sampling as a sampling of networks the similarities are to be expected. Horwitz–Thompson estimators are more complex since they involve computing the inclusion probabilities of edge-units. Since edge units have to be searched they are viewed as part of the sample. Edge units can be selected if they are either intercepted by a sample plot or because they are an edge to one or more networks intercepted by the initial sample. Horwitz–Thompson estimators are said to be less sensitive to the spatial distribution of population elements than the HH estimators (Salehi 1999, 2003; Christman 2000, 2002; Felix-Medina 2003). In adaptive cluster sampling the number of basic units included in the final sample is at least the same as in the initial random sample. The expected sampling variances in adaptive cluster sampling are therefore less than the expected sampling variance in SRS. This argument extends naturally to stratified adaptive sampling. If the population is truly clustered with a sizeable within-cluster variation of the attribute of interest and the extra costs associated with delineating and searching for networks are modest compared with the cost of interplot travel then the adaptive sampling approach can be very efficient (Christman 2002; Brown 2003). It should be mentioned though that the efficiency can fluctuate widely as a function of the spatial distribution of attribute values (Acharya et al. 2000; Hanselman et al. 2003; Smith et al. 2003).

An example of adaptive cluster sampling will hopefully clarify the sampling protocol and the estimators. A survey to determine the number of rare orchids within a 110.2-ha forested area is to be conducted. From ecological studies of the orchids we know that they are clustered in a few suitable locations scattered throughout the forest. Most orchid clusters have 11–27 individuals with a mean cluster size of 16 and a standard deviation of 3.8. A cluster occupies between 25 and 270 m^2 (median 200 m^2). It is estimated that the orchids occupy less than 1% of the forested area. The survey designer decided that adaptive cluster sampling with square 5-m×5-m plots would be a suitable approach. The sampling frame can be regarded as 44,531 basic units of 5 m×5 m located in a regular grid. According to the prior information about the orchids the expected network sizes would be between one and eleven 5-m×5-m units. Differential GPS will be used to stake out sample plots and the networks of connected 5-m×5-m units with an accuracy that warrants the assumption of no errors in the orchid counts. To safeguard with 99% probability against the possibility of an empty sample (see before) a sampling intensity of 2% or 891 is deemed adequate. If the orchid counts of 5-m×5-m units are distributed approximately as a Poisson distribution with a mean of $\lfloor 0.01 \times 25 \rfloor$ orchids per unit we would expect a relative standard error of 15% on the estimated population size. We introduce this detail to highlight that sampling for rare and elusive populations is a costly endeavor. What is not known to the surveyor but is listed here for the sake of completeness is that there are a total of 480 orchids in the population (equivalent to 4.4 per hectare) distributed across 31 networks occupying a total of 243 5-m×5-m units (0.56% of the total). A map of the orchid clusters is shown in Fig. 3.18.

In the initial SRS, a total of 13 orchids were found in four plots, whereas the remaining 887 plots had no orchids. The orchid counts in the six nonempty plots were 4, 1, 4, and 4. Without adaptive sampling around these nonempty plots the estimated population density would have been 5.8 per hectare with an estimated error of 2.7 per hectare. The adaptive search of networks around the four nonempty plots added a total of 27 new units to the sample, i.e., a total of 918 units were sampled. Orchid counts in the four networks were 8, 8, 8, and 7, and the sizes of these networks were 21, 13, 15, and 18 units (of 25 m^2). Inserting these figures in the previous modified HH estimators yields a density estimate of 3.9 orchids per hectare with a standard error of 2.0. For the extra effort of delineating and counting units in six networks we have obtained estimators that are clearly superior to what we would have obtained had we stuck with a SRS design. Four of the six networks intercepted by the initial sample are shown in Fig. 3.19. Edge units have not been highlighted, but there would be eight, ten, eight, and ten edge units surrounding the four intercepted networks. Note that three of the four networks joined one or more additional networks at a plot corner. By the adopted definition of connectedness they are not to be included in the sample.

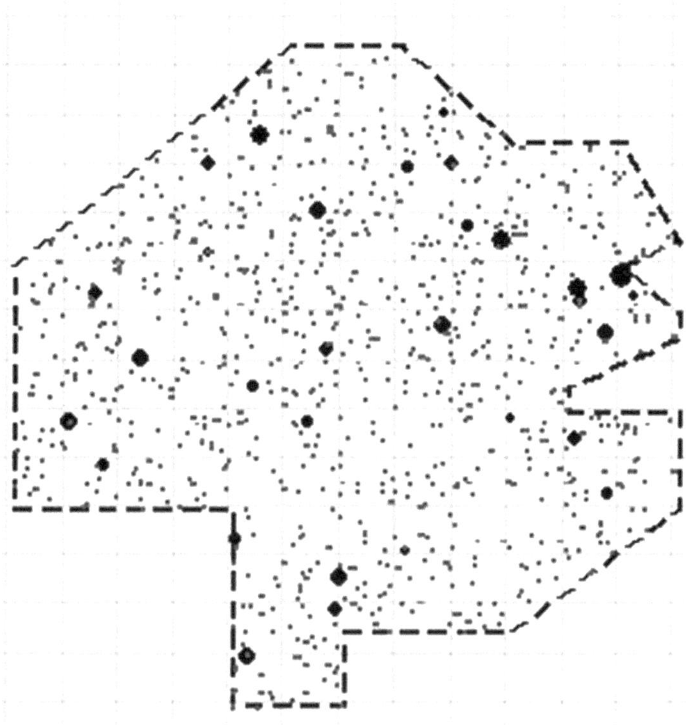

Fig. 3.18. Map of 31 orchid cluster locations (*dark circles*). Cluster sizes are proportional to the number of orchids in a cluster. Sample locations of 875 5-m×5-m sample plots are indicated by *light-gray squares*. The grid spacing is 100 m×100 m

We stated before that the efficiency of adaptive cluster sampling in terms of sampling variance relative to that of a SRS depends on the within-cluster variance of orchid density. For a fixed among-unit variance of orchid density the among-network variance in orchid density declines as the within-network variance increases and vice versa. In the previous example the within-network variance of orchid density was approximately 1.6 times as large as the among-network variance.

3.7.2 Sampling with Probability Proportional to Size

A concentration of sampling efforts in locations with a higher density of rare/elusive population units has a substantial potential for boosting the efficiency of sampling. Sampling with unequal probability is designed to give a

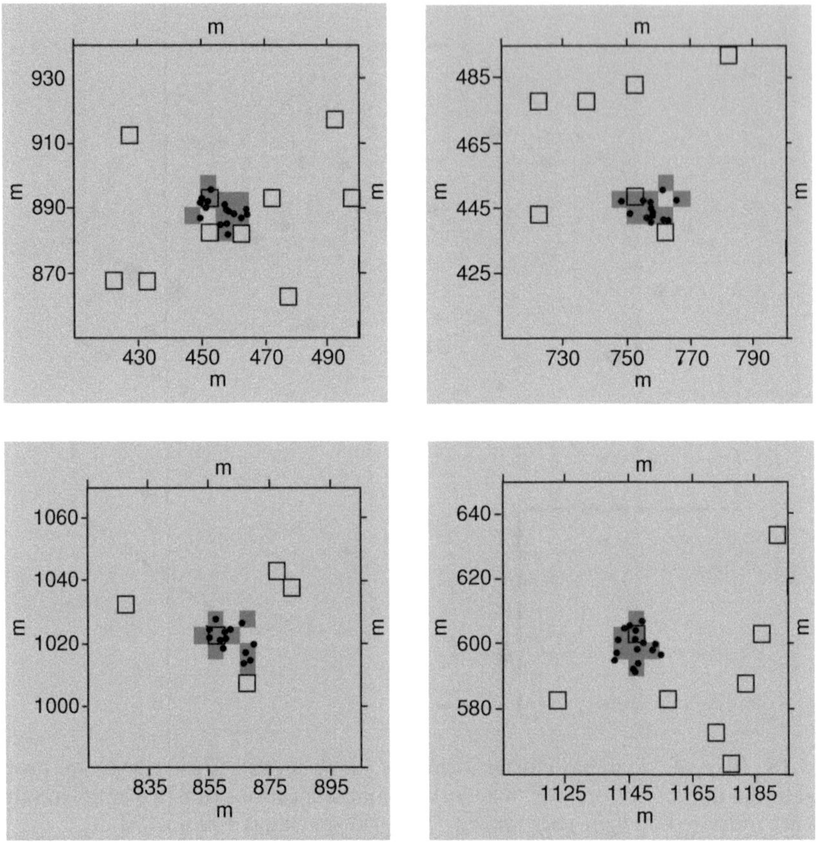

Fig. 3.19. Four intercepted networks of connected sample units with orchids (*dark-gray squares*). Initial random sample plots are indicated by *open squares*. Orchids are displayed as *black dots*

higher probability to sample locations where one expects the highest return (Brewer and Hanif 1983; Godambe and Thompson 1988; Särndal 1996). The challenge is to find an auxiliary attribute that is both known for the entire population and also is approximately proportional to the attribute of interest. Point sampling with an angle gauge is but one example of sampling attributes of trees with probability proportional to basal area. Other applications include, for example, volume sampling from known sample tree lists of basal areas (Schreuder et al. 1968, 1971, 1992; Gregoire and Valentine 1999; Magnussen 2000). Classified remotely sensed images of a population may also provide clues about the location and quantities of interest which can be used in the process of selecting samples (Ståhl et al. 2000; Williams 2001). The general theory and estimators for sampling with unequal probability can be found in many forestry textbooks, for example, de Vries (1986), Schreuder et al. (1993), and Shiver and Borders (1996). Estimators for sampling with and without replacement are distinctly different and the latter are usually quite complicated for all but the simplest cases (Brewer 2000) owing to nonnegative joint inclusion probabilities. Although sampling with replacement is less efficient than sampling without replacement owing to the potential of repeat samples with no new information, the computational challenges involved in the estimators of variance are such that we shall forgo this efficiency and present only the estimators for sampling with replacement.

Let x_i be the auxiliary attribute and y_i the attribute of interest for the ith population unit; x_i is known for all units in the population. The sum of x_i over the entire population is T_x. We are interested in estimating the population total Y. The draw-by-draw inclusion probability of the ith population unit is $\pi_i = n \times x_i \times T_x^{-1}$ with n equal to the desired sample size. When the sample is selected at random with these inclusion probabilities the unbiased Horvitz–Thompson estimator of the population total is (Brewer and Hanif 1983)

$$\hat{Y}_{\text{PPS}} = \sum\nolimits_{i \in s} y_i \times \pi_i^{-1},$$

where the summation is over the units in the sample (s). The unbiased sample-based estimator of sampling variance is

$$\hat{\text{var}}\left(\hat{Y}_{\text{PPS}}\right) = \frac{1}{n \times (n-1)} \sum_{i \in s} \left(\frac{y_i}{\pi_i} - \hat{Y}_{\text{PPS}}\right)^2 .$$

We illustrate the PPS methods using the same population of orchids as exemplified under adaptive cluster sampling. We assume that the orchids are typically associated with a certain group of tree species and that this group of species can be identified with reasonable success by interpreters of remotely sensed images. A classified grayscale remote-sensing image of the forest with $5 \times 10^4 \times 5 \times 10^4$ pixels is in Fig. 3.20. The grayscale levels are assumed to reflect the likelihood of the tree species group being associated with orchids. A darker tone reflects a higher belief in the occurrence of orchids and vice versa. Pixel values were generated synthetically from the following algorithm:

$$x_i = 0.5 \times \left[\log\left(y_i + 1\right) + \gamma_i + \tfrac{1}{4} \sum\nolimits_{j \sim i} \log\left(y_j + 1\right) + \gamma_j\right],$$

where

$$\gamma_i \sim \Gamma(0.4, 1),\ E\left(\gamma_i\right) = \text{var}\left(\gamma_i\right) = 0.4$$

and where the summation is over the four first-order neighbors to pixel j (to mimic scaling and sensor spread function). The average signal-to-noise ratio was 0.15 and the correlation between the feature of interest and the grayscale value was 0.16.

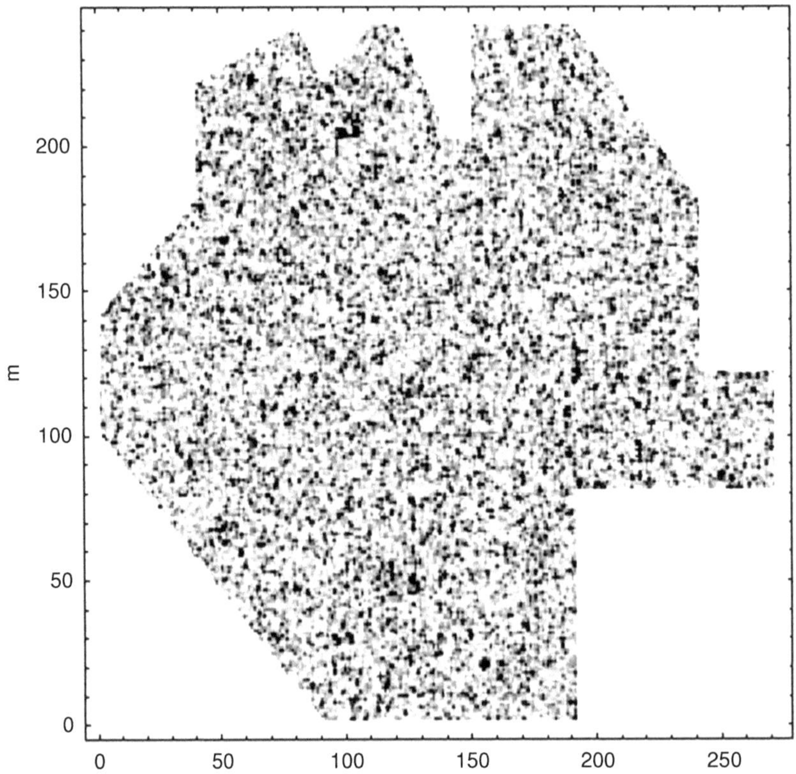

Fig. 3.20. Grayscale map of the spatial domain of the orchid population. A darker tone reflects a higher belief in the presence of orchids

Two 15-pixel×15-pixel windows taken from the grayscale map reveal the clustering of darker pixels with the presumed highest likelihood of orchid presence (Fig. 3.21).

The sample size of 891 was maintained and each unit was selected by forming a two-column list with the number (1,. . . .,44,531) of each population unit in the first column and the cumulative inclusion probabilities of these units in the second column. Now draw 891 random numbers uniformly distributed between 0 and 1. For each random number, find the unit number of the first cumulative inclusion probability that is larger than or equal to the random number. Select the population unit associated with that number. This selection protocol ensures that units are selected with probability proportional to π_i. For example, you have drawn the random number 0.447297. Excerpts of the numbered list of cumulative selection probabilities are in Table 3.5. The highlighted population unit with the number 19,839 should be selected.

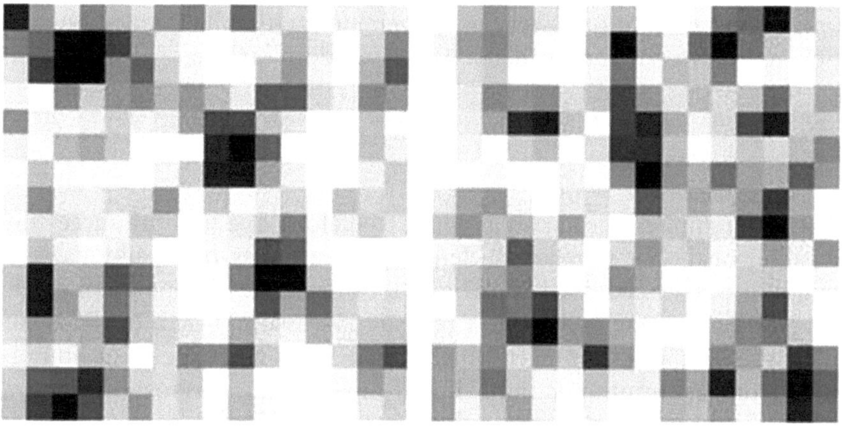

Fig. 3.21. Two randomly chosen 15-pixel×15-pixel windows of details taken from the grayscale map in Fig. 3.20

After selection we computed the average inclusion probability of the sample: it was 1.8 times higher than the average inclusion probability of a unit. The grayscale values of selected units were also slightly more strongly correlated with the actual orchid count (0.18) than seen across the entire population (0.16). A total of 30 orchids were recorded for the PPS sample (compare with 13 for SRS and 31 for the adaptive cluster sampling design). The PPS estimator of orchid density was 4.3 (actual is 4.4) with a standard error of 1.4, a clear improvement compared with the aforementioned alternatives. This PPS example was governed by realistic choices of the correlation and association between the auxiliary and the target attribute. Even with a modest relationship the possible gains in efficiency are striking. A note of caution is nevertheless warranted. If the assumptions about the auxiliary variable turn out to be

Table 3.5. Selection list of unit numbers and cumulative draw-by-draw inclusion probabilities. Random draw is 0.447297 and unit 19,839 should be selected

Unit no.	$Sum(\pi_{no.}, 1, no.)$
1	7.8×10^{-6}
⋮	⋮
19,837	0.447272
19,838	0.447282
19,839	**0.447303**
⋮	⋮
44,530	0.999949
44,531	1

wrong, or the association is very heterogeneous across the population, a PPS sample may perform poorer than a simple random sample.

3.7.3 Line Transect Sampling

As the name implies, the survey is conducted along one or more survey lines. In a survey of mobile population elements like animals, birds, and insects the population is assumed fixed in size and location of each population element during the time of the survey. How realistic this assumption is depends naturally on the time and manner in which the survey is conducted. For immobile elements this assumption is implicit and mostly quite reasonable. Line transect sampling does not depend on the existence of a sample frame, a feature it shares with point sampling designs. In a line transect an observer moves along the transect line(s) in an nonintrusive manner and records sightings of population elements and their distance to the survey line. The line can be staked out on the ground or can be a line on a photograph or some other medium and the observer can move on foot, in a vehicle, or in some elevated platform of observation, for example, an airplane. It is a strict requirement that no element is recorded more than once and that observations are mutually independent. That is, the sighting of one object does not in any way impact on the sighting of another. These are assumptions that can be difficult to justify in many surveys of elusive animals or birds. Observations made with an angle gauge (Ståhl 1997, 1998) are possible too and for specific purposes they are efficient.

A characteristic of most line transect surveys is the nonconstant probability of detection of population elements. In surveys with human observers visual obstacles along the transect lines, the possible elusive nature of the elements of interest, and our limited field of vision combine to make the sightings imperfect. Many elements cannot be seen and others are simply missed. This phenomenon has to be taken into account in the estimation of a desired population statistic obtained from an imperfect transect line survey. Unbiased estimates of, say, population totals are only attainable if we know the probability of detecting a given population element given our location of observation. This assumption is rarely satisfied in practice. Often the probability of detection is some function of the distance between the elements and the survey line. Detection could be perfect up to a critical distance, after which it declines rapidly or it is only nearly perfect for elements on the survey line and then decreasing monotonically with distance (Ramsey and Harrison 2004). In repeat surveys within a single fixed region the surveyors may obtain solid information about the detection function (f), in others an estimate must be derived from the survey itself. When the detection function is derived from the survey data the estimated population statistics will only be approximately unbiased (Thompson 1992).

There are a plethora of line transect methods reflecting requirements and adaptations to the specific nature of the survey population and the environment in which it exists. Only the most common line transect survey design will be presented in this section. Other popular alternatives have been detailed by, for example, Ramsey et al. (1988).

The design we have chosen for detailing has a random selection of survey lines, with the survey lines selected with probability proportional to their length (L). The objective of the survey is to estimate a population total (N), viz., density N per hectare, of an attribute of interest, for example, the number of beetle-infested trees within a known fixed area representing a population or a natural stratum of host trees for the beetle. First a fixed baseline (B) for the reference and orientation of the transect lines is constructed. It is customary to let the baseline run parallel to one of the axes in an orthogonal reference coordinate system defining the outline of the population and all its elements. A number (n_L) of transect lines running orthogonal to the baseline and extending across the entire population are now chosen by simply generating n_L random locations within the population of interest. Since the number of points on a transect line is proportional to its length the procedure will automatically generate transect lines with probability proportional to their length. The baseline need not be a single line. For some populations with a very irregular outline or a very large spatial domain it is often advantageous to slice the populations by a system of parallel baselines. In that case the population is viewed as a series of disconnected slices each defined relative to their baseline. Selection of transect lines proceeds as for the case of a single baseline. Along the entire length of the ith survey line $\left(L_i, i = 1, \ldots, n_L\right)$ the surveyor records the shortest distance $\left(x_{ij}, j = 1, \ldots, n_i\right)$ from the survey line to each of the n_i sighted elements along the ith line. Alternatively, the surveyor records a sighting angle θ_{ij} and a distance r_{ij} and converts these two measurements to x_{ij} via $x_{ij} = r_{ij} \times \sin\theta_{ij}$. If sightings were perfect up to a distance of w_{100} with no sightings possible beyond this distance then the obvious (unbiased) density estimator for the ith survey line would be $D_i = n_i \times \left(2 \times w_{100} \times L_i\right)^{-1}$ and we would proceed to a length-weighted estimate of the population density and the sampling variance. Given that the area of the population is known, an estimate of the population totals is obtained by a simple multiplication of the density estimate and PA. However, we shall assume that detection is an unknown function of distance but no element at distance zero would be missed. With these assumptions the density estimator obtained from the ith transect line becomes

$$\hat{D}_i = \frac{n_i \times \hat{f}(0)}{2 \times L_i},$$

where $\hat{f}(0)$ is the estimated value of the detection function at a distance of zero. We detail the estimation of the detection function in the following. The density estimator is clearly a ratio of two random variables (n_i and L_i) and is consequently biased. A jackknife estimator of the population density is therefore often preferable to a direct estimator (Efron 1982). A jackknife estimator reduces the first-order bias by taking the average of n_L leave-one-out estimates of the population density. Specifically we have

$$\hat{D}_{jk} = \frac{1}{n_L} \times \sum_{i=1}^{n_L} \hat{D}^{(i)} = \frac{1}{n_L} \times \sum_{i=1}^{n_L} \frac{\hat{f}(0) \sum_{i' \neq i} n_{i'}}{2 \times \sum_{i' \neq i} L_{i'}},$$

where $\hat{D}^{(i)}$ is a population density estimate obtained after excluding data from the ith transect line. The corresponding jackknife estimator of the sampling variance is

$$\hat{\text{var}}\left(\hat{D}_{jk}\right) = \frac{(n_L - 1)}{n_L} \times \sum_{i=1}^{n_L} \left(\hat{D}^{(i)} - \hat{D}_{jk}\right)^2.$$

Note, this variance estimator does not account for the uncertainty in the estimate of the detection function at distance zero nor the covariance between this estimate and the random variables n_i and L_i (Shenk et al. 1998). The omission is intentional since reliable estimates of these quantities require an intensive sampling of transect lines (n_L>30) since we would need a separate estimate of $\hat{f}(0)$ for every transect line. Estimates derived from a small number of survey lines tend to be erratic. Confidence intervals for the population densities are obtained as outlined in Sect. 3.3.1.3.

The detection function can be estimated in a number of ways. A subjective but quick method derives $\hat{f}(0)$ from a histogram of the sighting distances in the survey. If there is a sharp drop in the number of sightings beyond a distance of, say, w_{100}, and one is willing to assume that no element was missed at shorter distances, $\hat{f}(0)$ would be estimated as $1/w_{100}$; conversely $\hat{f}(0)$ is estimated as the scaled height of the first class in a histogram with class intervals chosen in some suitable way (Wand 1997). Estimation by kernel smoothing (Izenman 1991) would convey attractive properties to $\hat{f}(0)$ but kernel-based estimation of the lower endpoint of a density function restricted to the domain of positive real numbers remain problematic. We opt for an estimation via the Fourier series method (pp. 67–70 in Silverman 1986). The Fourier series method of estimating $\hat{f}(0)$ is

$$\hat{f}(0) = \frac{1}{w^*} + \sum_k \hat{a}_k, k = 1, \ldots, n_{\text{obs}},$$

where w^* is the maximum distance at which an element can be sighted, n_{obs} is the total number of sightings in the n_L transect lines, and $\hat{a}_k$ are the Fourier coefficients given by

$$\hat{a}_k = \frac{2}{n_{\text{obs}} \times w^*} \sum_{i=1}^{n_L} \sum_{j=1}^{n_i} \cos\left(\frac{k \times \pi \times x_{ij}}{w^*}\right).$$

The number of Fourier coefficients $\hat{a}_k$ to include is determined by the first value of k for which

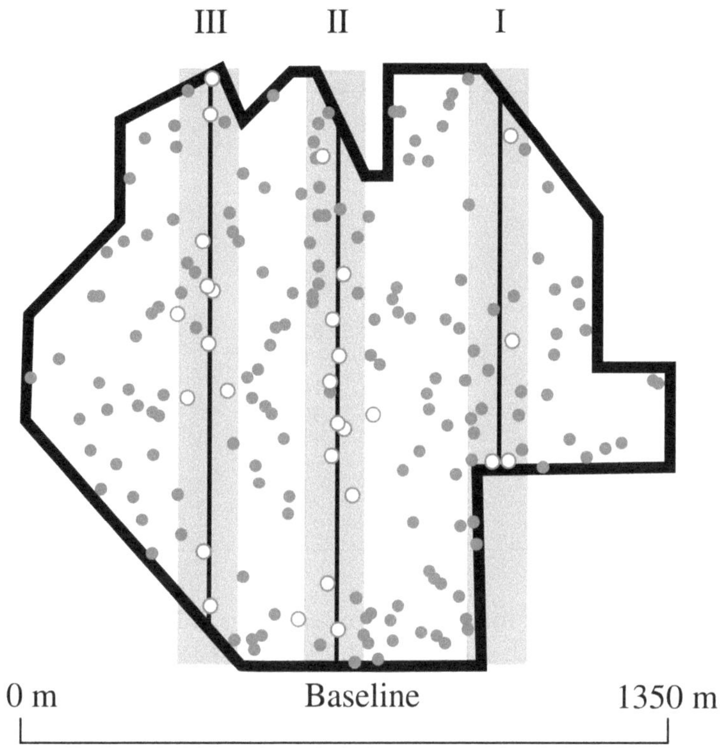

Fig. 3.22. Population outline with baseline and three transect survey lines. Population elements are indicated with *dots* and those observed from any of the three lines with *circles*. The *shaded area* around each survey line gives the true 95% detection interval

$$\frac{1}{W^*}\sqrt{\frac{2}{n_{\text{obs}}+1}} \geq |\hat{a}_k|$$

is true (Burnham et al. 1980). In practice, the number is rarely above 2.

A simple example will illustrate the estimation process. The population density of bark-beetle-infested trees within the 100.3-ha population domain shown in Fig. 3.22 is to be estimated by a transect survey with three random survey lines selected with probability to their length. The 1,350-m-long baseline and the three selected survey lines of length 756, 1,102, and 1,114 m and orthogonal to the baseline are indicated in the figure. The surveyor(s) move along the entire length of each survey length and record every infested tree (recognizable by resin exuding on the stem and possibly by a reddish needle discoloration) they can spot and then record the distance between the tree spotted and the survey line. Observations are not perfect: some trees will be missed and there is a natural (unknown) limit to the distance an infestation can be ascertained

Table 3.6. Number of sights on survey lines

j	Distance x_{ij} (m)		
	x_{1j}	x_{2j}	x_{3j}
1	17	1	3
2	15	25	3
3	16	8	37
4	22	15	11
5		70	4
6		30	2
7		82	67
8		1	17
9		14	6
10		10	4
11		12	42
12		2	
13		34	
Mean	17.5	23.4	17.8

from the survey line. The total number of sightings is 28. Table 3.6 gives the number of sightings on each survey line (4, 13, and 11) and the distances to each sighting from the survey line.

Sighted and nonsighted trees are indicated by different point signatures in Fig. 3.22. A histogram of the observed distances of elements from the survey lines is in Fig. 3.23. The sharp drop-off in the frequency of distances beyond

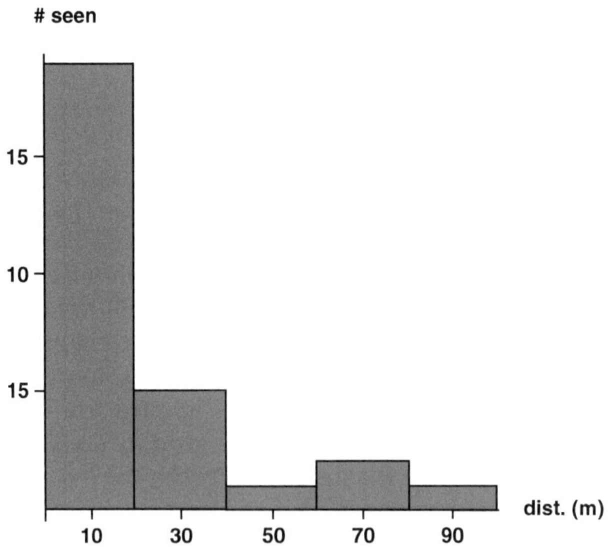

Fig. 3.23. Histogram of distances between seen objects and the nearest survey line

20 m is noticeable. We should expect w_{100} to be about 20 m. As an estimate of w^*, we took the 95% quantile of the recorded distances, which was 70.2 m. Recall that only sighted elements are recorded; hence, a 95% sample quantile corresponds to a higher quantile in the population of distances, in our case the 98% quantile. One should avoid the use of an observed maximum distance for w^* as it is a highly variable statistic; the three line-specific maxima bear this out. The number of Fourier series coefficients to use according to the previous rule is 1 and $\hat{a}_1 = 0.0153$, which gives us $\hat{f}(0) = 0.0295$ as the estimated fraction of perfect sightings. Delete-one jackknife estimates of density were accordingly 1.60, 1.18, and 1.35 with an average $\hat{D}_{jk} = 1.38$ and a relative standard error of 17.5%. A nominal 95% confidence interval runs from 0.34 to 2.42. The actual population contained 200 infested trees with a density of 1.81 trees per hectare.

A density estimate obtained from a single transect line does not have a design-based estimator of variance. Only an analytical estimate of the variance is possible, and only if the surveyor is willing to make assumptions about the distribution of elements and the arrangement of all possible survey lines inside the population. The correctness of these assumptions can be difficult to ascertain and the resulting estimates of variance can be quite poor. In the same vein, observed distances from either a single line or multiple lines can be used together with other predictors to generate spatial predictions of occurrence (Hedley and Buckland 2004). The quality of all estimators obtained from a line transect survey rests with the homogeneity of the detection function. If the probability of detection depends on more than distance then these factors must be incorporated into a generalized detection function (Marques and Buckland 2003; Ramsey and Harrison 2004). Finally, it is also critical that distance recordings are without errors. The impact of measurement errors can be serious and should be assessed whenever possible (Marques 2004). In a finite spatial domain of a population with a finite number of elements, any elements close to a population boundary will have a lower likelihood of detection than an element further away from the boundary. The reasons are the same as discussed for point sampling. The area of average detection is smaller for points close to the boundary because they can only be detected from locations inside the populations. The integral of all possible detection distances multiplied by the probability of detection is smaller owing to the restrictions imposed by nearby boundaries with nontrivial detection probability. When the area associated with points closer than w^* from a boundary only represents a small fraction of the total area, the boundary effect will be small. To gauge the potential of bias the surveyor can compute the reduction in the survey area due to boundaries. In our example, 95% of the recorded distances were within the gray bars in Fig. 322. The search area "lost" owing to boundary effects can be obtained by drawing six lines parallel to the baseline and going through the starting and end points of the three survey lines I, II, and III. Six triangles form between these six lines, the population boundary, and the outside of the shaded

95% detection limits. In the example, these areas account for 0.5% of the nominal (95%) search area in the absence of any boundary. It is, then, probable than the number of observed elements is biased downwards by this amount. In a larger survey with hundreds of observations it might be reasonable to add a matching fraction of the total count to the observations and assign an average distance to these "pseudo-observations."

3.7.4 Capture–Recapture Sampling

Capture–recapture sampling is primarily used for estimation of the population size of mobile population elements. Applications extend to estimation of the probability of detection in line transect surveys with a fixed survey width along the survey line (Borchers et al. 1998). In its simplest form a number of population elements (n_1) are captured at time 1 according to a chosen sample design and capture method, marked, and then released. At time 2 a new sample of n_2 elements is captured, of which n_{20} were unmarked and n_{21} were marked at time 1 $\left(n_2 = n_{20} + n_{21}\right)$. If it can be assumed that the total population size N is fixed during the time of the survey, that the first sample is representative of the population, that the n_1 marked elements distribute themselves uniformly across the population domain after the first capture, that the probability of catching a population element at time 2 is unaffected by the outcome of the first sample, and that the second sample is also representative of the population, then the minimum biased estimator of the population size is (Seber 1982)

$$\hat{N} = \left(n_1 + 1\right) \times \frac{\left(n_2 + 1\right)}{\left(n_{21} + 1\right)} - 1.$$

We prefer this estimator to Petersen's estimator of $n_1 \times n_2 \times n_{21}^{-1}$ (Seber 1982) as it is undefined for $n_{21}=0$. Implicit in this estimator is the assumption that the ratio of recaptured elements in the second sample extends to the population at large; an assumption that only holds if the two samples are truly representative of the population at large. Large sample sizes are needed before the assumptions can be fully justified. It should be noted that there is no unbiased estimator of N. Considerable effort has been spent on devising sample designs and capture methods that mitigate potential sources of bias (Seber 1986; Knight 2003; Wintle et al. 2004). An approximate unbiased estimator of the variance of this model-based estimator is

$$\hat{\text{var}}\left(\hat{N}\right) = \frac{\left(n_1 + 1\right)\left(n_2 + 1\right)\left(n_1 - n_{21}\right)\left(n_2 - n_{21}\right)}{\left(n_{21} + 1\right)^2\left(n_{21} + 2\right)}.$$

One of the most persistent problems of capture–recapture surveys is the potential for interactions between the population elements and the capture process. Models describing the effect of differential probabilities of capture at time 1 and time 2, birth and mortality processes, emigration, and immigration have been developed (Burnham et al. 1980; Norris and Pollock 1998; Efford 2004) in order to obtain model-based estimates of population size at the first, or second, survey time. Estimators of population sizes and their sampling variances for sampling on more than two occassions are given by, for example, Cormack (1993).

A maximum likelihood or a Bayesian estimation of N is possible if one is willing to make assumptions about the distribution of n_{12}, the only unknown random variable in the estimation problem. The distribution of n_{12} is usually assumed to be of hypergeometric, binomial, or Poisson form. In the binomial and Poisson model, N is a random variable, not fixed. A likelihood function can be associated with each of these models, which, in turn, would allow a likelihood-based estimation of n_{12}. In many cases some prior knowledge exists about what the distribution of n_{12} might be. Earlier surveys or subject knowledge could forward a prior distribution of n_{12} which would open up the possibility for a Bayesian estimation procedure (Poole 2002).

Shiver and Borders (1996, p. 333, example 11.4.1) illustrate a capture–recapture estimation problem with n_1=125, n_2=100, and n_{21}=44. The estimated population size using the previous estimator was 282 (rounded) with a standard error of 25 (rounded). Had we made the assumption of a hypergeometric distribution for n_{12} the estimated population size would have been 284 with a standard error of 29 (rounded). The variance of the maximum-likelihood estimate is obtained from

$$\hat{\text{var}}\left(\hat{N}_{\text{MLE}}\right) = \left(-\left.\frac{\partial^2 \ell\left(n_{21} \mid n_1, N\right)}{\partial n_{21}^2}\right|_{N=\hat{N}}\right)^{-1} \times \left(\frac{n_2}{\hat{N}}\right)^2,$$

where $\ell\left(n_{21} \mid n_1, N\right)$ is the likelihood function of n_{21} and the last factor accounts for the scaling from the sample size (n_2) at time 2 to the estimated population size. Computation of the derivatives of the likelihood function is difficult and complex regardless of the model chosen. Software that can do symbolic calculations is needed for easy estimation.

3.7.5 Inverse Sampling

With modest sample sizes a low sample yield of marked elements (n_{21}) at time 2 puts the surveyor in a conundrum. Sample variation may simply have reduced n_{21} by chance but the ensuing estimator of N may be counterintuitive or apparently in error. Inverse sampling is a sample design in which the sample yield is fixed prior to sampling, which makes the sample size an unknown random variable (Panchapakesan et al. 1998; Cuzick 2001; Moore et al. 2003). The advantage is clear: a target yield is assured. The downside is equally clear:

there is no control over sample size and in practice the order of the sites to be visited until the target yield has been reached has to be random to avoid the sample being nonrepresentative of the population. For example, if the surveyor decides that n_{21}=44 then every single site to be visited before the target is reached has to be determined by a random draw of all possible sites. This may result in excessive travel times and cost and is only practical if the samples can arrive in random order at a fixed location of observation. While estimators under inverse sampling are generally close to or identical to the estimators under random sampling the estimated variances are typically much larger. The sampling distribution (of sample sizes) under inverse sampling at time 2 for recapture could be assumed to be of negative hypergeometric or negative binomial form (Johnson et al. 1992). To give an example of the variance inflation, suppose that the surveyor in the previous example had decided that a yield of 44 marked elements at time 2 would be desired and that this yield by chance was achieved after catching 100 elements. The maximum-likelihood estimate of N would again be 284 but now the standard error would be 44 (rounded), or almost 50% higher. Since there is no guarantee that the target yield can be obtained within the available time and with existing resources, and given that there is considerable risk of an inflated sampling variance, a cautionary approach to inverse sampling is prudent.

3.7.6 Double Sampling

Two independent surveys or a survey in combination with a registration system can be an efficient design for estimation of the total rare/elusive population. Let n_1 be the number of elements recorded during the first survey, n_2 be the number for the second survey, and n_{12} be the number identified in both surveys. Recorded objects must be identified clearly and uniquely in order to obtain n_{12}. Let N be the unknown population total that we wish to estimate. N is assumed constant from the onset of the first survey to the end of the second survey. For two independent surveys we expect to find $n_1 \times n_2 \times N^{-1}$ elements recorded in both surveys. Given the observed count n_{12}, we obtain a double-sampling estimate of N via

$$\hat{N}_{\mathrm{DS}} = \frac{n_1 \times n_2}{n_{12}}.$$

This estimator was first proposed by Chandra-Ssekar and Deming (1949). Owing to its simplicity it has found widespread applications in human surveys, and wildlife surveys. There is no need to have an exact estimate of the PA or for that matter an estimate at all in order to estimate the total, perhaps one of the main attractions of double sampling. No variance estimator has been forwarded for this double-sampling estimator of the total. When sampling is with well-identified units, like line transects, survey lines, city blocks, or forest stands, then a jackknife estimator of variance and sampling error is recommended (Shao 1996). Note that the estimator is undefined for n_{12}=0 and is generally very unstable for small counts of n_{12}. It is our experience that n_{12} should be around 5% of $n_1 \times n_2$ before estimates with an acceptable accuracy (relative error less than 20%) can be expected. This means that the sampling intensity has to be rather high in both surveys; if n_1/N or n_2/N sinks below 10% the chance that n_{12}=0 is nontrivial. A more specific assessment would require assumptions about both survey design and population statistics, such as total and distribution across the spatial domain.

Double sampling in forest inventory could be an option for the estimation of, for example, the total number of trees of a rare species, the number of stems logged at a logging site, the number of trees fallen owing to windthrow, or the number of diseased trees. Counting could be done along random survey lines with markings of all observed elements falling on the line(s) or in close proximity to the line(s) or it can be done on remotely sensed images. We shall illustrate the double-sampling design for the estimation of the number of windthrown trees in the same forest we used for demonstrating line transect sampling, capture–recapture, and sampling with PPS.

A severe storm felled 1,688 trees in the 110.2-ha forest. The damage was mostly concentrated in 11 areas (12% of total) of size 0.1–4.2 ha (average 1.2 ha) but scattered fall downs were observed throughout the forest. The number of trees downed in the hardest-hit areas ranged from 41 to 59 per hectare with a mean of 50 stems per hectare. Tall trees were predominantly hit by the storm. The stem length of the fallen trees was 51 m, with a standard deviation of about 5 m. The surveyor decides to assess the damage by laying out two independent surveys, each with 30 random survey lines, random with respect to location and length. The orientation was random within a limited range of angles. The average length of a survey line is approximately 275 m but individual lines range from 30 to 890 m. A map of the fallen trees and the two sets of survey lines is shown in Fig. 3.24.

In the first survey 266 fallen stems crossed the survey lines (n_1), while 200 crossed the lines of the second survey (n_2). A total of 41 stems were common to both surveys (n_{12}). From this we get an estimated total of 1,298 fallen trees. A jackknife estimator of the total was 1,309, which indicates a bias of 11 (1%) in the double-sampling estimator. The jackknife estimator of the standard error was 122. To obtain the jackknife estimators we deleted one survey line from the first survey and one survey line from the second survey number; hence, 900 delete-one estimates were obtained. The distribution of these delete-one estimates of the number of fallen trees is shown in the histogram in Fig. 3.25. Notice the skewed distribution and the appearance of a mixture of two distributions arising from the spatial heterogeneity of the intensity of

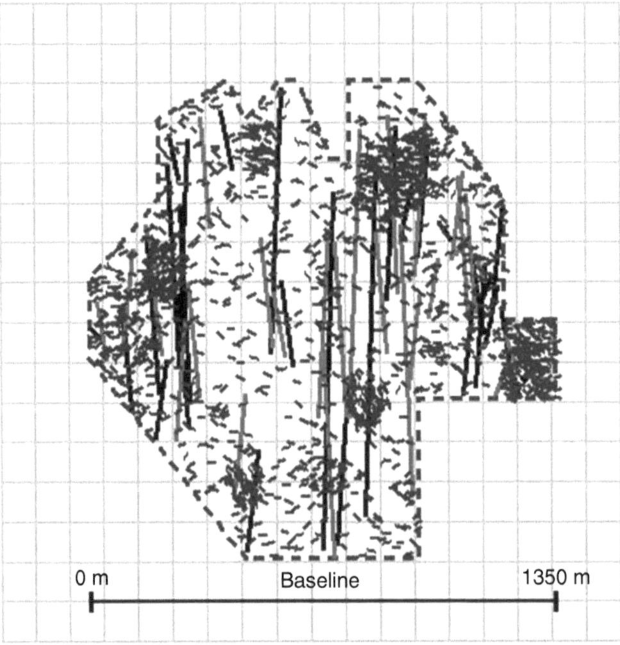

Fig. 3.24. Map of fallen trees and survey lines in the first survey (*black*) and in the second independent survey (*gray*). The grid spacing is 100 m × 100 m

windthrow across the forest. The fact that the nominal 95% confidence interval for the sample-based estimator based on the assumption of a normally distributed sample estimate does not include the true number is a direct consequence of this skewed distribution. A confidence interval with a coverage closer to the nominal value should, therefore, be obtained from the quantiles of the bootstrap distribution of sample estimates (Shao 1996).

3.7.7 Composite Sampling

In sampling for rare/elusive population elements the time and cost to identify the presence/absence or to quantify the attribute of interest in a sample unit can be very costly. For a rare/elusive element most sample units will have a value of zero but will still carry the full cost of analysis. Soil sampling for rare contaminants, sampling containers of wood chips for the presence of a rare staining fungi or nematode, or landscapes for rare deforestation events are but a few examples with relevance to forest inventory. As the name implies, in composite sampling several sample units are joined into a single composite unit.

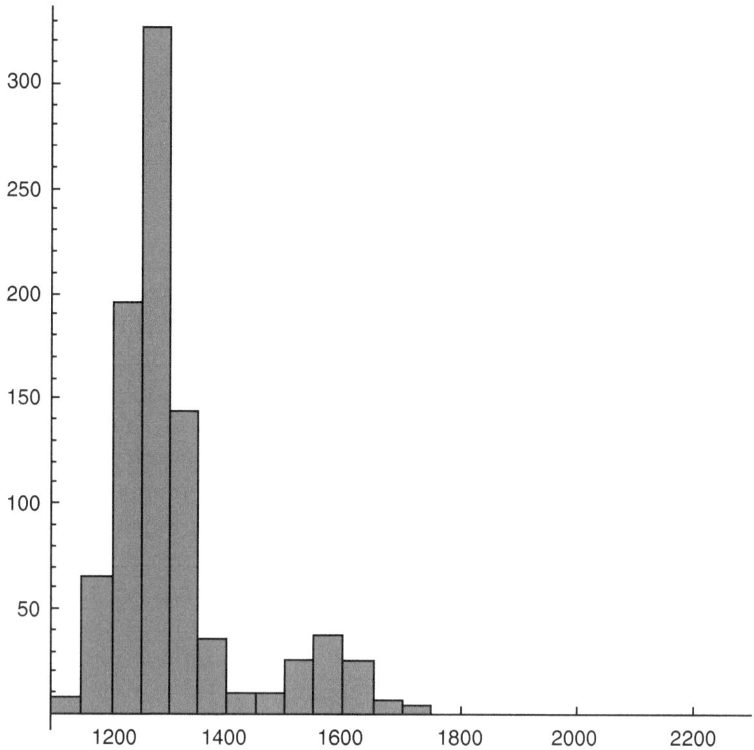

Fig. 3.25. Histogram of jackknife (delete-one) estimates of the number of fallen trees

The idea behind composite sampling is simple enough: the screening for the presence/absence of an elusive element is concentrated on fewer composite units. The only sample units to be examined individually are those for which the composite sample came up positive for the presence of the rare element. Lancaster and Keller-McNulty (1998) have reviewed the composite sampling method and they provide a succinct overview of the pros and cons of this method. Composite sampling is a process that involves defining and optimizing the compositing design, the measurement process, and the data analysis process. The observed composite response y for composite unit i can be represented by

$$y_i = f\left(x_{i1}, x_{i2}, \ldots, x_{in_i}; \mathrm{c}\right) + e_i, i = 1, \ldots, n,$$

where x_{ij} is the attribute value of the jth sample units in the ith composite unit, $f(\cdot;\cdot)$ is a function of the physical process of compositing, **c** is a set of weights that depend on assumptions made about the physical process, and e represents the measurement error. The utility of this expression may not be immediately clear or useful unless we know the function *f*, the vector **c**, and the measurement error. When the compositing process model and the errors are not fully known the inference about *y* becomes model-based and the uncertainty regarding model parameters must be accounted for in the estimators of sampling variance. In its simplest form, when the statistical objective is to estimate a population mean, the physical process is presented as

$$f\left(x_{i1},\ldots,x_{in_i};\mathrm{c}\right)=\sum_{j=1}^{n_i}c_{ij}x_{ij}.$$

If the sample units enter with equal amount then the expected value of *y* is the population mean with a variance of $\sigma_x^2/n+\sigma_e^2$. Compare this with the variance of $\sigma_x^2/n+\sigma_e^2/n$ that one would expect if the *n* measurements were done on *n* sample units. Hence, composite sampling is a trade-off: cost savings are achieved at the expense of precision and information. The trade-off will have to be weighed carefully in each case. More elaborate schemes are possible: sample units can themselves be sampled before they are combined into composite units and the composite units can, in turn, also be sampled. Finally, measurement units may be a small fraction of a composite unit. The process of mixing and subsampling is captured by the vector **c** and the error term. Lancaster and Keller-McNulty give a good example of how the sampling designs can be employed to illicit estimates associated with population features, such as, for example, row and column effects in spatial sampling. Estimation of prevalence is also possible from composite sampling but model assumptions that need to be verified are needed. Commonly one assumes that *x* is a Bernoulli random variable with probability π_x of taking the value of 1. If *r* sample units are pooled equally into a composite sample then the probability that y_i is 1 becomes $1-(1-\pi_x)^r$. A maximum-likelihood estimator of prevalence is

$$\hat{\pi}_x=1-\left(1-\frac{\sum_i y_i}{n}\right)^{\frac{1}{2}}.$$

A note of caution is in place: the estimator is not unbiased and it can be quite imprecise, especially for high values of *r*.

Classification of pixels sampled from a remotely sensed image is akin to composite sampling when the size of the pixel is a multiple of the sampling unit applied in forest inventory. What is observed is a composite of features in several individual inventory units. Attempts to obtain estimates at the scale of sample units by "unmixing" (Oleson et al. 1995; Bosdogianni et al. 1997; Grandell et al. 1998; Mertens 2003; Vikhamar and Solberg 2003) are in essence attempts to solve for *x* given *y* and a model for **c**. The problem is generally underdetermined (more variables than equations) but if one is willing to specify *f* and assume **c** to be invariant then one can use ordinary least-squares or mixed linear models for the estimation of **c** and ultimately estimators of *x*.

3.8 Small-Area Estimation

A forest inventory is designed to provide estimates of attribute values of interest for the entire population or for a set of strata identified during the design phase. After completion of an inventory it often happens that attribute estimates or an update of estimates for one or several small geographic areas of the sampling frame of the inventory is needed. Estimators pertaining to the forest in a particular county, district, or any other zone of interest to someone are therefore needed on a routine basis. Related postsampling estimation problems have surfaced in most large-scale surveys and spawned a search for effective, design-consistent estimators applicable to domains within a population and small geographically defined areas (Särndal et al. 1992; Rao 2003). A common feature in all small-area estimation problems is the small number of samples taken within the small area. A direct estimation based on only the samples taken inside the small area would in most cases provide estimates with low precision and that are possibly biased. The survey context, the particular features of the small area or domain, and the availability of auxiliary information determine in each case the most promising approach to estimation. A rich and diverse collection of estimators have been tailored to a wide spectrum of small-area estimation problems. The majority are model-based or at least model-assisted. The data for a small-area or domain are assumed to adhere to a model that we wish to estimate. Optimality of estimates in terms of minimum bias and minimum variance is the ideal that is pursued but it is rarely achieved. Most estimators are in some sense the "best possible" given the estimation problem posed. One of the first published forestry applications was given by Green et al. (1987). Timber volumes per hectare and by county were to be estimated from a regionwide inventory. They assumed that county-specific sample-based estimates (means) of timber volumes per hectare were estimates of the sum of a random county-specific effect and a random "error." Improved, in terms of mean squared error, county-specific estimates were then obtained as a weighted average of the county-specific sample means and a weighted mean of all counties. Related estimation problems are recurrent in forestry and with the increase in the use of remotely sensed data as auxiliary information we have greatly expanded our options for effective small-area estimation (Kangas 1996; Lappi 2001). Only a few of the most commonly applied estimators will be presented here. Rao (2003) provides a recent summary of small-area estimators and Pfeffermann (2002) offers a review of current trends, unresolved issues, and the future direction of small-area estimation research.

3.8.1 Direct Small-Area Estimators

Direct small-area estimation means that only the samples collected in the small area (U_i) are used for estimation purposes. The sample size n_i in U_i is a random variable with a possible value of zero. As long as the probability of n_i=0 remains low and n_i is close to its expected value, the estimators employed for estimation of population level attributes apply (Särndal et al. 1992; Rao 2003). Often, however, there is a risk that n_i=0 is nontrivial and this must be considered in estimating a small-area sampling variance.

Under SRS in a population occupying an area A, with n fixed-area plots each with an area A_{plot} the direct small-area estimator for U_i say a total Y, is

$$\hat{Y}_{i(\text{dir}|\text{SRS})} = \frac{A}{n \times A_{\text{plot}}} \sum_{j \in U_i} Y_j,$$

with an approximate variance of

$$\hat{\text{var}}\left(\hat{Y}_{i(\text{dir}|\text{SRS})}\right) = \left(\frac{A_i \times n}{A \times n_i}\right)^2 \left(\frac{A}{n \times A_{\text{plot}}}\right)^2 \left(1 - n \times \frac{A_{\text{plot}}}{A}\right)\left(\frac{n_i - 1}{n - 1}\right) \times \text{var}\left(Y_j | j \in U_i\right),$$

where A_i is the area of U_i and Y_j is the plot total in the jth plot in U_i. Direct variance estimation is obviously only possible when $n_i \geq 2$. When the area of U_i is not known or only known with error, the added uncertainty stemming from either predicting A_i from the ratio n_i/n or the error in A_i must be factored in. In that case a variance approximation based on a Taylor series approximation would be appropriate.

Direct estimates can be improved if the small area can be stratified into G groups based on an attribute value closely related to the attribute of interest. The sample size in each of the G groups must be larger than or equal to 2. In that case a poststratified direct estimate for the small area $\hat{Y}_{i\text{dir}(\text{SRS}|\text{poststrat})}$ will have less variance than a simple direct estimate. $\hat{Y}_{i\text{dir}(\text{SRS}|\text{poststrat})}$ is obtained as per stratified random sampling, but the approximate variance of $\hat{Y}_{i\text{dir}(\text{SRS}|\text{poststrat})}$ is

$$\hat{\text{var}}\left(\hat{Y}_{i\text{dir}(\text{SRS}|\text{poststrat})}\right) = \sum_{g=1}^{G} \left(\frac{n}{n-1} \times \frac{n_{ig} - 1}{n_{ig}}\right) \times \left(\frac{A_{ig}}{n_{ig} \times A_{\text{plot}}}\right)^2 \times \left(\frac{1}{n_{ig}} - \frac{n}{A \times n_{ig}}\right) \times \text{var}\left(Y_{ig}\right),$$

where n_{ig} is the sample size in the intersection of the gth group with U_i and Y_{ig} is the plot total of y in group g plots in U_i.

3.8.2
Synthetic Small-Area Estimators

A small-area estimator is called synthetic if it is obtained from a larger area, covering several small areas, under the assumption that the large population is homogenous with respect to the attributes of interest. In its simplest form the estimator of the mean value of an attribute in the ith small area becomes $\hat{\bar{Y}}_{i\,(\mathrm{syn})} = \hat{\bar{Y}}$, where $\hat{\bar{Y}}$ is the sample-based estimate of the population mean. Small-area totals are obtained by multiplying the mean by the area of the small area (assumed known without error). A synthetic estimator benefits from the (usually) low MSE of $\hat{\bar{Y}}$, but suffers from a potentially serious bias if the small area differs from the area of the rest of the population. Despite the obvious potential problem in applying a population mean to a small area, its precision makes it harder to ignore. Conversely an estimate based solely on data from the small area $\left(\hat{Y}_i\right)$ may appeal on the grounds of bias but not in terms of precision. As we shall see, a compromise in the form of a weighted average of a small-area estimate and global information can strike a good balance.

Auxiliary information in the form of a vector $\mathbf{x}$ related to y through the population model $y = \mathbf{x}' \cdot \boldsymbol{\beta}$ may be available in the form of known totals $\mathbf{X}_i$ for the small area. A synthetic regression estimator of the total is then $\hat{Y}_{i\mathrm{syn}_{\mathrm{reg}}} = \mathbf{X}'_i \cdot \hat{\boldsymbol{\beta}}$, where $\hat{\boldsymbol{\beta}}$ is an estimate of the population regression coefficients (c.f. Sect. 3.3.5.1). The bias of the regression estimator will be small if the small-area regression coefficients $\beta_i \approx \beta$, an assumption that can be examined more closely in a separate assessment of the model (Ronchetti et al. 1997; Zhang and Davidian 2001). A special case of the synthetic regression estimator is the synthetic ratio estimator $\hat{Y}_{i\,(\mathrm{syn}\,|\,\mathrm{ratio})} = X_i \times \hat{R}$, where X_i is the total of x in the ith small area and $\hat{R}$ is the ratio of the estimated population totals of y and x. Since the synthetic estimators can be biased an estimate of their MSE is needed (variance plus squared bias) to gauge precision. Design-consistent variance estimators of the synthetic estimators, $\hat{\mathrm{var}}\left(\hat{Y}_{i\,(\mathrm{syn})}\right)$, are obtained as per the design employed, but a reliable estimate of bias is harder to obtain. A common approach to estimate the MSE of a synthetic estimate is to obtain synthetic and direct sample-based estimates for a group of m "similar" small areas and then compute

$$\widehat{\mathrm{MSE}}\left(\hat{Y}_{i\,(\mathrm{syn})}\right) = \hat{\mathrm{var}}\left(\hat{Y}_{i\,(\mathrm{syn})}\right) + \mathrm{bias}^2\left(\hat{Y}_{i\,(\mathrm{syn})}\right),$$

$$\mathrm{bias}^2\left(\hat{Y}_{i\,(\mathrm{syn})}\right) = \overline{\mathrm{MSE}}_m\left(\hat{Y}_{i\,(\mathrm{syn})}\right) - \frac{1}{m} \sum_{j=1}^{m} \hat{\mathrm{var}}\left(\hat{Y}_{j\,(\mathrm{syn})}\right),$$

and

$$\overline{\mathrm{MSE}}_m\left(\hat{Y}_{i\,(\mathrm{syn})}\right) \approx \frac{1}{m}\left(\sum_{j=1}^{m} \left(\hat{Y}_{j\,(\mathrm{syn})} - \hat{Y}_j\right)^2 - \sum_{j=1}^{m} \hat{\mathrm{var}}\left(\hat{Y}_j\right)\right).$$

Estimators applicable to small-area means are obtained in a similar way after the appropriate scaling. Provided that sampling errors dominate bias the previous estimator of MSE is robust and asymptotically design-consistent for $m, n_j (j = 1, \ldots, m) \rightarrow \infty$. Variance estimators based on resampling or leave-one-out jackknifing are often preferred to design-based estimators of variance. Synthetic estimators can, just like direct estimators, frequently be improved by poststratification into G groups as outlined for the direct estimators.

3.8.3 Composite Small-Area Estimators

We saw that a synthetic estimate could be seriously biased if the small area was distinctly different from the general population and that a local area estimate could be very imprecise. As a compromise the composite estimator provides a weighted average of two available estimators for the ith small area, say $\hat{Y}_{i1}$ and $\hat{Y}_{i2}$ of totals

$$\hat{Y}_{i(\text{comp})} = \phi_i \times \hat{Y}_{i1} + (1 - \phi_i) \times \hat{Y}_{i2},$$

where $0 \leq \phi \leq 1$. Many small-area estimators have the composite form. Composite estimators of means are obtained in a similar way. The MSE of the composite estimator of a small-area total is given by

$$\widehat{\text{MSE}}\left(\hat{Y}_{i(\text{comp})}\right) = \phi_i^2 \times \widehat{\text{MSE}}\left(\hat{Y}_{i1}\right) + (1 - \phi)^2 \times \widehat{\text{MSE}}\left(\hat{Y}_{i2}\right) + 2\phi_i(1 - \phi_i)\,\text{E}_p\left(\hat{Y}_{i1} - Y_i, \hat{Y}_{i2} - Y_i\right),$$

where $E_p\left(\hat{Y}_{i1} - Y_i, \hat{Y}_{i2} - Y_i\right)$ is the expected mean-square cross-product of the two estimators taken over all possible samples under the employed design (p). By choosing the weights that minimize the MSE one obtains

$$\widehat{\text{MSE}}_{\text{opt}}\left(\hat{Y}_{i(\text{comp})}\right) = \frac{\widehat{\text{MSE}}\left(\hat{Y}_{i1}\right) \times \widehat{\text{MSE}}\left(\hat{Y}_{i2}\right) - \widehat{E}_p^{\,2}\left(\hat{Y}_{i1} - Y_i, \hat{Y}_{i2} - Y_i\right)}{\widehat{\text{MSE}}\left(\hat{Y}_{i1}\right) \times \widehat{\text{MSE}}\left(\hat{Y}_{i2}\right) - 2\,\widehat{E}_p\left(\hat{Y}_{i1} - Y_i, \hat{Y}_{i2} - Y_i\right)}$$

The difficulty with this estimator is that only approximate estimates of the MSE of cross-products between $\hat{Y}_{i1}$ and $\hat{Y}_{i2}$ can be obtained by application of some form of data resampling consistent with the design (Efron and Tibshirani 1993; Shao 1996; Lahiri 2003; Shen et al. 2004). If one is willing to assume that the MSE of the cross-product is negligible compared with the MSEs of Y_{i1} and Y_{i2}, the approximately optimal weight becomes

$$\varphi^*_{\text{opt}} \approx \text{MSE}\left(Y_{i2}\right) \times \left[\text{MSE}\left(Y_{i1}\right) + \text{MSE}\left(Y_{i2}\right)\right]^{-1}$$

and

$$\widehat{\text{MSE}}_{\text{opt}}\left(\hat{Y}_{i\,(\text{comp})}\right) \approx \frac{\widehat{\text{MSE}}\left(\hat{Y}_{i1}\right) \times \widehat{\text{MSE}}\left(\hat{Y}_{i2}\right)}{\widehat{\text{MSE}}\left(\hat{Y}_{i1}\right) \times \widehat{\text{MSE}}\left(\hat{Y}_{i2}\right)} = \varphi^{*}_{\text{opt}} \times \widehat{\text{MSE}}\left(Y_{i1}\right)$$

from which we see that the MSE of the composite estimate is less than the smallest of the component MSEs. The maximum reduction is 50% below the smallest value which is achieved when the components receive equal weights. Under a SRS design in a homogenous population the composite estimator that combines an estimate based on samples inside the *i*th small area with one for the population at large becomes

$$\hat{Y}_{i\,(\text{comp}\,|\,\text{SRS})} = \frac{n_i}{N_i} \times \hat{Y}_i + \left(1 - \frac{n_i}{N_i}\right) \times \hat{Y},$$

which is also the best linear unbiased prediction (BLUP). Under similar circumstances and assuming that the variance of *y* is proportional to *x* the composite ratio estimator becomes

$$\hat{Y}_{i\,(\text{comp}\,|\,\text{ratio})} = \frac{X_i}{X} \times \hat{Y}_{i\text{ratio}} + \left(1 - \frac{X_i}{X}\right) \times \hat{Y}_{\text{ratio}}.$$

Multivariate composite estimators are obtained by simple extensions of the univariate estimators (Gregoire and Walters 1988).

3.8.4 Model-Based Small-Area Estimation

A model describing an attribute value of a population element as a linear combination of fixed large-scale effects and random local effects offers the most general and flexible approach for small-area estimation. Fixed effects are constant for all population elements, while the local effects are specific to a small area. The population or a large part of the population is viewed as an ensemble of several small areas and estimation is done for all members of the ensemble simultaneously. A model-based simultaneous small-area estimation approach offers the advantage that the estimate for a specific small area can be improved by “borrowing” information from either the entire ensemble of small areas or a specific subset of small areas with certain attributes in common. The model and the associated model assumptions detail the communality of attribute values between population elements within a single small area and between population elements in different small areas. A set of nested models is often necessary to succinctly describe the relationship between observed sample values in various parts of the population. Our ability to obtain robust design-consistent and asymptotically unbiased estimates of local random effects have improved dramatically over the last 2 decades and continue to do so (Pfeffermann 2002). Two versatile yet simple models, the area-level linear mixed model and the element linear mixed model, will be given as examples. Checking of model adequacy, model fit, and model assumptions is incumbent upon the analyst whenever a model-based inference is deemed appropriate (Ritz 2004).

In the area-level model we assume that a small-area effect θ_i, $i = 1,\ldots,m$ is a known function $g\left(\bar{Y}_i\right)$ of a small-area attribute value, say the mean $\bar{Y}_i$, and furthermore is related to p area-specific auxiliary values $\boldsymbol{X}_i$ through the linear model

$$\theta_i = X'_i\beta + b_i \times v_i,$$

where the b_i are known positive constants and β is the $p\times 1$ vector of population-specific regression coefficients, and the v_i are area-specific random effects assumed to be independent and identically distributed with an expected value of zero and a variance σ_v^2. The function g can be the identity function, a linear function, or a nonlinear function, while the constants b_i are introduced to allow for heterogeneity in the variance of random effects. Note that the expectations of the random effects are with respect to the model, an important issue since it can be difficult to justify for areas for which n_i is small (less than 10).

We are interested in obtaining the BLUP of θ_i, which means that we seek a design-consistent minimum-variance estimator of $\boldsymbol{\beta}$ and a BLUP of v_i. Preliminary estimates of θ_i can be obtained directly from the sample as $\hat{\theta}_i = g\left(\hat{\bar{Y}}_i\right)$ but they are not the BLUP. We can write our direct sample-based estimates as an observational equation as

$$\hat{\theta}_i = \theta_i + e_i \text{ with } E_p\left(e_i \middle| \theta_i\right) = 0 \text{ and } \operatorname{var}\left(e_i \middle| \theta_i\right) = \psi_i \text{ known}$$

Combining the model with the observational equation, we get

$$\theta_i = X'_i\boldsymbol{\beta} + b_i \times v_i + e_i,$$

which is a special case of a linear mixed model. The mix of sampling errors (e_i) and random model effects (v_i) makes the model rather unique and introduces inferential complexity. Especially, the assumption of known area-specific sampling variances may be viewed as restrictive, and typically a direct estimate or some smoothed estimate $\hat{\psi}_i$ is used in place of ψ_i.

Since we must rely on estimated variance components our estimators are no longer the BLUP but the empirical best linear unbiased prediction (EBLUP) (Wolter 1985). The EBLUP of θ_i is

$$\hat{\theta}_{i\,\mathrm{EBLUP}} = \hat{\gamma}_i \times \hat{\theta}_i + \left(1 - \hat{\gamma}_i\right) X'_i\,\tilde{\boldsymbol{\beta}},$$

where

$$\hat{\gamma}_i = \hat{\sigma}_v^2 \times b_i^2 \times \left(\psi_i + \hat{\sigma}_v^2 \times b_i^2\right).$$

We recognize in $\hat{\theta}_{i\,\mathrm{EBLUP}}$ a composite estimator of a direct design-consistent estimate $(\hat{\theta}_i)$ and a synthetic estimate $(X'_i\tilde{\boldsymbol{\beta}})$ for the *i*th small area with weights (γ_i) determined by the strength of the among-area variation (σ_v^2) relative to that of the total random variation $(\psi_i + \sigma_v^2)$. More weight is given to a direct local estimate when the data point to strong local effects and vice versa, an intuitively appealing attribute. Only area-level auxiliary variables $(\mathbf{X}_i)$ are used for the estimation, which makes the estimate $\hat{\theta}_{i\,EBLUP}$ valid for any statistically valid sampling design. When $\theta_i = \mathbf{X}'_i\tilde{\boldsymbol{\beta}} + b_i \times v_i$ holds, the average bias will be zero. Estimators of $\boldsymbol{\beta}$ depend on an available estimate of σ_v^2 and vice versa; therefore, an iterative estimation process is needed. A current estimate of $\boldsymbol{\beta}$ is obtained from a current estimate of σ_v^2 and so on until convergence is achieved. Current estimates of $\boldsymbol{\beta}$ and σ_v^2 are

$$\tilde{\boldsymbol{\beta}} = \left[\sum_{i=1}^{m} X_i \hat{\theta}_i \times \left(\psi_i + \hat{\sigma}_v^2\right)^{-1}\right]\left[\sum_{i=1}^{m} X_i X'_i \times \left(\psi_i + \hat{\sigma}_v^2\right)^{-1}\right]^{-1}$$

and

$$\hat{\sigma}_v^2 = \hat{\sigma}_v^2 \left| \sum_{i=1}^{m} (\hat{\theta}_i - X'_i\tilde{\boldsymbol{\beta}})^2 \times \left(\psi_i + \hat{\sigma}_v^2\right)^{-1} - m + p \cong 0, \right.$$

where $\hat{\sigma}_v^2$ is a method of moments estimator of σ_v^2. Alternative estimators each relying on a set of specific assumptions are abound. The specifics of the data at hand and the experience of the analyst decide the choice.

Estimators of the MSE of $\hat{\theta}_{i\,\mathrm{EBLUP}}$ are approximate only since we rely on model-based estimates of model parameters and a sample-based estimate of error variances. The estimates are generally also biased. It is important to note that the estimation of MSEs should be tailored to the estimation procedure used for the fixed and random effects (Rao 2003). A slightly conservative MSE estimator that is valid for the previous estimate is

$$\widehat{\mathrm{MSE}}\left(\theta_{i\,\mathrm{EBLUP}}\right) = \hat{g}_{1i}\left(\hat{\sigma}_v^2\right) + \hat{g}_{2i}\left(\hat{\sigma}_v^2\right) + 2\hat{g}_{3i}\left(\hat{\sigma}_v^2\right),$$

where

$$\hat{g}_{1i}\left(\hat{\sigma}_v^2\right) = \hat{\gamma}_i \psi_i,$$

$$\hat{g}_{2i}\left(\hat{\sigma}_v^2\right) = \left(1 - \hat{\gamma}_i\right)^2 X'_i X_i \left[\sum_{i=1}^{m} X'_i X_i \times \left(b_i \hat{\sigma}_v^2 + \psi_i\right)^{-1}\right],$$

and

$$\hat{g}_{3i}\left(\hat{\sigma}_v^2\right) = \left[b_i^4 \psi_i^2 \times \left(b_i \hat{\sigma}_v^2 + \psi_i\right)^{-4}\right]\left(\hat{\theta}_i - X'_i\, \tilde{\boldsymbol{\beta}}^2 \times \hat{\mathrm{v}}\mathrm{ar}\left(\hat{\sigma}_v^2\right)\right),$$

where $\hat{\mathrm{v}}\mathrm{ar}\left(\hat{\sigma}_v^2\right)$ is an estimate of the variance of an estimated variance component. A jackknife estimate of $\hat{\mathrm{v}}\mathrm{ar}\left(\hat{\sigma}_v^2\right)$ can be obtained by repeating the previous estimation procedures m times, each time with one different small area excluded from the analysis. Alternatively one can approximate this variance by

$2\hat{\sigma}_v^4 \times (n - m - p)^{-1}$. Datta et al. (1991) detailed a multivariate extension of the area-level linear mixed model.

Our second small-area model is the element linear mixed model. In this model the attribute value (y) of the jth individual population elements within the ith small area is modeled as a linear combination of p fixed $\left(x'_{ij}\boldsymbol{\beta}\right)$ effects known for every element in i and two random effects $\left(v_i + b_{ij}e_{ij}\right)$

$$y_{ij} = x'_{ij}\boldsymbol{\beta}' + v_i + b_{ij}e_{ij},\ j = 1,\ldots,n_i, i = 1,\ldots,m,$$

where b_{ij} are known constants and e_{ij} is assumed to be a random variable with an expected value of 0 with respect to the model and a variance of σ_v^2. Again, $\boldsymbol{\beta}$ is a vector of population-level design-consistent regression coefficients. For estimation purposes it is often assumed that the distribution of the random variables is normal. We assume that a sample of size n_i has been taken from the N_i elements in the ith small area $\left(n = \sum_{i=1}^{m} n_i\right)$ and that this sample is consistent with the model. We wish to estimate, say, $\bar{Y}_i$, the mean of y in the ith small area. SRS from the ith small area or a sample selection based on $\mathbf{x}_{ij}$ both satisfy an appeal to validity of the generic model (Rao 2003) Under the element linear model the EBLUP estimator of the ith small-area mean can be written as a composite estimator of the survey regression estimator and the regression synthetic estimator:

$$\hat{\bar{Y}}_{i\,\mathrm{EBLUP}} = \hat{\gamma}_i + \left[\hat{\bar{Y}}_i + \left(\bar{\mathbf{X}}_i - \hat{\bar{\mathbf{X}}}_i\right)'\tilde{\boldsymbol{\beta}}\right] + (1 - \hat{\gamma}_i) \times \bar{\mathbf{X}}_i'\tilde{\boldsymbol{\beta}},$$

where $\hat{\bar{Y}}_i$ is the direct small-area estimate of $\bar{Y}_i$, $\bar{\mathbf{X}}_i$ is the $p \times 1$ vector of known small-area means of the auxiliary variables, $\hat{\bar{\mathbf{X}}}_i$ is the small-area sample estimate of $\bar{\mathbf{X}}_i$, and γ_i is the weight given to the survey regression estimator. Note, when the constants b_{ij} are not all 1 the sample means become

$$\hat{\bar{Y}}_i = \frac{\sum_{j=1}^{n_i} b_{ij}^{-2} y_{ij}}{\sum_{j=1}^{n_i} b_{ij}^{-2}},\quad \hat{\bar{\mathrm{X}}}_i = \frac{\sum_{j=1}^{n_i} b_{ij}^{-2} x_{ij}}{\sum_{j=1}^{n_i} b_{ij}^{-2}}.$$

The best linear unbiased estimator of the population-level regression coefficients is

$$\tilde{\boldsymbol{\beta}} = \left(\sum_{i=1}^{m}\sum_{j=1}^{n_i} b_{ij}^{-2}\mathrm{x}_{ij}\mathrm{x}'_{ij} - \hat{\gamma}_i b_{i.}^{-2}\hat{\bar{\mathrm{X}}}_i\hat{\bar{\mathrm{X}}}_i'\right)^{-1}\left(\sum_{i=1}^{m}\sum_{j=1}^{n_i} b_{ij}^{-2}\mathrm{x}_{ij}y_{ij} - \hat{\gamma}_i b_{i.}^{-2}\hat{\bar{\mathrm{X}}}_i\hat{\bar{Y}}_i\right),$$

where $\mathrm{b}_{\mathrm{i.}}$ is the sample sum of fixed variance constants b_{ij} for the ith small area. The weight given to the direct survey regression estimate is

$$\hat{\gamma}_{\mathrm{i}} = \hat{\sigma}_{\mathrm{v}}^2 \times \left(\hat{\sigma}_{\mathrm{v}}^2 + \hat{\sigma}_{\mathrm{e}}^2 \Big/ \sum_{j=1}^{n_i} b_{ij}^{-2}\right).$$

Provided we have large small-area sample sizes and $b_{ij} \equiv 1 \forall \{i,j\}$, the survey regression estimator is approximately design-unbiased but the synthetic regression estimator $\overline{\mathbf{X}}'_i \tilde{\boldsymbol{\beta}}$ may be a biased for $\hat{\overline{Y}}_i$. Estimation of the variance components σ_v^2 and σ_e^2 can proceed in different directions depending on the assumptions made and the preferences of the analyst. Maximum-likelihood and restricted-maximum-likelihood estimation requires assumptions about the distribution of the random effect. If warranted, these methods lead to more efficient estimates but only if the distributional assumptions actually hold. They also provide a generic framework for estimation regardless of the values chosen for the b_{ij}. As done for the area-level model we shall present the method of moment estimation procedures under the assumption of random sampling in small areas. First, we obtain ordinary least squares (OLS) estimates of the element residuals e_{ij} as

$$\hat{e}_{ij\,\mathrm{OLS}} = y_{ij} - \hat{\overline{Y}}_i - \left(\mathbf{X}_{ij} - \hat{\overline{\mathbf{X}}}_i\right)\hat{\boldsymbol{\beta}}_{\mathrm{OLS}}$$

where $\hat{\beta}_{\mathrm{OLS}}$ is the OLS estimate of the regression coefficient $y_{ij} - \hat{\overline{Y}}_i$ regressed on $\mathbf{X}_{ij} - \hat{\overline{\mathbf{X}}}_i$ (no intercept). Only residuals for nonzero values of $\mathbf{X}_{ij} - \hat{\overline{\mathbf{X}}}_i$ are computed. From these residuals we estimate

$$\hat{\sigma}_e^2 = \frac{1}{(n - m - n_0)} \sum_{i=1}^{m} \sum_{j=1}^{n_i} \hat{e}_{ij\mathrm{OLS}},$$

where n_0 is the number of zero-valued $\mathbf{x}$ residuals. Second, we estimate the OLS residuals (u) from a regression of $y_{ij} \times b_{ij}^{-1}$ on $x_{ij} \times b_{ij}^{-1}$ (no small-area effects), i.e.,

$$\hat{u}_{ij\mathrm{OLS}} = b_{ij}^{-1}\left(y_{ij} - \mathbf{X}'_{ij}\hat{\boldsymbol{\beta}}_{\mathrm{OLS}}\right),$$

and obtain an estimate of σ_v^2 from

$$\hat{\sigma}_v^2 = \frac{1}{n^*}\left(\sum_{i=1}^{m} \sum_{j=1}^{n_i} \hat{u}_{ij\mathrm{OLS}}^2 - (n - p)\hat{\sigma}_e^2\right),$$

where

$$n^* = \sum_{i=1}^{m} n_i \hat{\overline{X}}_i \left[1 - n_i \left(\sum_{i=1}^{m} \sum_{j=1}^{n_i} \mathbf{X}_{ij}\mathbf{X}'_{ij}\right)_i^{-1} \hat{\overline{\mathbf{X}}}_i\right].$$

Summation in these expressions should be limited to small areas with $n_i > 1$. Ghosh and Rao (1994) proposed the following estimator for the MSE:

$$\widehat{\mathrm{MSE}}\left(\overline{Y}_{i\mathrm{EBLUP}}\right) = \left(1 - \frac{n_i}{N_i}\right)^2 \left[\hat{g}_{1i}\left(\hat{\sigma}_v^2, \hat{\sigma}_e^2\right) + \hat{g}_{2i}\left(\hat{\sigma}_v^2, \hat{\sigma}_e^2\right) + 2\hat{g}_{3i}\left(\hat{\sigma}_v^2, \hat{\sigma}_e^2\right)\right],$$

where

$$\hat{g}_{1i}\left(\hat{\sigma}_v^2, \hat{\sigma}_e^2\right) = \hat{\gamma}_i \left(\frac{\hat{\sigma}_e^2}{n_i}\right) + \left(1 - \frac{n_i}{N_i}\right)^2 \frac{(N_i - n_i)}{N_i^2},$$

$$\hat{g}_{2i}\left(\hat{\sigma}_{v}^{2},\hat{\sigma}_{e}^{2}\right)=\hat{\sigma}_{v}^{2}\left[\mathbf{X}_{i}^{C}-\hat{\gamma}_{i}\hat{\bar{\mathbf{X}}}_{i}\right]\hat{\Lambda}^{-1}\left[\mathbf{X}_{i}^{C}-\hat{\gamma}_{i}\hat{\bar{\mathbf{X}}}_{i}\right]',$$

and

$$\hat{g}_{3i}\left(\hat{\sigma}_{v}^{2},\hat{\sigma}_{e}^{2}\right)=\frac{1}{n_{i}^{2}\times\hat{\gamma}_{i}^{3}}\times\left(\frac{2\hat{\sigma}_{e}^{8}}{n-m-n_{0}-2}+\frac{2\hat{\sigma}_{v}^{8}}{n^{*}-2}\right)+\frac{4(m-1)}{n^{*}}\hat{\sigma}_{e}^{6}\hat{\sigma}_{v}^{2},$$

where $\mathbf{X}_i^C$ is a matrix of nonsampled $\mathbf{x}_{ij}$ values in the *i*th small area and

$$\hat{\Lambda}=\sum_{i=1}^{m}\sum_{j=1}^{n_i}\mathbf{X}'_{ij}\mathbf{X}_{ij}-\hat{\gamma}_{i}\times n_{i}\hat{\bar{\mathbf{X}}}'_{i}\hat{\bar{\mathbf{X}}}_{i},$$

Flores and Martínez (2000) entertained the unit level mixed linear model for the estimation of crop areas under irrigation in 53 small areas in the Duero river basin in northwestern Spain. Auxiliary variables were estimates obtained from remotely sensed images and the element of sampling was a 500-m×500-m ground area (25 ha). A random sample of 158 elements was taken (0–17 per small area). The use of the auxiliary information resulted in a reduction of the MSE of small-area estimates by 30–70%. Kangas (1996) used the mixed element level model for estimating the timber volumes in eight Finnish municipalities and found it efficient (as opposed to direct or synthetic estimation) even in the absence of auxiliary information. Wang and Fuller (2003) recently suggested some improvements to the MSE estimation procedures of mixed linear models; the improvement makes the MSE more robust when the among-area variation is strong. Interested readers are referred to their text for details.

3.8.5 Small-Area Estimation by Block Kriging

The spatial distance between two population elements can be an indicator of the expected similarity of their attribute values. Within a forest stand, for example, one would expect that the basal area in a 100-m² unit, on average, would be more similar to the basal area in units that are spatially close than to the basal area in more distant units. This phenomenon of distance-dependent similarity, if manifest, can be exploited in certain small-area estimation problems. Samples taken in the vicinity of the small area can be used efficiently to predict, via a spatial model (Cressie and Chan 1989), the average attribute value in the small area. Simple kriging is a basic form of spatial prediction for a location with unknown attribute values (Goovaerts 1997). A simple kriging prediction is a linear combination of known attribute values observed in locations within a neighborhood of the location for which we seek a prediction The weight given to an observation depends on the strength of the expected covariance between the observed value and the value to be predicted.

Small-area prediction by block kriging illustrates one of the simplest yet most powerful spatial models for small-area estimation problems. As before, we have a population sample of size n with attribute y and we wish to estimate, say, the mean $\bar{Y}_i$ for small area i. We have n_i sample records available for the small area i $(n_i \geq 0)$, which we consider as a "block" in the context of kriging. Let us assume that there is no significant spatial trend in the observed y values in the small area and its vicinity but y values in locations separated by less than a relatively short distance of a few hundred meters do tend to be significantly and positively correlated with each other. Furthermore, we also have a function $C(y_k, y_l)$ that predicts, without bias, the expected covariance between two y values observed in locations k and l. Normally the covariance is governed by the distance between locations k and l. Issues surrounding the selection, estimation, and validation of the spatial covariance function (variogram) are beyond the scope of this text. Suffice to say that there are many complex statistical issues one must consider before accepting a spatial model, as the risk of inadvertently introducing a serious bias is nontrivial (Cressie 1991; Chilès and Delfiner 1999; Atkinson and Lewis 2000; Diblasi and Bowman 2001; Zhao and Wall 2004).

In our chosen variant of block kriging the n_i sample records from the small area are used only for a direct estimate $\hat{\bar{Y}}_i$, which is then combined with a block kriging prediction $\hat{\bar{Y}}_{i\,\mathrm{bkrig}}$ in the form of a composite estimator. Attribute values, both sampled and nonsampled, for the small area i (SA_i) are denoted by y_i. The first step towards obtaining $\hat{\bar{Y}}_{i\,\mathrm{bkrig}}$ is to choose a number of sample records $Y_k, k = 1, \ldots, N_i^{OK}$ taken outside the small area but close enough to the small area to justify the expectation that their attribute values would be significantly correlated with attribute values inside the small area. The covariance function can guide the cutoff distance for N_i^{OK} since $C(y_k, y_i) \times \mathrm{var}(y)^{-1}$ can be viewed as a crude predictor of the expected correlation of y values. Since block kriging computations increase with the square of N_i^{OK} there are good reasons to keep the number as low as possible but high enough to take advantage of spatial correlation. In practice, sample locations with an expected correlation below 0.2 can be excluded with only a minimal impact on the predictions and their estimated variance. The set of outside sample values included for block kriging prediction is $\Im_i^{OK}$. From the selected outside sample points we obtain the block kriging prediction for the small area as

$$\hat{\bar{Y}}_{i\mathrm{bkrig}} = \sum_{k=1}^{N_i^{OK}} \hat{\lambda}_k \times Y_k, Y_k \in \Im_i^{OK},$$

where $\hat{\lambda}_k$ is the estimated block kriging weight assigned to Yk in $\Im_i^{OK}$. Estimated block kriging weights are obtained as solutions to the following system of block kriging equations (Goovaerts 1997):

$$\sum_{k=1}^{N_i^{OK}} \lambda_k \times C(Y_k, y_l) = \overline{C}(y_l, y_i), y_i \in SA_i, \{Y_k, y_l\} \in \Im_i^{OK},$$
$$\sum_{k=1}^{N_i^{OK}} \lambda_k = 1,$$
$$l = 1, \ldots, N_i^{OK},$$

where $\overline{C}(y_h, y_l)$ is the average expected covariance between Y_k from $\Im_i^{OK}$ and an element attribute value y_l in SA_i. We can obtain a good approximation of $\overline{C}(Y_k, Y_l)$ by computing the expected covariance between elements in $\Im_i^{OK}$ and a series of elements y_l, $l = 1, \ldots, M_{SA_i}$ distributed evenly over SA_i and then take the average of these expected covariances (Goovaerts 1997); hence,

$$\overline{C}(y_l, y_i) \simeq \frac{1}{M_{SA_i}} \sum_{i=1}^{M_{SA_i}} C(y_l, y_i), y_l \in \Im_i^{OK}, y_i \in SA_i,$$

The number of locations M_{SA_i} to be included in the average depends on the rate of convergence. At one point increasing the number further will have only a minor impact on the average. Around 16 is probably a reasonable choice. An estimator of the variance of $\hat{\bar{Y}}_{i\,bkrig}$ is

$$\hat{\text{var}}\left(\hat{\bar{Y}}_{i\,bkrig}\right) = \overline{\overline{C}}(Y_k, y_j) - \sum_{k=1}^{N_i^{OK}} \hat{\lambda}_k \times \overline{C}(Y_k, y_l), \{y_i, y_k\} \in SA_i,$$

where $\overline{\overline{C}}(y_i, y_j)$ is the expected block-to-block covariance which we approximate by the average covariance between elements in SA_i, i.e.,

$$\overline{\overline{C}}(Y_k, y_l) - \sum_{i=1}^{M_{SA_i}} \sum_{l=1}^{M_{SA_i}} C(Y_k, y_l),$$

The block kriging predictor $\hat{\bar{Y}}_{i\,bkrig}$ and the direct estimate $\hat{\bar{Y}}_i$ can now be combined to a composite estimator as outlined in Sect. 3.8.3. To compute the MSE of the composite estimator we need an estimate of the expected covariance between $\hat{\bar{Y}}_{i\,bkrig}$ and $\hat{\bar{Y}}_i$. They cannot be assumed to be independent since $\hat{\bar{Y}}_{i\,bkrig}$ implicitly generates M_{SA_i} pseudo-observations for SA_i. The expected covariance is approximated by the average expected covariance between the n_i sample locations in SA_i and the N_i^{OK} sample locations in $\Im_i^{OK}$. Alternatively, a single prediction $\hat{\bar{Y}}_{i\,bkrig}$ could have been obtained by including the n_i sample points in SA_i in the set $\Im_i^{OK}$ with no other change to the previously outlined procedure. We chose a composite estimator as it is more transparent and computationally easier to optimize. The choice will have only a minor impact on the results.

When y values display a spatial trend the block kriging procedure has to be expanded to include the prediction of local trend values. While the extension is technically straightforward the presence of a trend nevertheless complicates matters. First, the trend has to be estimated precisely to avoid introduction of potentially serious bias. A precise estimation is often not possible from typical survey data. Also, only few surveys will have enough data to support a thorough model selection and validation process and a high enough sampling density to ensure that there are a sufficient number of suitable predictors available in the area around SA_i (Lappi 2001). Still, there are situations in forest inventory where spatial-model-based small-area predictions are attractive (Mandallaz 1993, 2000). They are identified by the presence of a significant distance- or location-dependent correlation, viz., covariance, between sampled attribute values.

3.8.6 Empirical Bayesian Methods for Small-Area Estimation

The Bayesian framework uses the posterior distribution for inference (Ghosh and Meeden 1997; Gaudard et al. 1999; Congdon 2001). The posterior distribution is the product of a likelihood and a prior distribution, and as such it is entirely model-based. A likelihood function f with parameters $\boldsymbol{\theta}_i$ can be postulated for the data sampled in a small area i (SA_i) and then combined with prior expectations with regard to the probability distribution function of the parameters $\boldsymbol{\theta}_i$ in order to obtain the posterior distribution p of $\boldsymbol{\theta}_i$. When prior distributions are estimated from samples taken outside SA_i p is said to be the empirical Bayesian (EB) posterior for SA_i (Singh et al. 1998; Pfeffermann 2002). The EB approach offers a very flexible and rich framework for small-area estimation. In applications with a Gaussian-likelihood function and a conjugate prior (a conjugate prior produces a posterior distribution of the same type as implied by the likelihood), the posterior estimates will be similar to the composite estimator (Congdon 2001).

An example with a continuous real-valued positive attribute y and one with counts of a categorical attribute value illustrate the flexibility and power of the EB approach. In our first EB example, suppose we have from SA_i four (n_i=4) sample values $Y_i = \{Y_{i1}, Y_{i2}, Y_{i3}, Y_{i4}\} = \{156, 220, 181, 185\}$ with a mean $\hat{\bar{Y}}_i$ of 185.5 and a variance $\hat{\text{var}}\left(\hat{\bar{Y}}_i\right)$ of 173.4. A larger sample of size $n-n_i$=100−4=96 from outside of SA_i produced $\hat{\bar{Y}}_{\supset SA_i} = 200$ and $\hat{\text{var}}\left(\hat{\bar{Y}}_{\supset SA_i}\right) = 4.59$. The small-area likelihood f is a Gaussian with $\boldsymbol{\theta}_i = \{\theta_{i1}, \theta_{i2}\} = \{\bar{Y}_i, \text{var}(\bar{Y}_i)\}$ and we seek to estimate $\hat{\boldsymbol{\theta}}_i^{EB}$ the EB posterior of $\boldsymbol{\theta}_i$. We assume a Gaussian prior with parameters $\hat{\omega} = \left\{\hat{\bar{Y}}_{\supset SA_i}, \hat{\text{var}}\left(\hat{\bar{Y}}_{\supset SA_i}\right)\right\}$ for the mean and a Γ (gamma)distribution prior with parameters $\hat{\lambda}$ = {96.00, 0.048} for the variance. The Γ distribution prior was chosen so that its expected value would be $\hat{\text{var}}\left(\hat{\bar{Y}}_{\supset SA_i}\right)$ and its variance $2 \times \hat{\text{var}}^2\left(\hat{\bar{Y}}_{\supset SA_i}\right) / (n - n_i - 2)$, which is the variance of a variance when y is normally distributed (Snedecor and Cochran 1971). From these preliminaries we obtain the posterior p of $\boldsymbol{\theta}_i$ as

$$p\left(\theta_i \middle| \hat{\bar{Y}}_i\right) \propto f\left(y_{ij} \middle| \theta_i\right) \times \varphi\left(\theta_{i1} \middle| \hat{\omega}\right) \times \Gamma\left(\theta_{i2} \middle| \hat{\lambda}\right).$$

The maximum posterior log-likelihood was –10.87 with $\theta_i^{EB} = \left\{\hat{\bar{Y}}_{i\,EB}, \hat{\text{v}}\text{ar}\left(\hat{\bar{Y}}_{i\,EB}\right)\right\}$ = {192.8, 9.7}. In this example we gave the sample from outside SA_i maximum weight, in that our priors were tailored to the large-sample results. The EB posterior mean is almost a perfect average of $\hat{\bar{Y}}_{\neg SA_i}$ and $\hat{\bar{Y}}_i$ (an estimated weight of 0.49 is given to the direct estimate) and the posterior variance is almost 18 times smaller than the variance of the direct estimate, but also about 18 times as large as the variance of the $\hat{\bar{Y}}_{\neg SA_i}$. We are of course free to change the priors if only a part of the large sample is deemed representative as a prior for SA_i or we have other information that warrants a change. In any case, the choice of an informative "prior" must be decided carefully and should be justified explicitly in the same fashion as one would justify a model choice. The EB framework is extended easily to deal with regression and ratio estimators and multivariate attribute values (Ghosh and Meeden 1997; Green and Valentine 1998; Elliot and Little 2000; Denison et al. 2002; O'Brien and Dunson 2004).

Our second example show the flexibility of the EB appraoch to handle binary data. We have done a survey of a beetle infestation. In each plot n_t=16 trees are examined for the presence (y=1) or absence (y=0) of a certain beetle species. We have a total of n=47 plots, of which n_i=7 are inside SA_i. We wish to estimate the proportion of trees infested with the beetle in the small area (P_i) and a variance of this estimate. At the plot level, the likelihood of observing, say, n_{bj} beetle-infested trees in the jth plot is

$$\Pr\left(n_{bj} \middle| n_t, P_i\right) = \binom{16}{n_{bj}} \times P_i^{n_{bj}} (1 - P_i)^{n_t - n_{bji}}$$

as per the binomial distribution. The results from the 40 "outside" plots are used to form prior expectations of the proportion P_i. The survey produced the following estimates:

$$\hat{P}_{\supset SA_i} = 0.172, \hat{\text{v}}\text{ar}\left(\hat{P}_{\supset SA_i}\right) = 5.196 \times 10^{-4}$$

for the outside area and $\hat{P}_i = 0.223$, $\hat{\text{v}}\text{ar}\,(\hat{P}_i) = 2.949 \times 10^{-3}$ for SA_i, where 25 of the 112 sample trees were infested with the beetle. We assume conveniently a beta distribution as a prior for the parameter P_i in the small-area data likelihood. The beta distribution has two parameters, α and β and a mean of $\bar{P} = \alpha \times (\alpha + \beta)^{-1}$ and a variance of $\bar{P}\,(1 - \bar{P}) \times (1 + \alpha + \beta)^{-1}$. From the 40 outside plots and by maximum-likelihood methods, we obtained $\hat{\alpha} = 1.005$ and $\hat{\beta} = 4.843$. The convenience in the choice of the prior is that the posterior distribution of the small-sample estimate of P is also a beta distribution (Congdon 2001) but with parameters $\left\{\hat{\alpha} + \sum_{i=1}^{n_i} n_{bi}, \hat{\beta} + n_i \times n_t + 1\right\} = \{26.00, 117.84\}$, from which we obtain $\hat{P}_{iEB} = 0.221$ and $\hat{\text{v}}\text{ar}\,(\hat{P}_{iEB}) = 1.44 \times 10^{-3}$. While the posterior mean is close to the direct sample mean, the EB estimate of the posterior variance is only half of the variance of the direct estimate.

EB extensions to multinomial data are straightforward: instead of a beta distribution as a prior one conveniently chooses instead the Dirichlet distribution (Santner and Duffy 1989; Green and Clutter 2000), which generates a Dirichlet posterior with parameters determined as the sum of small-area counts and the prior parameters (pseudo-counts).

3.9 k Nearest-Neighbor Prediction

When an auxiliary attribute(s) is known for all N population elements and a functional relationship can be assumed to exist between them and the attribute of interest, available for n $(n \ll N)$ sampled elements only, then the predictive power of the auxiliary attribute values can be exploited in several ways for the purpose of improving the precision of an estimated mean or total of a population or a stratum. This was illustrated in Sect. 3.3.5 for two-phase sampling. While global and strata estimates of totals and means are useful in their own right, the management of natural resources often requires attribute values to be provided for all population elements within specified areas. Essentially a map showing the spatial distribution of attribute values is desired. A naive prediction of local attribute values from a population-level regression model can produce unacceptable local artefacts because the predictions ignore any spatial correlation among the predictors and because application of a single population-level model may produce biased results when applied to spatial subsets of the population (Rao 2003).

A forest can be viewed as a composition of a finite set of distinct compositions of auxiliary attribute values. If, furthermore, we assume that the value of the attribute(s) of interest is fixed for a given distinct composition of the auxiliary value(s) then, if the assumption holds, perfect predictions would be possible when the distinct set of auxiliary compositions matches that of the entire population. The predicted attribute value would naturally be the value recorded for the sample with matching auxiliary values. In practice a perfect match is rarely possible because the sample simply does not exhaust the natural variability in the auxiliary attribute(s). To make our predictions we could relax our requirement of a perfect match and assume that similar auxiliary attribute values means similar values of the desired attribute. The *k*NN method of prediction is based on this relaxed assumption and was first developed for the purpose of replacing within-item missing attribute values (Rubin 1987). In the *k*NN method a prediction is derived from the *k* sample records that match most closely the auxiliary values of the element we wish to predict.

The *k*NN method of prediction is intuitively appealing and conceptually simple; however, a successful application demands complex and demanding analyses and computations. The numbers of neighbors, the auxiliary traits to include, and the definition of similar values are all nontrivial issues in need of careful analysis (Moeur et al. 1995; Katila and Tomppo 2001; McRoberts et al. 2002).Otherwise, carefully calibrated *k*NN predictions will be biased and the result could be worse than predictions based on global expectations (Holmgren et al. 2000; Franco-Lopez et al. 2001). *k*NN is now used routinely to provide local estimates of national forest inventory attributes from local auxiliary attribute values obtained from remotely sensed images (Gjertsen et al. 1999; Katila et al. 2000; Katila and Tomppo 2001; Tomppo and Halme 2004).

All *k*NN methods require large sample sizes to ensure that similar matches are indeed found. It is difficult to make specific recommendations, the natural variability in attribute values is the decisive factor, but even for rather homogeneous forests of northern climes sample sizes over 200 seem to be required for *k*NN methods to be even modestly successful (Haara et al. 1997; Franco-Lopez et al. 2001; Holmström 2002). As the number of auxiliary attributes increase it becomes increasingly difficult to find a good match, a paradigm known as the curse of dimensionality (Scott 1992). Predictions derived from the single most similar set of auxiliary attribute values are asymptotically unbiased and they will preserve the sample variability in the desired attribute value(s) (Moeur et al. 1995; McRoberts et al. 2002). When more than one similar sample record is used for prediction, then it is a common observation that predictions at the extreme tend to be biased in opposite directions (Moeur et al. 1995; McRoberts et al. 2002).

The local kNN prediction of the attribute y from the auxiliary variables $\mathbf{X}$ for the ith nonsampled population element is

$$\hat{Y}_i = \sum_{j \in NN_k(\mathbf{x}_i)} w_{ij} \times y_j,$$

where y_j is the attribute value of the jth sample, w_j is the weight given to this value, and j is one of k NNs to the ith population element in terms of the auxiliary attribute values, $j \in NN_k(\mathbf{X}_i)$. In the multivariate case a scalar would be replaced by the appropriate vector notation. The weights are chosen to reflect the degree of similarity in the auxiliary attribute values between the ith nonsampled and the jth sampled population element. Weights are usually based on an index (d_{ij}) of similarity, viz., distance, between the auxiliary attribute values of the ith and jth elements. Subject knowledge, prior beliefs, ecology, spatial distance, and statistical consideration guide the choice of weight function. For example, a close match in $\mathbf{X}$ could still receive a low weight if population elements i and j are located on different soil types, on different aspects, in different vegetation zones, or are separated by a large spatial distance. Prior knowledge can be used to guide the search towards locations with the highest chance of a suitable match (van Lieshout and Baddeley 2002). It is customary to choose weights such that

$$w_{ij} = \frac{1}{d_{ij}^t} \times \left(\sum_{j \in NN_k(x_i)} \frac{1}{d_{ij}^t} \right)^{-1}, t = 0,1,2,\ldots.$$

t=0 implies equal weighting of the k neighbors, whereas t=1 weights with the inverse of distance and higher t values ensure a more rapid decline in the weight is given to the less similar values in the group of k most similar neighbors. The choice of t is intimately connected to the number k of most similar neighbors. It makes little sense to have a high k and a high t as most neighbors would then contribute little towards a prediction. Conversely, a lower k would argue for a lower t. A t value of 1 seems to be the most popular choice.

The similarity index d_{ij} should reflect the impact that discrepancies in the auxiliary variable(s) have on a local prediction. An ideal index is linear in the square of the absolute prediction errors (Barbieri and Berger 2004). The index is inevitably a function of the auxiliary attributes included as predictors, their scale, and predictive power. Finding an optimal index or distance metric is the crux of the kNN method and is often a very time consuming step. A generic index takes the form

$$d_{ij} = (\mathbf{x}_i - \mathbf{x}_j)\Omega_{x'x}^{-1}(\mathbf{x}_i - \mathbf{x}_j)',$$

where $\mathbf{x}$ is a $p\times 1$ vector of auxiliary attribute values, $\Omega_{x'x}$ is a weight matrix, and $\mathbf{x}'$ is the transpose of $\mathbf{x}$. In the case of $\Omega_{x'x} = I$, the identity matrix, the similarity index is equal to the Euclidian distance in the feature space of $\mathbf{x}$. A Euclidian distance weighting disregards the predictive power of individual auxiliary attributes and distances are strongly influenced by scale differences in the auxiliaries. The choice of $\Omega_{x'x} = D(\sigma_j^2)$, where $D(\sigma_j^2)$ is a diagonal matrix of the variances of the auxiliary variables, removes the scale effect on the distance measure but does not reflect a possible correlation among the predictors. Disregarding the correlation can lead to biased predictions. Choosing $\Omega_{x'x} = \Sigma_{x'x}$, where $\Sigma_{x'x}$ is the variance covariance matrix of the auxiliary attributes, leads to a similarity index based on Mahalanobis distances (Rencher 2002) and removes both scale effects and correlation between the auxiliary attributes, but their predictive powers is not taken into account. These choices of the weight matrix result in nonparametric kNN predictions. The predictive power of the auxiliary attributes can be incorporated by assuming that predictions of y are linear in $\mathbf{x}$, i.e., $\hat{y} = \hat{\boldsymbol{\beta}}\mathrm{x}'$, or are linear in a set of mutually independent (orthogonal) variables $\mathbf{z}$ obtained by premultiplying $\mathbf{x}$ by the Cholesky decomposition of $\Sigma_{\mathrm{x}'\mathrm{x}}$ (Rencher 2002). In the former case we get

$$d_{ij} = \left(\mathbf{x}_i - \mathbf{x}_j\right)\hat{\boldsymbol{\beta}}\Omega_{x'x}^{-1}\hat{\boldsymbol{\beta}}'\left(\mathbf{x}_i - \mathbf{x}_j\right)',$$

where $\hat{\boldsymbol{\beta}}$ is either an ordinary or a generalized least-squares estimate of regression coefficients, and in the latter case we get

$$d_{ij} = \left(\mathbf{x}_i - \mathbf{x}_j\right)\Gamma\Lambda^2\Gamma'\left(\mathbf{x}_i - \mathbf{x}_j\right)',$$

where Γ is a matrix of canonical correlation coefficients and $\Lambda_{z'z}$ a diagonal matrix of canonical correlation coefficients (Rencher 2002). The two distance measures are identical if all p of the transformed variable $\mathbf{z}$ are used as predictors. If only a subset q ($p>q$) with a significant correlation to y (or $\mathbf{y}$) is used then the two will differ. Further details on the canonical approach can be found in Moeur et al. (1995).

The expected error of a kNN-predicted value of y is usually estimated by some leave-one-out cross-validation procedure (Franco-Lopez et al. 2001; McRoberts et al. 2002; Rao 2003; Efron 2004). The procedure is relatively simple but time-consuming. The one-by-one procedure makes a kNN prediction $\hat{y}_{k\mathrm{NN}}$ for one of the n sampled elements by withholding this observations from the calculations of similarity indices, weights, and ranking of indices. The mean of the errors made in these n predictions is the cross-validation estimate of error:

$$\mathrm{RMSE}_{\mathrm{CV}}\left(\hat{Y}_{NN_k}\right) = \sqrt{\frac{\sum_{i=1}^{n}\left(Y_i - \hat{Y}_{NN_k}^{(i)}\right)^2}{n}},$$

where $\hat{Y}_{NN_k}^{(i)}$ is a kNN prediction of the ith sample value derived independently from Y_i. Bootstrapping offers an alternative method for estimating this error. Instead of the delete one-at-a-time procedure of cross-validation, a sample of size n is sampled with replacement from the original sample and a kNN prediction rule is obtained from the bootstrap sample and is applied to the original sample. By repeating this process a large number of times (more than 500) one obtains a distribution of prediction errors from which statistics such as the mean, mode, and quantiles are easily obtained. Estimates of the prediction variance are only approximations, possibly biased since predictions are based on order statistics with a nonsmooth distribution function (Chen and Shao 2001).

Individual kNN prediction errors can be assumed to depend on the index of similarity values d_{ij} for the kNNs used for a prediction; hence, a regression model with the square of the prediction errors obtained during the cross-validation process as the dependent variable, and the k d_{ij} index values as the predictors could be used to estimate the *k*NN error of individual predictions (Moeur et al. 1995).

3.10 Resampling for Nonlinear Inventory Statistics

Users of forest inventory information are often interested in estimates that go beyond the mean, the total, and associated estimates of their sampling variance. Estimates of, for example, population percentiles (e.g., the median and the lower and upper 2.5 percentiles), ratios of estimates involving two or more inventory attributes (e.g., percentage change during a given period of time or the proportion of area in plantations), number of species in a population, or simply a model-based transformation of one or several inventory estimates into another attribute of interest (e.g., composite estimators, transformation of volume to biomass or carbon content, small area estimates, estimates of non-sampling errors) are demanded on a routine basis from the analyst. Let $\hat{T}$ be such an estimate obtained from one or several inventory estimates $\hat{Z}$ via some function g as in $\hat{T} = g\left(\hat{Z}\right)$, where $\hat{Z} = \left\{\hat{Y}_1, \ldots, \hat{Y}_k, \left(\hat{\text{cov}}\left(\hat{Y}_i, \hat{Y}_j\right)(i,j) = 1, \ldots, k\right)\right\}$. If g is linear in the inventory estimates (as in a weighted average with fixed and known weights), the variance of $\hat{T}$ is estimated via a first-order Taylor series linearization-substitution method (Rao 1988):

$$\hat{\text{var}}\left(\hat{T}\right) = \hat{g}'(Z)^T \hat{\Omega}(Z) \hat{g}'(Z)$$

where $g'(Z)$ is the vector of derivatives with respect to the inventory attributes and $\hat{g}'(Z)$ is $g'(Z)$ evaluated at $\hat{Z}$, $\hat{\Omega}(Z)$ is the estimate of the variance covariance matrix of **Z**, and superscript T denotes the transpose of a vector or matrix.

For g linear in all the parameters **Z** the estimate $\hat{T}$ and the estimate of the variance of T will exhibit properties that are a linear function (g) of the elements of **Z**. If $\hat{\mathbf{Z}}$ is design-unbiased so is $\hat{T}$, and if the variance estimates for **Z** are all design-unbiased and consistent so is the estimated variance of T. However, when g is nonlinear, or possibly nonsmooth (derivatives do not exist everywhere, as in a discrete distribution or when g embodies a series of hierarchical functions or the output of g is constrained), the statistical properties of $\hat{T}$ are no-longer predictable from g and the properties of $\hat{\mathbf{Z}}.\hat{T}$ may be biased and the Taylor series method may produce a poor approximation to the variance of T since higher moments of the sampling distribution of T do not vanish (as they do in a normal distribution).

When g is nonlinear or nonsmooth the analyst may chose to adopt a resampling scheme as an alternative to the Taylor series method. Research has shown that estimates of $g\left(\hat{Z}\right)$ and $\hat{\text{var}}\left[g\left(\hat{Z}\right)\right]$ with g nonlinear or nonsmooth derived from a correct application of the bootstrap resampling technique are at least as good as those based on the Taylor series method; often they are better (Shao 1996, 2003; Shao and Chen 1998; Hall et al. 2001; Van Hees 2002; Lahiri 2003; Shen et al. 2004; Zhu and Morgan 2004). At times the function g is so complex as to preclude an analytical estimator that the bootstrap or another resampling alternative (e.g., jackknife, balanced replicate resampling, or Pòlya-urn) offers the only practical option (Shao 1996; Meeden 1999). Estimation of the MSE of composite estimators as exemplified by small-area estimation problems (see also Sect. 3.8) becomes almost straightforward with bootstrap resampling. Bootstrap resampling is also attractive for estimation problems when missing data are imputed at random (hot-deck) or model-based (Saho and Sitter 1996; van Deusen 1997; Shao and Steel 1999; McRoberts 2001; Lahiri 2003; Shao 2003). Although computer-intensive, the bootstrap computations are simple.

3.10.1 The Bootstrap

Efron (Efron and Tibshirani 1993) introduced the bootstrap resampling method for the study of the properties of no-linear and nonsmooth statistics. The bootstrap simulates the estimated sampling distribution of a statistic estimating a population attribute by generating a large number (B) of bootstrap estimates $\hat{T}_1^*, \ldots, \hat{T}_B^*$ of T from which a mean, a variance, and an approximation to the distribution function $\hat{\Pr}\left(\hat{T}^* \leq T\right)$ are obtained by standard techniques.

In the simplest (naive) implementation of bootstrap resampling a single bootstrap estimate $\hat{T}_l^*$, $l = 1, \ldots, B$, is obtained by a SRS *with replacement* from n observed values of Y_i, $i = 1, \ldots, n$. The resampling yields a bootstrap sample Y_j^*, $j = 1, \ldots, n$ from which $\hat{T}_l^*$ is estimated. The naive implementation requires that the observed Y_i are identically and independently distributed (iid), which is only possible if the data are collected by SRS. Sample selection with unequal probabilities, however, invariably introduces a complex correlation structure which makes the development of a theoretical valid bootstrap method challenging (Lahiri 2003). For sample surveys, bootstrapping methods have been validated under randomization theory.

In forest inventories sampling is commonly from a finite population and without replacement to avoid sampling the same element (unit) more than once. Even point-sampling locations are usually chosen amongst a finite set of possible locations. Consequently variance estimators include a correction factor for the sampling fraction (f) in a finite population and the variance-effective sample size under sampling without replacement is n–1 not n as in sampling with replacement (Thompson 1992). These differences, if not carefully identified and accounted for in the bootstrap resampling procedure, can lead to a problem of bias in bootstrap estimates of variance and percentiles (Lahiri 2003; Shao 2003). Schreuder and Williams (2000) found conventional 95% confidence intervals for the mean under SRS and sample sizes of 20, 40, and 60 to be slightly superior in terms of actual coverage of the true mean than corresponding naive bootstrap confidence intervals.

A large number of modified bootstrap procedures have been proposed to account for the sampling procedure and finite-population corrections (Shao 1996). The bootstrap can adapt to any sampling design with the provision that resampling is done at the unit level (h) at which the *iid* assumption is still valid given that the unit was sampled. In stratified multi-stage cluster sampling, for example, bootstrap resampling would occur at the level of clusters within strata. Attribute values in a cluster can not be assumed *iid* conditional on inclusion of the cluster in the sample.

Common features of modified bootstrap procedures are (1) the resample size m_h is less than the size of the available sample n_h at unit level h of resampling, (2) a scaling of **y**, and (3) resampling without replacement from a synthetic "complete" population with the sample records multiplied from n_h to N_h. Only the flexible rescaling bootstrap proposed by Rao and Wu (1988) will be detailed here for the case where the actual sample units were selected by SRS or unequal inclusion probabilities.

Under the SRS scenario at unit level h one obtains a bootstrap sample $Y^*_{hj}, j=1,\ldots,m_h < n_h$, by SRS with replacement which is then rescaled according to

$$\tilde{Y}^*_{hi} = \overline{Y}_h + \sqrt{\frac{m_h(1-f_h)}{n_h-1}}\left(Y^*_{hj} - \overline{Y}_h\right), j = 1,\ldots, m_h,$$

where $\overline{Y}_h$ is the mean of the actual sample at unit level h and f_h is the actual sample fraction in unit level h (by count or area). This process is repeated for all unit levels h (h=1,....,H). Then, one obtains a design-based estimate $\tilde{T}^*_l$ of T as if $\tilde{Y}^*_{hi}, j=1,\ldots m_h, h=1,\ldots,H$, was an actual sample. B replications of this process produce the rescaled bootstrap estimate of the sampling distribution of T. Under an unequal probability sampling design the jth element in unit level h is given a weight w_{hj} in order to expanded it to an (unbiased) estimate of the population total. The bootstrap resampling is done as under SRS but instead of rescaling Y^*_{hj} one rescales the sampling weights

$$w^*_{hj} = w_{hj}\left(1 - \sqrt{m_h \times (n_h - 1)}^{-1} + \sqrt{m_h \times (n_h - 1)}^{-1} \times \frac{n_h}{m_h} \times r^*_{hj}\right),$$

where r^*_{hj} is the number of times Y_{hi} is included in the bootstrap resample. After completing the bootstrap resampling across all H unit levels the desired sample estimate is obtained from Y^*_{hj} using weights w^*_{hj} in place of w_h (Rao et al. 1992).

Modified bootstrap procedures extend naturally to estimation with missing data replaced by random or model-based imputations (Rubin 1987). The bootstrap sample is taken from the complete nominal sample. Missing sample records are simply added as empty records with a label that identifies them as missing. After each round of modified bootstrap resampling the missing values are imputed from the drawn bootstrap sample (only) by applying the exact same protocol as would be used for the actual sample. If one did the imputations before the bootstrapping the variance estimates would be downward biased (Shao 1996).

A major advantage of the bootstrap of multivariate data is the effortless provision of measures of multivariate associations within and between sampling units since these are estimated by standard procedures from the replicated bootstrap samples. These associations are almost always needed for the estimation of variances of complex survey estimators. Needless to say, conventional methods for obtaining estimates of these quantities can be exceedingly difficult.

3.10.2 The Jackknife Resampling

The jackknife is a delete n units at a time resampling technique (Efron 1982) used to obtain first-order approximations to estimates of bias, and sampling variance. When the function g is linear, the jackknife estimates will be equivalent to the design-based estimates. When g is nonlinear or nonsmooth, a jackknife variance estimator may be inconsistent (Shao 1996). As for the bootstrap, the rationale for using a jackknife resampling procedure for estimation is not statistical but rather convenience when design-based or model-based estimators are exceedingly complex or nonexistent.

When the data are a simple random sample of $y_i, i = 1, \ldots, n$, the ith leave-one-out jackknife sample is

$$y_{(i)} = \{Y_1, Y_2, \ldots, Y_{i-1}, Y_{i+1}, \ldots Y_n\},$$

from which the ith jackknife replication of $\hat{T}_{(i)} = g\left(Y_{(i)}\right)$ is obtained as for the actual sample. After obtaining possible distinct all n jackknife replication estimates of T the jackknife estimate of bias in the estimate $\hat{T}$ obtained from the actual sample is

$$\text{bias}_{\text{jack}} = (n-1)\left(\hat{T}_{(\cdot)} - \hat{T}\right) \text{with } \hat{T}_{(\cdot)} = n^{-1}\sum_i T_{(i)}$$

and the jackknife estimate of the standard error is

$$\hat{\text{var}}_{\text{jack}}\left(\hat{T}\right) = \frac{n-1}{n}\sum_{i=1}^{n}\left(\hat{T}_i - \hat{T}_{(\cdot)}\right).$$

In finite-population sampling a correction for the sample fraction must be reflected in the jackknife variance estimator. In multistage inventory one has to decide what the unit to delete is. As for the bootstrap, one should delete the highest unit level unit for which the iid assumption is valid given the unit is included in the actual sample. Also, under unequal probability sampling the effect of deleting one unit on the expanded (weighted) sample observations must be addressed. We illustrate the jackknife procedure for unequal probability stratified cluster sampling. Notation is as per the bootstrap example given before. After deleting the j th unit in the h th unit level we obtain the $h\,j$ th jackknife replication of Y from

$$\hat{Y}_{(h'j')} = \sum_{h \neq h'} \sum_{j \neq j'} w_{hj} \times Y_{hj} + \frac{n_{h'}}{n_{h'} - 1} \sum_{j \neq j'} w_{h'j} \times Y_{h'j},$$

from which we obtain $\hat{T}_{(h'j')}$ as before. We repeat this delete-one process across all units j' in a unit at level and across all unit levels h' (h'=1,...,H) and obtain

$$\hat{T}_{(h')} = n_{h'}^{-1} \sum_{j'=1}^{n_{h'}} \hat{T}_{(h'j')}$$

and finally the jackknife variance estimator for $\hat{T}$ from

$$\hat{\text{var}}_{\text{jack}}(\hat{T}) = \sum_{h=1}^{H} \frac{(1 - f_h)(n_{h'} - 1)}{n_{h'}} \sum_{j'=1}^{n_{h'}} \left(\hat{T}_{(h'j')} - \hat{T}_{(h')}\right)^2 .$$

3.10.3 The Pòlya-Urn Resampling Scheme

The ease of implementation of the flexible and simple Pòlya-urn resampling scheme and the fact that Pòlya-urn estimators of means, totals, and variances are design-consistent and asymptotically equivalent to design-based estimators (Ghosh and Meeden 1997) makes Pòlya-urn resampling an attractive alternative to the bootstrap. Pòlya-urn resampling generates a posterior distribution of the statistic of interest. It is a predictive joint distribution for the unobserved or unseen units in the population conditional on the seen sample values, similar to the Bayesian bootstrap of Rubin (1981). Problems associated with the sampling process, unequal inclusion probabilities, and finite populations are not encountered in the Pòlya-urn resampling scheme.

The basic Pòlya-urn resampling scheme is very simple. Let us assume that we have n sample observations Y_i, $i=1,\ldots,n$, from a finite population of size N. To implement the Pòlya-urn sampling we place the n sample records in a virtual urn. We draw one sample record at random from the urn, and return the record and one additional copy of this record to the urn. There are now n+1 sample records in the urn. We repeat this drawing scheme a total of N–n times. After the last draw, the urn contains N sample records, which we interpret as one posterior prediction of a population census. From the N sample records, we compute the attribute of interest by standard techniques, just as in a bootstrap. A large number K ($K > 400$) of posterior predictions are obtained in this manner. As for the bootstrap, the number K is determined by the among-replicate variability of the posterior predictions.

The Pòlya-urn resampling scheme adapts well to more complex designs. In one-stage cluster sampling the resampling scheme is unchanged; a sample record is the data from a cluster (Magnussen et al. 2004). Under a stratified random sampling design, a Pòlya-urn resampling scheme such as the one just described is implemented for each strata (Magnussen and Köhl 2002). In multistage cluster sampling (Meeden 1999) the resampling is done in a nested sequence. For example, we have under SRS sampled n first-stage clusters out of a total of N, and we have sampled m out of M ultimate units within each cluster. We would then do an urn resampling of the n first-stage units until the urn contained N such units. Then, we would take each first-stage unit in the urn and conduct a second-stage round of urn resampling until the chosen unit contained M ultimate units. After one completion of the nested urn resampling scheme we have one posterior prediction of a population census and we can proceed to compute the statistics of interest. And we continue until additional replications of the nested resampling only produce minimal gains.

Remote Sensing 4

4.1 Introduction

Remote-sensing data play a growing role in studies of natural and seminatural environments, a role that stretches from a visual interpretation to sophisticated extraction of information by advanced image analysis and statistical algorithms. Today, the synergy between traditional data types and remotely sensed data is widely recognized as essential and critical for extensive "on-time" environmental studies and monitoring at the local, regional (landscape), and national levels.

The technologies and methods of remote sensing have evolved dramatically over the last few decades. The spectrum of satellite and airborne sensors available provides a suite of imaging scales and information content of interest and importance to planners and land managers. Easy access to historical remotely sensed data, a continued reduction in data cost, and improved resolution from satellite platforms all point towards the increased role and importance of the remote-sensing technology for planning agencies and land management initiatives involved in monitoring forestry, land-cover, and land-use change at a variety of spatial scales.

Current remote-sensing technology offers collection and analysis of georeferenced data from ground-based, atmospheric, and Earth-orbiting platforms. It offers linkages to Global Positioning System (GPS) data, geographic information systems (GIS) data layers and functions, and emerging modeling capabilities. The ease with which remotely sensed data can be integrated with other sources of information and other data structures is cardinal to its success in land cover and land-use information projects. Globally, the demand for landscape- and regional level information on the state of the environment continues to grow as does the demand for information content. Continued improvements in remote sensing are needed to meet these demands.

The benefits of satellite-based remote sensing in mapping, monitoring, and management of actual forest cover was recognized long ago by several authors

(Dutt et al. 1994; Woodcock et al. 1994; Trotter et al. 1997; Achard et al. 2002; Mollicone et al. 2003). Moreover, the temporal frequency at which new satellite data for a particular region become available allows changes in the extent of forest cover at national and regional levels to be monitored in a reliable and efficient manner. Forestry planning is, to a certain degree, already assisted by remote sensing, mainly in North America and Europe: there have been some good "showcases" after a hurricane, after flooding, after a snowstorm (freezing rain), fire, etc. when major disturbances makes it pivotal to have up-to-date maps and for management of disturbances both during and after the event. The goal of the applications of remote sensing to forestry is to add value (utility) in the planning process (Holmgren and Thuresson 1998), which should mean one of the following: the information retrieved from satellite remote sensing should be less expensive than that collected by other means; it should contain better information at the same cost; or it should be both cheaper and better in some way. Even, if detailed data were available – like high-resolution images and field training sets – it would be already possible to optimize responses to the continuous changing market demands. The need for a less expensive and more comprehensive approach has prompted much research into extracting forest information from aerial and satellite remote sensing (Trotter et al. 1997); Woodcock et al. (1994) analyzed the costs which can be saved, both in the long term and per unit area. The conversion of satellite image format to GIS is easy, making map production possible within a short time and also allowing the fast update of computer-based maps, a particularly useful feature in areas where resource information is either limited or where rapid developments quickly date information collected by conventional methods (Roy et al. 1985). In the years since the launch of the Landsat Multispectral Scanner (MSS) in 1972, the first optical satellite, technology has developed and improved towards better spectral, spatial, and radiometrical resolutions: with newer sensor bands (green, red, and near IR are usually available) being extended to blue and two mid-IR bands.

The principal applications of remote sensing and GIS in forestry domains, the forest geomatics discipline that includes all the sciences for monitoring and surveying the land and the environment (Gomarasca 2004), are in short the following:

1. *Forest inventory*: restocking and assessment of tree cover/density/number per hectare; identification of vegetation and forest types; production forecasting through the estimation of tree height, diameters, and volumes; natural revegetation; management issues in continuous-cover forestry; monitoring changes in woodland cover.
2. *Forest health and nutrition*: vegetation stress; disease and pest infestations; abiotic hazards, windthrow, fire and snow damage; pollution.

3. *Forest sustainability*: criteria and indicators for sustainable forest management international processes. Satellite imagery is considered critical to criteria and indicator reporting. The ability of remote sensing to measure the direction of change and to compare measurements over time is essential for determining and controlling progress towards sustainable management and for aiding sustainable development. Information previously generated in separate localities from estimates derived by extrapolation (rather than actual field measurement) may now be produced by a few centers of excellence that can efficiently treat large amounts of data and disseminate the results effectively: satellite imagery could be used to monitor different indicators in a substantial or partial capacity (Table 4.1) (Peterson et al. 1999).
4. *Forest growth*: physiological processes, photosynthetic processes, and water-use efficiency; leaf water potential/content; variations of growth rate and carbon sequestration.

Table 4.1. Potential Applications of Space Imagery for Ministerial Conference on the Protection of Forests in Europe indicators

Substantial role for these indicators	Partial role for these indicators
1. Percent and extent of forest types relative to historical condition and to total forest area	1. Area and severity of disease infestation
2. Percent and extent of area by forest type and age	2. Percent of area successfully naturally regenerated and artificially regenerated
3. Area, percent, and representativeness of forest types in protected areas	3. Mean annual increment by forest type and age
4. Level of fragmentation and connectedness of forest ecosystem components	4. Tree biomass volume
5. Area and severity of insect attack	5. Participation in the climate change conventions
6. Area and severity of fire damage	6. Annual removal of forest products relative to the volume of removals determined to be sustainable.
7. Percent and extent of area by forest type and age	7. Distribution of and changes in the land base available for timber production
8. Area of forest converted to non-forest land use	8. Availability of habitat for selected wildlife species of economic importance
9. Percent of forest managed primarily for soil and water protection	9. Area and percent of protected forest by degree of protection
10. Area, percent, and representativeness of forest types in protected areas	
11. Percent canopy cover	
12. Percent biomass volume by general forest type	
13. Area of forest depletion	
14. Area of forest permanently converted to non-forest use	
15. Semi-permanent or temporary loss or gain of forest ecosystems, such as grasslands, agriculture	
16. Surface area of water within forested areas	

Source Goodenough et al. (1998)

5. *Forest ecology*: monitoring biodiversity; landscape ecology; population dynamics. Remote-sensing techniques offer also potential for assisting in the analysis of large forest tracts for identification of ecosystem classes or aggregations of ecologically similar classes. However, satellite images are acquired at predetermined spatial resolutions, designed primarily for general land-cover and land-use analysis and mapping. Although airborne systems are capable of acquiring data at a variety of resolutions (i.e., spatial, spectral, and temporal), optimal resolutions for specific terrain analyses are still determined on an ad hoc basis. This problem was presented by Woodcock and Strahler (1987) in a paper discussing the *scale dependence of prediction in remote sensing*. Remote-sensing data are generally collected at a single spatial resolution, in contrast to the many scales at which nature's units and processes exist. It is therefore difficult to identify a single spatial resolution of remote-sensing data that will provide the most suitable level of information for extracting forest ecosystem characteristics. Multiresolution remote-sensing data can be expected to provide suitable information at a variety of levels for inventories and forest ecosystem classifications (Fig. 4.1).

For remote sensing of forest ecosystems to become operational, the spatial resolutions of remote-sensing data must be appropriate for the application: the chosen resolution will determine the information content and the measurement error (Atkinson 1993; Atkinson et al. 1996). For instance, to discriminate forest ecosystems at a landscape scale, the best spatial resolution for the spectral reflectance of each forest type should be determined a priori.

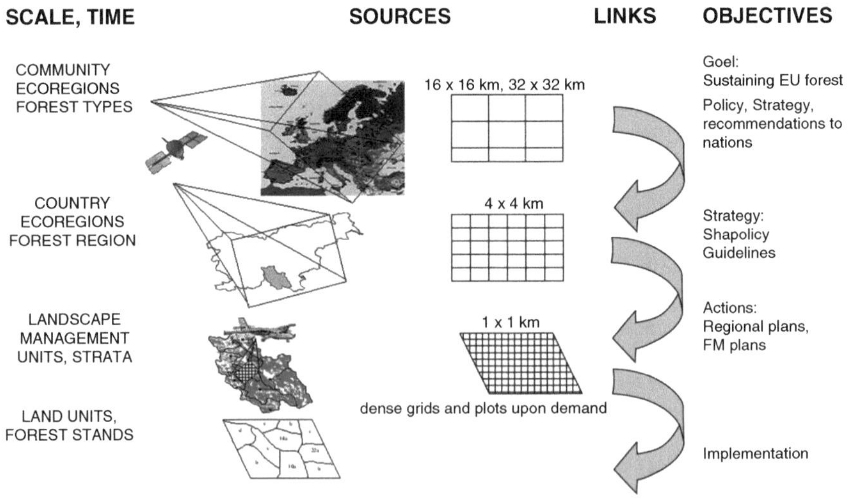

Fig. 4.1. Multiresolution scheme for forest monitoring (Courtesy of M. Kovac 2004)

Remote-sensing data provide a means of monitoring the rate of change with respect to land-cover conversion or systematic change in health or productivity. Integration of various datasets allows the land-use planner to make decisions based on existing information within the digital database, as well as to create new information through spatial analysis. Timely and spatially consistent remote-sensing data for systematic analysis of landscape change (i.e., local to regional scales) over space and time are the main focus of remote sensing. Remote-sensing data, image access solutions, and GIS provide opportunities for integrated analysis of spatial data and product development. The interactions of these components have been summarized by Wilkinson (1996):

1. Remote-sensing data can be used as input data for analysis within a GIS.
2. GIS can provide ancillary data for improved remote-sensing analysis for discrimination of forest types and land-cover and land-use classes.
3. The application of remote sensing and other spatial data within a GIS provides capabilities for modeling and scenario analysis.

4.2 Basic Concepts

"Remote sensing is the science of acquiring information about the Earth's surface without actually being in contact with it. This is done by sensing and recording reflected or emitted energy and processing, analysing, and applying that information" (Canada Centre for Remote Sensing 2003). The efficiency of Earth observation (EO) has been evident when GIS and information technology have overtaken traditional methods in forest assessment and management: this is *forest geomatics*. In the EO process an interaction between incident radiation and the targets of interest is involved. This is exemplified in imaging systems where the following seven elements are involved:

1. Energy source or illumination – the first requirement for remote sensing is to have an energy source which illuminates or provides electromagnetic energy to the target of interest.
2. Radiation and the atmosphere – since the energy travels from its source to the target, it will come in contact with and interact with the atmosphere it passes through. This interaction may take place a second time as the energy travels from the target to the sensor.
3. Interaction with the target – once the energy makes its way to the target through the atmosphere, it interacts with the target depending on the properties of both the target and the radiation.

4. Recording of energy by the sensor – after the energy has been scattered by or emitted from the target, we require a sensor (remote – not in contact with the target) to collect and record the electromagnetic radiation.
5. Transmission, reception, and processing – the energy recorded by the sensor has to be transmitted, often in electronic form, to a receiving and processing station where the data are processed into an image (hard copy and/or digital format).
6. Interpretation and analysis – the processed image is interpreted, visually and/or digitally or electronically, to extract information about the target which was illuminated.
7. Application – the final element of the remote-sensing process is achieved when we apply the information we have been able to extract from the imagery about the target in order to better understand it, reveal some new information, or assist in solving a particular problem (Chap. 6).

These seven elements comprise all aspects of a remote-sensing process. We shall detail each of these seven elements, focusing on forest applications. Note, however, that in general remote sensing also involves the sensing of emitted energy and the use of nonimaging sensors.

4.2.1 Electromagnetic Radiation

As already noted, the first requirement for remote sensing is to have an energy source to illuminate the target (or the energy being emitted by the target), in the form of electromagnetic radiation. All electromagnetic radiation has fundamental properties and behaves in predictable ways according to the basics of wave theory: it consists of an electrical field which varies in magnitude in a direction perpendicular to the direction in which the radiation is traveling, and a magnetic field oriented at right angles to the electrical field. Both these fields travel at the speed of light. Two characteristics of electromagnetic radiation are particularly important for understanding remote sensing. These are the *wavelength* and the *frequency* (Fig. 4.2).

The wavelength is the length of one wave cycle, which can be measured as the distance between successive wave crests. Wavelength is usually represented by the Greek letter lambda (λ). Wavelength is measured in meters or some factor of meters such as nanometers, micrometers, or centimeters. Frequency refers to the number of cycles of a wave passing a fixed point per unit of time. It is normally measured in hertz, and various multiples of hertz.

Wavelength and frequency are related by the following formula:

$$c = \nu\lambda,$$

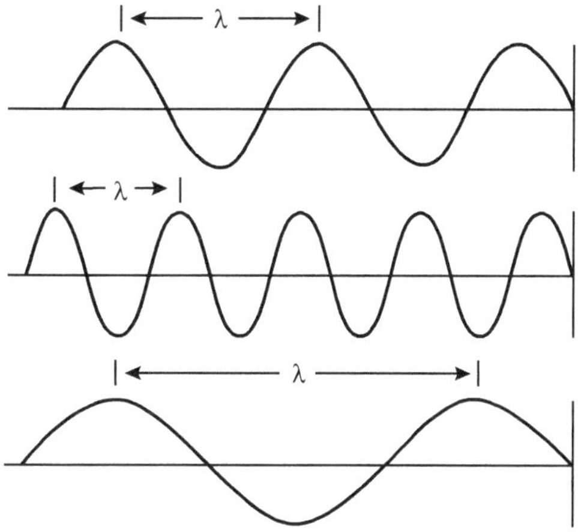

Fig. 4.2. Electromagnetic radiation with different wavelengths (modified after Canada Centre for Remote Sensing 1998)

where c is the speed of light (3×10^8 m/s), ν is the frequency (cycles per second, hertz), and λ is the wavelength (meters).

From this equation we see that wavelength and frequency are inversely related to each other. The shorter the wavelength, the higher the frequency and vice versa. Understanding the radiation in terms of wavelength and frequency is crucial for assessing the information to be extracted from images.

4.2.2 The Electromagnetic Spectrum

The electromagnetic spectrum covers the range from shorter (including γ-rays and X rays) to longer wavelengths (including microwaves and broadcast radio waves). Several regions of the electromagnetic spectrum are useful for remote sensing (Fig. 4.3).

For most purposes, the UV portion of the spectrum represents the shortest wavelengths with a practical significance in general remote sensing. The UV region is just beyond the violet region of the visible interval of wavelengths. The Earth's substratum, rocks and minerals, fluoresces or emits visible light when illuminated by UV radiation.

The light which our eyes (our remote sensors) sense is a very small part of the electromagnetic spectrum. There is a lot of radiation around us which is "invisible" to our eyes, but which can be detected by other remote-sensing instruments and used to our advantage. The visible wavelengths cover a range from approximately 0.4 to 0.7 μm. The longest visible wavelength is red and the

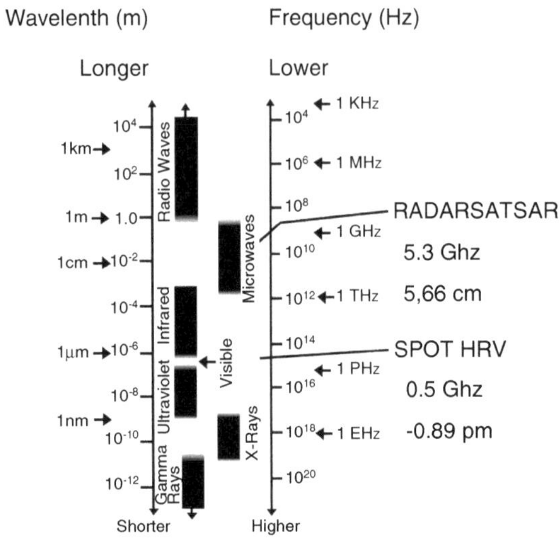

Fig. 4.3. Subdivision of the electromagnetic spectrum (after Canada Centre for Remote Sensing 1998)

shortest is violet. What we perceive as a color in the visible portion of the spectrum is given next: it is important to note that this is the only portion of the spectrum associated with the concept of color.

- *Violet*: 0.4–0.446 µm
- *Blue*: 0.446–0.500 µm
- *Green*: 0.50 –0.578 µm
- *Yellow*: 0.578–0.592 µm
- *Orange*: 0.592–0.620 µm
- *Red*: 0.620–0.7 µm

Blue, green, and red are the primary colors of the visible spectrum. They are defined as such because no single primary color can be created from the other two, but all other colors can be formed by combining blue, green, and red in various proportions. Although we see sunlight as a uniform or homogeneous color, it is actually composed of UV, visible and IR wavelengths. The visible portion of radiation is divided into its component colors when sunlight is passed through a prism, which diffracts light according to the specific wavelengths of the light.

The next portion of the spectrum of interest for remote sensing is the IR region, which covers the wavelength range from approximately 0.7 to 100 µm – more than 100 times as wide as the visible portion! The IR region can be divided into two categories – *reflected* IR, and *emitted* or *thermal* IR. Radiation in the reflected IR region is used in a very similar manner to radiation in the visible region. It covers wavelengths from approximately 0.7 to 3.0 μm. The thermal or emitted or *mid*-IR region is quite different, with this energy being essentially the radiation emitted in form of heat from the Earth's surface. It covers wavelengths from approximately 3.0 to 15 μm. Then, the *far*-IR region follows, with wavelength longer than 15 μm. It is useful to compress multispectral remote-sensing information into a low number of synthetic bands that contain the highest total information level and that can be visualized with additive synthesis, to take advantage of the sensor information in a reduced number of channels.

The portion of the spectrum of more recent interest to remote sensing (especially in the tropics) is the microwave region, from about 1 nm to 1 m. The microwave region covers the longest wavelengths used in remote sensing. The shorter wavelengths have properties similar to those of the thermal IR region, while the longer wavelengths approach the wavelengths used for radio broadcasting.

4.2.3 Interactions with the Atmosphere

Radiation used for remote sensing has to travel through the atmosphere before it reaches the Earth's surface. Particles and gases can modify the properties of incoming radiation. These modifications are caused by scattering and absorption of photons (the particles of light) on their way through the atmosphere.

Scattering occurs when particles or large gas molecules present in the atmosphere interact with and redirect the electromagnetic radiation from its original path. The extent of scattering depends on several factors, including the wavelength of the radiation, the abundance of particles or gases, and the distance the radiation travels through the atmosphere. There are three major types of scattering.

1. *Rayleigh scattering* occurs when radiation interacts with particles that are small compared with the wavelength of the light. Small specks of dust or nitrogen and oxygen molecules when illuminated will produce this scattering. Shorter wavelengths are scattered much more than longer wavelengths. It is the dominant scattering mechanism in the upper atmosphere. The fact that the sky appears "blue" during the day is a direct consequence of Raleigh scattering. As sunlight passes through the atmosphere, the shorter wavelengths (blue) of the visible spectrum are scattered more than the other visible wavelengths. At sunrise and sunset the light travels a longer distance through the atmosphere than at midday and the scattering of the shorter

wavelengths is more complete; this leaves a greater proportion of the longer wavelengths to penetrate the atmosphere.

2. *Mie scattering* occurs when the particles are just about the same size as the wavelength of the radiation. Dust, pollen, smoke, and water vapor are common causes of Mie scattering, which tends to affect wavelengths longer than those affected by Rayleigh scattering. Mie scattering occurs mostly in the lower portions of the atmosphere, where larger particles are more abundant. It dominates in overcast conditions.
3. Nonselective scattering occurs when radiation collides with particles larger than the wavelength. Water droplets and large dust particles can cause this type of scattering. Nonselective scattering got its name from the fact that all wavelengths are scattered about equally. It causes fog and clouds to appear white: blue, green, and red light are all scattered in approximately equal quantities (blue plus green plus red gives white light).

Absorption is the other main process at work when electromagnetic radiation interacts with the atmosphere. In contrast to scattering, this phenomenon causes molecules in the atmosphere to absorb energy at various wavelengths. Ozone, carbon dioxide, and water vapor are the three main atmospheric absorbents.

Ozone absorbs the harmful (to most living things) UV radiation from the sun. Without a protective layer of ozone in the atmosphere our skin would burn when exposed to sunlight.

Carbon dioxide is an efficient absorbent of radiation in the far-IR region of the spectrum, and is associated with thermal heating. The absorption by carbon dioxide basically traps heat, otherwise emitted to outer space, inside the Earth's atmosphere. You may have heard carbon dioxide referred to as a greenhouse gas. The trapping of heat is the working principle of a greenhouse. Water vapor in the atmosphere absorbs much of the incoming longwave IR and shortwave microwave radiation (between 22 m and 1 nm). The presence of water vapor in the lower atmosphere varies greatly from location to location and over time. Because these gases absorb electromagnetic energy in very specific regions of the spectrum, they influence where we can "expect" to find useful information. Regions not severely influenced by atmospheric absorption are the most useful for remote-sensing applications: they are generally known as *atmospheric windows* (Fig. 4.4).

By comparing the characteristics of the two most common energy/radiation sources (the sun and the Earth) with the atmospheric windows available to us, we can define the wavelengths that are a priori expected to be most useful for remote sensing. The visible portion of the spectrum corresponds to both an atmospheric window and the peak energy level of the sun. Heat energy emitted by the Earth corresponds to a window around 10 μm in the thermal IR region. A large window at wavelengths beyond 1 nm is associated with the microwave region.

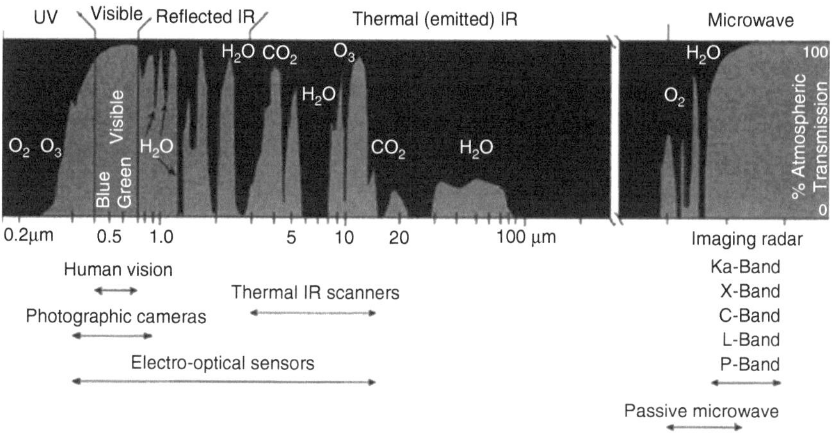

Fig. 4.4. Distribution of atmospheric windows throughout the electromagnetic spectrum. Sensor types suitable for each window are indicated (after Holton et al. 1995)

4.2.4 Radiation–Target Interactions

Radiation that is neither absorbed nor scattered in the atmosphere can reach and interact with the Earth's surface. There are three forms of interaction that can take place when energy strikes or is incident upon the surface: absorption; transmission; and reflection (Fig. 4.5). The total incident energy will interact with the surface through one or more of these processes. The proportion of absorption, transmission, and reflection generated by the interaction will depend on the wavelength of the energy and the features of the material and its conditions.

Absorption occurs when radiation (energy) is absorbed by a target, while transmission occurs when radiation passes through a target. Reflection occurs when radiation "bounces" off the target and is redirected. In remote sensing, we are most interested in measuring the radiation reflected from targets.

Two types of reflection from a target are distinguished: specular and diffuse. Radiation hitting a smooth surface produces a specular or mirrorlike reflection,

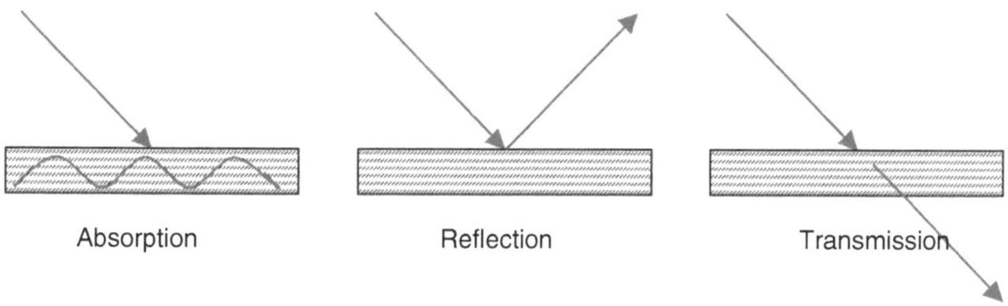

Fig. 4.5. Interactions between radiation and objects

where the energy (or almost all) is directed away from the surface in a single direction. Diffuse reflection occurs when radiation hits a rough surface, with the energy being reflected almost uniformly in all directions. Most of the Earth's surface generates reflections that are somewhere between perfectly specular and perfectly diffuse. Whether a particular target reflects specularly or diffusely, or somewhere in between, depends on the surface characteristics (roughness) of the feature and the wavelength of the incoming radiation. For wavelengths much smaller than the surface variations or the sizes of the particles that make up the surface, a diffuse reflection will dominate. Two examples of how radiation at visible and IR wavelengths interacts with leaves and water at the Earth's surface can illustrate the main features and complexities of reflection.

1. *Leaves*: Chlorophyll – a chemical compound found in all (living) leaves – strongly absorbs red and blue wavelengths but reflects green wavelengths; leaves appear "greenest" therefore to us in the summer, when chlorophyll content is a maximum. During senescence the amount of chlorophyll in leaves drops significantly, which means less absorption and proportionately more reflection of the red wavelengths, making the leaves appear red, yellow, or brown (yellow is a combination of red and green wavelengths). The striking fall color of deciduous forests in temperate and boreal regions is a perfect example of this phenomenon. The internal structure of healthy leaves acts as excellent diffuse reflectors of near-IR wavelengths. If our eyes were sensitive to near-IR radiation, trees would appear extremely bright to us at these wavelengths. In fact, measuring and monitoring the near-IR reflectance is one way that scientists can determine how healthy (or unhealthy) vegetation may be (Marchetti and Castagnoli 1989; Marchetti et al. 1998).
2. *Water*: Longer-wavelength visible and near-IR radiation is absorbed more than shorter-wavelength radiation by water; thus, water typically looks blue or blue-green and darker if viewed at red or near-IR wavelengths. If there are suspended sediments in the upper layer of a water body, then more light is reflected and the water will appear brighter. They are easily confused with shallow (but clear) water, since their reflection is quite similar. Water with a high algae concentration will appear greener than algae-free water: as for leaves, it depends on the differential absorption by chlorophyll. The shifting topography of a water surface (rough, smooth, floating materials, etc.) leads to complications in the interpretation of a reflection pattern observed over water bodies or wetlands. A mixture of specular and diffuse reflections modifies color and brightness in complex ways.

From these two examples we see that, depending on the complex make-up of the target that is being looked at and the wavelengths of radiation, very different "outcomes" of the three processes involving absorption, transmission, and/

reflection, that are quite difficult to predict, can be observed. By measuring the energy reflected (or emitted) by specific targets across a variety of different wavelengths, a spectral response for that target can be obtained. For example, water and vegetation may show similar reflection in the visible wavelengths but not in the IR part of the energy spectrum (Fig. 4.6). The spectral response of a specific target can vary over time and space, as illustrated for leaves and water bodies. Knowing where to "look" spectrally and understanding the factors that modify the spectral response of a feature is of critical importance for the correct interpretation of the interaction of electromagnetic radiation with the surface of the target.

The main part of the incoming radiation in the visible domain is absorbed by the leaf pigments; roughly 10% is reflected (Sims and Gamon 2003). Chlorophylls, caroteniods, and anthocyanins, with strong absorption in the blue (0.35–0.50 μm) and red (0.62–0.70 μm) regions, dominate the reflectance properties in this domain. The reflectance in the green region (0.50–0.62 μm) is slightly higher. Owing to the low reflectance in the visible portion of the spectrum, reflectance from only the uppermost portion of the canopy is detected and measured by the sensor (Williams 1989). In dense canopies, the influence from soil, understory, and lower parts of the canopy is small. This influence increases when the leaf cover is reduced.

In the near-IR the high reflectance of leaves is caused by the internal leaf structure, i.e., the cell wall to air interface within the leaves (Gausman 1977; Peterson and Running 1989). Approximately 50% of the incoming radiation is reflected by healthy canopies, the absorption is low, and the transmittance is high (Fig. 4.6); therefore, incoming solar radiation can penetrate through and reflect back up through multiple layers of leaves within the canopy. An increase

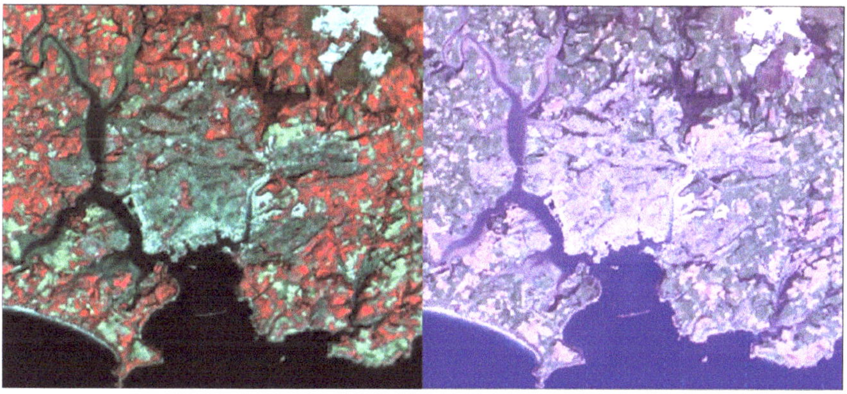

Fig. 4.6. Landsat Thematic Mapper (*TM*) images. Near-IR *left*) shows a very marked distinction between water (*black*), bare soils (*blue-green*), and vegetation (*red*). Bare soils and vegetation are less distinguishable in a simulated true color image *right*)

in the number of leaf layers significantly increases the near-IR reflectance (Williams 1989). Deviations from this reflectance increase with greater canopy closure and have been attributed to shadows, occurring if the increased vegetation cover is accompanied by more shadows.

In the mid-IR the optical properties are mainly affected by the water content of the target absorbing a large part of the incident energy. The strongest absorption occurs near 1.4 and 1.9 μm, with reflectance peaks around 1.65 and 2.2 μm (Guyot and Riom 1988; Williams 1989). As the moisture content decreases, the mid-IR reflectance increases. The 1.55–1.75-μm region (Thematic Mapper, TM, band 5) has been proposed as the best suited wavelength interval for satellite remote sensing of canopy water status in the mid-IR region (Tucker 1980). Owing to the strong relationship between leaf moisture content, stomatal conductance, respiration, and the rate and efficiency of photosynthesis, assessment of the relative photosynthetic activity can be performed using remote-sensing data (Williams 1989). Furthermore, compared with the visible range, in the mid-IR natural surfaces show a wider dynamic range that makes it easy to identify different land covers; in this spectral window atmospheric noise is limited and does not affect understanding of spectral information.

4.2.5
Passive and Active Sensing

So far, the sun has been considered the implicit source of energy and radiation. The sun provides a very convenient source of energy for remote sensing. Its energy is either reflected, as it is for visible wavelengths, or absorbed and reemitted in thermal IR wavelengths. Remote-sensing systems, which measure the energy that is naturally available, are called passive sensors (Fig. 4.7): they

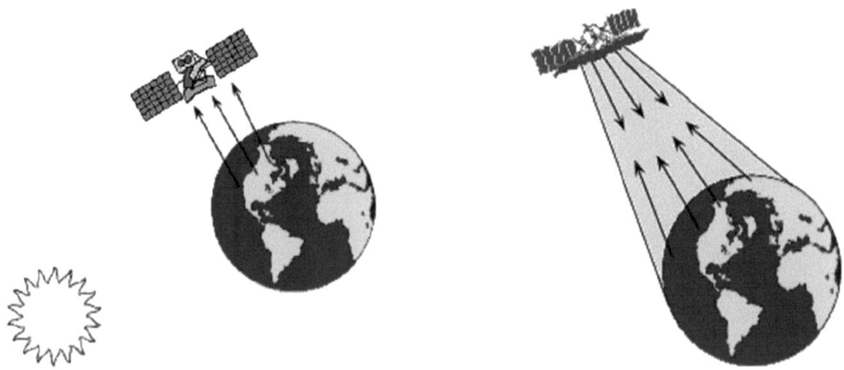

Fig. 4.7. Working principles of passive (*left*) and active (*right*) sensors (modified after Canada Centre for Remote Sensing 1998)

can only be used to detect the presence of energy that is available only during the time of solar illumination (no reflected energy from the sun at night). Otherwise, the energy naturally emitted (such as thermal IR) can be detected during the day or the night, as long as the amount of energy exceeds the sensitivity threshold of the sensor.

Active sensors, in contrast, generate an independent source of energy for illumination. The sensor emits radiation towards a selected target. The radiation reflected from the target is detected and measured by the sensor. The advantages of active sensors include the ability to obtain measurements anytime, regardless of the time of day or season. Active sensors can also use wavelengths that are not provided by the sun in sufficient quantities, such as microwaves. They also provide improved control over the way a target is illuminated. However, active systems require the generation of substantial amounts of energy for adequate illumination of distant targets. The most common active sensor types in forest applications are laser fluorosensors and synthetic aperture radars (SARs).

4.2.6 Characteristics and Analysis of Images

It is useful to define and understand a few fundamental terms and concepts associated with remote sensing before we look more in detail at sensors and their characteristics. EO involves electromagnetic energy that may be detected either photographically or electronically. The photographic process uses chemical reactions on the surface of light-sensitive film to detect and record energy variations. It is important to distinguish between the terms images and photographs in remote sensing. An *image* refers to any pictorial representation, regardless of what wavelengths or remote-sensing device has been used to detect and record the electromagnetic energy. A *photograph* refers specifically to images that have been detected as well as recorded on a film. Photographs are normally recorded over the wavelength range from 0.3 to 0.9 μm – visible and reflected IR. On the basis of these definitions we can say that all photographs are images but not all images are photographs.

A photograph can be represented and displayed in a digital format by subdividing the image into small areas of equal size and shape, called picture elements – *pixels*. The brightness of each pixel is represented by a numeric value or digital number. Using the previous definitions, the digital representation of the photograph is actually a digital image of the photograph! After a scanning process, where a special device assigns a number to each pixel in the film, the computer transforms the digital values of the pixels to brightness values in the display of the digital image. A multispectral shot of a scene gives back the spectral properties of the objects in accordance with the different spectral bands.

Electronic sensors record the energy as an array of numbers directly in digital format (as in digital cameras); resolution, scanning process, and digital spectrum (8 bit, 16 bit, 32 bit, etc.) affect the quality of the acquisition process. It is important to note that the digital format of capturing data is not necessarily the same as is used for displaying data.

These two different ways of displaying remote-sensing data, either pictorially or digitally, are increasingly interchangeable as they convey the same information, although some detail may be lost when converting back and forth.

The visible portion of the spectrum contains colors. We can see a color because our eyes can detect the entire visible range of wavelengths and our brain provides the context of human interpretation and recognition. The information from a narrow range of wavelengths is gathered and stored in a *channel*, the sensor's container for that portion of the spectrum, also often referred to as a *band*. The data stored in a channel are usually represented as one of the primary colors (blue, green, and red) and, depending on the relative brightness (i.e., the digital value) of each pixel in each channel, the primary colors can be combined in varying proportions to represent different colors. Using this method to display a single channel or range of wavelengths, we display the channel through all three primary colors. Because the brightness level of each pixel is the same for each primary color, they combine to form a black-and-white image, showing various shades of gray from black to white. A display of more than one band each as a different color, a combination of different bands (brightness levels may be different for each band), forms a color image.

4.2.6.1 Image Resolution

The digital image consists of discrete pixels: associated with each pixel is a value represented as a digital number, which depicts the average radiance of that area within that scene. The size of a pixel affects the reproduction of details within a scene: you normally have the possibility to choose the resolution – the pixel space of an image. detail in high resolution in a small scene or in low detail in a large scene since the more the pixel size is reduced the more scene detail is preserved in the digital representation. Remote-sensing imagery must always be expressed toghether with information on the spatial resolution, but in EO we can distinguish three types of resolution: the resolution depends both upon the sensor and the platform (Fig. 4.8).

1. *Spatial (geometrical) resolution.* This is the number of pixel per surface unit. In remote sensing it is referred to in terms of pixel real dimension on land and is expressed in ISO units of meter (Fig. 4.9).

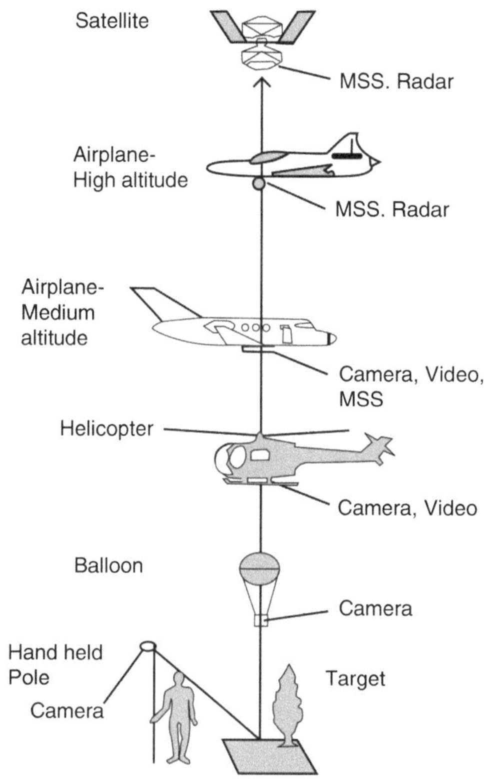

Fig. 4.8. Platforms with sensors on board. *MSS* Multispectral Scanner

2. *Radiometric resolution.* This is the finest distinction that can be made among objects viewed in the same part of the spectrum. Digital number values are displayed by image data, and vary from 0 to selected power of 2. The range corresponds to the number of bits used for coding numbers in binary format. Each bit records an exponent of power 2. The maximum number of brightness levels available depends on the number of bits used. If a sensor uses 8 bits to record the data, there would be 2^8=256 levels.
3. *Temporal resolution*, for satellite data only, is also important: it is the revisit period and refers to the length of time it takes for a satellite to complete one entire orbit cycle, expressed in days. The Landsat mission has, for example, a revisit time of 16 days, the Système Pour l'Observation de la Terre (SPOT) mission 26 days, and OrbView 3 days.

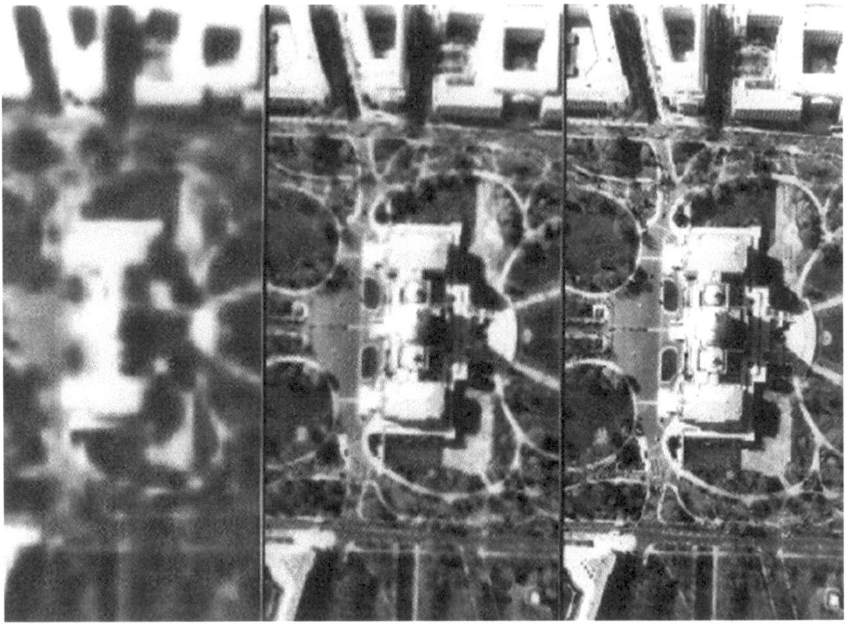

(a) (b) (c)

Fig. 4.9. Comparison between three different spatial resolution images. **a** SPOT (10 m); **b** OrbView-1 (3 m); **c** OrbView-1 (1 m)

4.2.6.2
Image Processing

Remote-sensing images are recorded in digital form and are processed by computers to produce images for interpretation in two forms: photographic film and digital format. Variations in the scene characteristics are represented as variations in brightness on photographic film. The parts of a scene that reflect a relatively high amount of energy will appear brighter and parts that reflect relatively less energy will appear darker. Digital image processing is a collection of computing techniques and algorithms for the manipulation of digital images: it encompasses operations such as noise removal, geometric and radiometric corrections, enhancement of images, information extraction, and image data manipulation and management. Image processing methods may be grouped into three functional categories:

1. *Geometric and radiometric corrections*: the correction of errors, noise, and geometric distortions introduced during flight, scanning, recording, and playback operations.

2. *Image Enhancement*: two distinct procedures are applied: linear contrast enhancement (LCE) and spatial filtering (SF). In LCE the full sensitivity brightness range of the detectors is utilized. The low end is put at 0 and the high end as 1: brightness values between the two fixed endpoints are linearly stretched, improving the contrast for most of the original values. SF is instead a pixel-by-pixel transformation of the image, which depends on the graylevel of the pixels concerned as well as the graylevel of the neighboring pixels. It is a procedure in which the graylevel of a pixel is altered according to its relationship with respect to the neighboring pixels. The neighborhood of the pixels to consider in SF is determined by experience or is optimized for specific applications.
3. *Information extraction and classificaton*: dedicated computing algorithms support the process of identification and extraction of specific pieces of desired information. Many approaches are currently available and used for the classification of remote-sensing data (Franklin 2001; Nyerges and Green 2000). New generations of multilevel "hybrid" classifiers (neural network, fuzzy, nonparametric Bayesian classifiers, expert systems) have significantly improved the practical utility of satellite imagery. From unsupervised to supervised classifiers relying on a single source or multiple sources of information a wide gamut of procedures and protocols have been developed. The results achieved vary dramatically in terms of precision and accuracy. The utility of a classifier depends intimately on definitions of pursued targets and the intrinsic properties of the images used. However, pixel-based common methods are characterized by systems of nomenclature not detailed enough to be used as useful tools supporting forest planning. Supervised methods are based on the manual acquisition of a certain number of pixels, *training sets*, from all the channels of the scanner, for each forest/land-cover class. *Principal component analysis* (PCA) is traditionally employed in remote sensing as a data reduction and decorrelation technique. Principal components are decorrelated (independently to the second order) and thus become, for example, available to Bayesian methods which require independent data (to the first order). Most of the variance is in the first component, while the last components contain mainly noise. In many landscapes, for example, the seven bands of Landsat TM have an inherent information content of two to three components. In the case of multitemporal data most of the variance due to seasonality is often present on the second and higher components (Fabbro 2000). The images resulting from PCA are, however, difficult to interpret. On the other hand, Kauth and Thomas (1976) proposed the *Tasseled Cap Transformation* (a spectral compression tool for multispectral images, producing the *greenness* for biomass, *wetness* for moisture, *brightness* for soil,

yellow staff for vegetation condition) and vegetation indices provide reduced data which can easily be interpreted. Tasseled cap transformation has been extensively used with high-resolution data for forest and land-cover change mapping (Rogan et al. 2002; Seto et al. 2002) and pastureland surveying (Todd et al. 1998).

Supervised image classification techniques are widely used (Campbell 1996). With such a technique, each pixel is classified according to the class that has the greatest spectral similarity with the available set defined in the training stage of the classification (Foody 2000). These types of classifications are called "hard", because the decision about the class to which each pixel belongs is finally and exclusively based on a single criterion. Hard classifiers are suitable when signature data are gathered from training sites (the ground truth) composed of single "pure" target type. Training sites satisfying this criterion may be difficult to find, especially when the pixel size is a multiple of the size of an object of interest (Eastman 1999). A hard classification may be inappropriate when most pixels contain more than one target type, where the target could be, for example, a land-cover class (Campbell 1996). Pixels with a mixture of target types dominate in images from coarser spatial resolution satellite sensors (Foody 1996). In these cases where one should expect a majority of the pixels to be mixtures of target types, a mixture analysis approach to classification may be more rewarding. *Spectral mixture analysis* assumes that the reflectance of each pixel is a linear combination of contributing subpixel components. The spectral signature of these components, end members, resulting from spectral unmixing (via PCA and parallel coordinates representation), provides reduced data which can easily be interpreted. Spectral mixture analysis approaches, widely used with hyperspectral data, for mapping remote-sensing images can take the form of "soft classifications," like mixture modeling (Shimabukuro and Smith 1991; Settle and Drake 1993; Oleson et al. 1995; Sohn and McCoy 1997; Faraklioti and Petrou 2000) or fuzzy supervised classification (Wang 1990; Foody 1996; Ricotta et al. 2003a).

Classification of large amounts of data invariably favors an automatic or semiautomatic approach to classification. The sheer cost of alternatives is prohibitive. Various methods for automatic segmentation of images have therefore seen the light in recent years. Most are based on a mixture of supervised and unsupervised classifiers. In particular, "soft"-computing-based approaches as in neural networks (e.g., multilayer perceptron, self-organizing maps), support vector machines, fuzzy inference schemes, PCA and/or independent component analysis, and hidden Markov models appear to hold promise. For complex and difficult classification problems both supervised neural networks (multilayer Perceptron; Benediktsson et al. 1990; Bischof et al. 1992; Azimi-Sadjadi et al. 1993; Paola and Schowengerdt 1995; Atkinson and Tatnall 1997; Dejhan et al. 2000), and nonsupervised linear vector quantization (Hara et al. 1995; Ito and Omatu 1998) ought to be considered as options. Results show that neural networks operate better than statistic classifiers when deductive information is lacking. Neural networks seem to produce fewer errors when the results are compared for pixels with a known target type. Neural networks also often produce images of a better visual quality. Particularly, in active remote sensing, the results from neural network classification appear to be more consistent. Maximum-likelihood classifiers have better performance when a reliable model for the image data can be found. The difficulties encountered in formulating a model for image data often dissuade one from using likelihood-based classifiers. For this reason also in the neural network models the integration of the statistical information is studied. The ease of dealing with heterogeneous informations (also merging data) is another advantage of neural networks (Atkinson and Tatnall 1997); weaving information can be also integrated, giving a group of pixels and not only one pixel in network input (Serpico and Roli 1995). The computational costs of classification by neural algorithms can be high and good training data are pivotal for the performance of a neural network. Most of the work published in the field of neural network classification resorts to pixel-by-pixel classification, followed by clustering/smoothing procedures. In this way information such as texture or regularity (smoothness) of land cover is only used (if ever) in a postprocessing phase.

A comprehensive approach to classification exploits attractive features of the most popular methods in a self-organizing and self-learning system with procedures that take a priori knowledge explicitly into account. At the same time, the spatial structure of the data (e.g., local correlation) and nonlinearities and noise properties of the sensing system should be considered from the first stages of processing, both from a statistical point of view (e.g., by blind deconvolution methods, such as independent component analysis); they have to be based on the correlation matrix of the data by resorting to spatially organized nonlinear filtering architectures such as cellular neural/nonlinear networks.

Given the importance of spatial characteristics, spatial organization of information and the limits of a per pixel classification, the need for new and better classification systems is widely recognized. The use of more complex legends makes the classification errors of traditional approaches greater, usually over standard acceptable limits (Chirici et al. 2003a, b). A new suite of classifications systems supposed to meet these needs are currently in various stages of development. This implies higher classification errors, usually over standard acceptable limits. Such problems still limit the operative diffusion of thematic maps produced by supervised and unsupervised pixel-based classifications of remotely sensed images. The last very high resolution generation of sensors and the development of *object-oriented classification* techniques seem to be able to by-pass such problems. In object-oriented *recognition* it is not single pixels that are classified but image objects are extracted/recognized in a previous image segmentation step. Object-oriented classification is bound to become more widespread as high-resolution and very high resolution sensors are employed in remote sensing. Such an approach is also likely to reduce the "salt-and-pepper" artefacts often seen in images classified on a per pixel basis. In fact, the traditional classifiers tend to produce salt-and-pepper classification and have difficulties to extract patches of interest. The new type of classification process is now turning into a polygon base approach. Furthermore they permit addition of textual or contextual information to the segments that must be describable in an appropriate way to achieve improved classification results. The object-oriented approach is based on the concept that important semantic information necessary to interpret an image is represented in meaningful segments and their mutual relationships. The first step in object-oriented classification is an automatic segmentation of the image. The segmentation produces segments with an overall internal homogeneity and taking spatial continuity (texture, topology) into consideration. The formatted objects will be labeled with values and statistic information of the pixels of which they consist and spatial features as well as their position within the hierarchical network. By adopting this paradigm, object-oriented classification can achieve a higher efficiency and accuracy of polygon-based classification than those possible with traditional classification procedures (Chirici et al. 2003a, b; Fig. 4.10). Segmentation tends to minimize the *spectral heterogeneity* (h_s) of each patch derived from the digital number value of their pixels and the *geometrical heterogeneity* (h_g), dependent on the shape of the new patches (Baatz and Schäpe 1999; Baatz et al. 2001):

$$h_s = \sum_{c=1}^{q} w_c \sigma_c,$$

where q is spectral bands; σ_c is the standard deviation of digital numbers of the c-band in the polygon considered, and w_c is the weight of the c-band;

$$h_{g\_smooth} = \frac{l}{\sqrt{n}};$$

$$h_{g\_compact} = \frac{l}{b}.$$

For each patch h_{g_smooth} is the fractal factor, l is the perimeter, n is the number of pixels, $h_{g\_compact}$ is the compactness factor, and b is the length of the minor side of the smallest square parallel to the raster datum where the patch is inscribable.

The algorithm works starting from each pixel and merges neighboring polygons till the change of heterogeneity between the original patches and the new polygon does not exceed a defined threshold.

Dealing with object-oriented techniques facilitates the manipulation of new generations of remote-sensing products:

Fig. 4.10. Landsat TM 5 image referred to Rondonia, Brazilian Amazonian rainforest. This is a color composite of bands 4, 5, and 3. Primary forests and converted land uses are evident

- The fusion technique makes it possible to obtain good geometrical resolution and spectral properties, combining different channels, often including thermal or very high resolution panchromatic bands. It can be performed using, for instance, the multisensor multiresolution technique (Zhukov et al. 1999; Minghelli-Roman et al. 2001). If higher spectral resolution provides better classification of the land cover, for instance, the spectral content of low-spatial-resolution sensors can be conserved within the geometrical resolution of high-resolution and very high resolution missions.
- The differences due to seasonal variation of phenological features of vegetation reflected in the spectral data, throughout the composition of various passages (e.g., two Medium-Resolution Imaging Spectrometer, MERIS, images taken on different dates consisting of 14 bands produce a single image consisting of 28 spectral bands).

4.2.6.3
Visual Computer-Aided Interpretation

Humans are able to detect and recognize as many as 10,000 distinct objects (Biederman 1985) under varying viewing conditions, while state-of-the-art remote-sensing classification can recognize relatively few objects. In EO also traditional instruments, such as aereo-photogrammetry and high-definition film, black and white, color, or color IR (CIR), have also improved dramatically. They now permit the user to obtain photographic data at convenient costs and with excellent spatial resolution capacity for very wide areas. Today's film is more stable and permits a resolution down to 10–30 cm. Very high resolution films and high-altitude flights could achieve lower prices than comparable satellite missions and can be employed cost effectively on regional areas on the scale between 1: 25,000 and 1:50,000. They produce a high information content at affordable prices. Image scales around 1:5,000–1:10,000 are also frequently used and a large body of experience has been accumulated with this type of material/image. Within Europe, the use of CIR photographs for the assessment of ecosystems initality gathered from representative transects or by sampling photoplots has been adopted in an operational way since the 1990s (Marchetti and Castagnoli 1989; EC 1992). Related applications have used a grid of photopoints georeferenced through GPS. Results from these applications including analytical restitution and interpretation are usually stored in a data base management system and are integrated with auxiliary information for the sake of landscape or regional assessments (Marchetti and Castagnoli 1989).

Both for traditional on-flight films and sensors, in the case of satellite missions, the basic concepts and paradigms of computer-aided interpretation of images are valid. Remote-sensing data are best viewed as wavelength intensity information and the information needs decoding before any sensible message can be extracted. Any decoding process relies on our knowledge of the properties of electromagnetic radiation. In order to extract "meaningful" information, the image interpreter has to exercise his or her judgement, scientific knowledge, general knowledge of the area and the phenomena as well as experience, so that he or she will be able to make truthful assumptions about the object/feature under investigation.

The first stage in a visual image interpretation is known as the *detection stage*. The detection stage is followed by a *recognition and identification* stage in which the image interpreter has to exercise general, local, as well as specific levels of reference to allocate objects into known categories. The general reference level is simply the interpreter's knowledge of the phenomena and processes to be investigated, the local reference level is the interpreter's intimacy with his or her own local environment, and the specific reference level is the interpreter's deeper understanding. In recognition and identification, the nongeometric image characteristics and location normally give a series of clues. The clues may be internally consistent but this is not always the case. The result of identification is a list of objects and features in the area, forming the basis of delineation of areas having homogeneous patterns and characteristics. This occurs in the *analysis stage*. Each delineated area has to be classified through a process of induction (general inference from particular cases) and deduction (particular inference from general observations). Accuracy is then controlled by field checks.

The final stage of the interpretation is the *classification*: labeling spatial data which can be displayed as maps, or incorporating them into a GIS. In this process it is helpful to recall the basic image features useful for classification: tone, color, shape, size, texture, pattern, shadow, and association.

1. *Tone* is the continuous grayscale varying from white to black (or bright to dark in color images) of a pixel. It is the most important feature: objects are often distinguished by tone.
2. *Color* is the chromatic value of a pixel. In a photograph it corresponds to spectral reflected values.
3. *Shape* refers to the general form, structure, or outline of individual objects. It could be regular or irregular. In both cases it can be distinguished into subcategories that help us to understand the process the objects are subjected to. Straight edge shapes typically represent urban or agricultural (field) targets, while natural features, such as forest edges, are generally irregular.
4. *Size* of objects in an image is a function of scale.
5. *Texture* (smoothness or roughness) of the image's elements is related to the amount of tonal change in a portion of the image. It is produced by elements that are too small to identify.
6. *Pattern* is a regular usually repeated shape with respect to an object.
7. *Shadow* is usually a visual obstacle for image interpretation. Shadows may permit height estimation for some objects.
8. *Association* takes into account the relationship between recognizable objects or features in the proximity of the target. The identification of expected associated features may provide information to facilitate identification.

Table 4.2 gives a list of land-cover classes and their main features in terms of visual characteristics.

Table 4.2. Main parameters to identify land-cover classes from visual interpretation

Land-cover class	Land-cover subclass	Characteristics	Parameters
Artificial surfaces	Other artificial lands	Tone	Variable from black to white
		Color	Variable but dark red and gray values are very common
		Shape	Regular
		Size	Variable. Anyhow objects are quite well interpreted at any scale
		Texture	Smooth or slightly coarse
		Pattern	Regular
		Shadow	It can be a marked problem with tall artificial buildings
		Association	Linear and geometric objects
	Green spaces	Tone	Dark
		Color	Mainly green
		Shape	Regular
		Size	Hundreds of square meters to a few hectares
		Texture	Coarse
		Pattern	Punctual or subcircular element next to other ones
		Associacion	Inside artificial surfaces
Agricultual lands	Other agricultural lands	Tone	Medium in vegetation season. Dark or bright, depending on soil feature, in postharvest seasons
		Color	Green in in vegetation season. Brown shadings in other sesons
		Shape	Regular
		Size	Variable depending on agricultural development of the area
		Texture	Smooth
		Pattern	Mainly quadrangular juxtaposed elements
		Shadow	Mainly absent
		Association	Linear borderline woody plants, water storages
	Tree plantations	Tone	Mainly dark. Variable in winter for deciduous trees
		Color	Mainly green. Variable in winter for deciduous trees
		Shape	Mainly quadrangular
		Size	A few hectares
		Texture	Coarse
		Pattern	Regular distribution of subcircular elements (trees)
		Shadow	It can help to distinguish them from other agricultural lands
		Association	Inside agricultural land
Natural environ-ments	Wooded lands	Tone	Mainly dark. Variable in winter for deciduous forest
		Color	Mainly green. Variable in winter for deciduous forest

Table 4.2. (Continued)

Land-cover class	Land-cover subclass	Characteristics	Parameters
		Shape	Mainly irregular. Often it is possible to have a regular shape for forests next to previous classes
		Size	Extremely variable, but more than 0.5 ha
		Texture	Coarse
		Pattern	Irregular and difficult to identify
		Association	Sometimes in rugged geomorphological sites
	Prairie	Tone	Medium in vegetation periods. Bright in dry periods
		Color	Mainly green. In therophytic-dominated prairie it is yellow or gold in dry periods
		Shape	Mainly irregular
		Size	Extremely variable
		Texture	Smooth or slightly coarse
		Pattern	Irregular and difficult to identify
		Association	Often next to wooded land
	Wetlands	Tone	Medium
		Color	Mainly blue-green
		Shape	Mainly irregular
		Size	Extremely variable
		Texture	Smooth
		Pattern	Irregular and difficult to identify
		Association	Often next to inner water or sea
	Unvegetated surfaces	Tone	Bright
		Color	Mainly bright gray or white
		Shape	Irregular
		Size	Variable
		Texture	Variable from coarse to smooth (ice or snow)
		Pattern	Irregular but often well delineated by linear and serpentine elements
		Association	Often located above the altitude limit of vegetation, even if it could be included in wooded lands
Water	Inner water and seas	Tone	Mainly dark. It can vary depending on reflection
		Color	Blue or blue-green
		Shape	Presence of rounded margins
		Size	Extremely variable
		Texture	Smooth or, in rare cases such as sea, slightly coarse
		Pattern	Absent
		Association	Linear water or vegetation elements ending on it, beaches, wetlands

4.3
The Instruments and Their Use

In the last 3 decades sensors operating at a wide range of imaging scales with potential interest and importance to forest planners have become available. Coupled with the ready availability of historical remote-sensing data, the reduction in data cost and increased "resolutions" from satellite platforms, remote-sensing technology appears poised to make an even greater impact on planning agencies and land management initiatives involved in monitoring land-cover and land-use change at a variety of spatial scales. Current remote-sensing technology offers collection and analysis of data from ground-based, atmospheric, and Earth-orbiting platforms, with linkages to GPS data, GIS data layers and functions, and emerging modeling capabilities (Franklin et al. 2000; Franklin 2001). Planning and forest management agencies have numerous and varied responsibilities and tasks (Jensen and Cowen 1999). Further, their ability to complete these tasks is hampered by the paucity of comprehensive information on the types and rates of land-use change and even less systematic evidence on the causes, distributions, rates, and consequences of those changes (Loveland et al. 2002). For example, at the forest/rural–urban fringe, large tracts of undeveloped rural land are rapidly converted to urban land use. This land-use dynamics makes it difficult for planners to obtain or maintain up-to-date land-cover and land-use information, where typical updating processes are on an interval scale of 5 years (Chen et al. 2001). Although the full potential of remote-sensing technology for change detection applications has yet to be completely realized, the need for remote-sensing information to help formulate policy and provide insight into future change patterns and trends is now recognized (Jensen and Cowen 1999). Otherwise remote-sensing technology often overlooks the needs of end users. Effective real-world operational examples of land-cover and land-use change still remain relatively rare (Loveland et al. 2002; Rogan et al. 2003).

In the near future, the field of remote sensing will change dramatically with the projected increase in the number of satellites of all types (Glackin 1998). This will further compound the problems already described. Although coarse spatial resolution meteorological satellite data have been available since the 1960s, civilian remote sensing of the Earth's surface from space at medium spatial resolutions (i.e., 250 m) only began in 1972, with the launch of the first of a series of Earth resource satellites (i.e., Landsat). Since 2000 a proliferation of satellite platforms with a large number of sensors (e.g., ENVISAT) and increasing spatial resolutions (e.g., IKONOS and Quickbird) can been seen. Indeed, the ever-expanding constellation of satellite platforms has acquired thousands of trillions of bytes of data that are often invaluable for applications (Jensen 2000). Furthermore, high-resolution airborne data acquisition technology has developed rapidly in recent years. A summary of the key characteristics of selected satellite sensors is presented in Table 4.3.

Table 4.3. Remote-sensing technology has been driven by three interrelated factors: (1) advancements in sensor technology and data quality, (2) improved and standardized remote-sensing methods, and (3) research applications (for more detail on these issues refer to Jensen 2000; Franklin 2001). High-resolution and very high resolution images are currently available (Landsat 7, 15 m; IRS-1C and SPOT-4, 6 m; Quickbird and IKONOS, 1–4 m) and they are characterized by an improvement of the signal-to-noise ratio and by even better informative content, up to 24-bit character

Sensor Mission	Organization	Oparation Period	Spatial Resolution (m)	Swath (km)	Spectral coverage (mm)	Number of channels
MSS (Landsat 1-5)	NASA, USA	1972-1983	79 (MS) 240 (TIR)	185	0.50-12.6	4
METEOSAT	EUMETSAT, Europe	1977	2500-5000		0.4–1.1, 5.7–7.1 10.5–1	3
AVHRR NOAA (6-15)	NOAA, USA	1978	1100	2700	0.58-11.5	5
TM (Landsat 4,5)	NASA, USA	1982	30 (MS) 120 (TIR)	185	0.45-2.35	7
HRV (SPOT 1,2,3)	SPOT Image, France	1986	10 (PAN) 20 (MS)	60	0.50-0.89	3
LISS-I (IRS-1A)	ISRO, Indian	1988	72,5	148	0,45-0.86	4
AVIRIS	JPL, USA		20	20	0.4-2.5	224
LISS-II (IRS-1B)	ISRO, Indian	1991	36,3	146	0,45-0-86	4
SAR, AMI (ERS-1)	ESA	1991-2000	26,0	102	NA	NA
SAR, OPS (JERS-1)	NASDA, Japan	1992	18,0	75	0.43-1.7	7
LISS-III (IRS-1C, 1D)	ISRO, Indian	1995	32, 70 188(WiFS)	142	0.52-1.7	4
SAR (RADARSAT-1)	Canada	1995	8-100	45-500	NA	NA
Panchromatic (IRS-1D)	ISRO, Indian	1997	5,8	70	0.50-0.75	1
Goes-8-10	NESDIS, USA	1994	1000 (VNIR) 8000 (SWIR) 4000 (TIR)	8	0.52-12.5	5
SAR, OPS (ERS-2)	ESA	1995	26	102	NA	NA
EarlyBird	DigitalGlobe, USA	1997	3 (PAN) 15 (MS)	6,30	0.45-0.89	3
HRVIR (SPOT4)	SPOT Image, France	1998	10 (PAN) 20 (MS)	60	0.50-1.75	3
Vegetation (SPOT4-5)	SPOT Image, France	1998	1150	2250	0.43-1.75	5
SeaWiFS (OrbView-2)	Orblmage, USA	1997	1130	2800	0.40-0.88	8
MODIS	NASA, USA	1999	250 (PAN) 500 (VNIR) 1000 (SWIR)	2300	0.620-2.155, 3.66-14.385	36

Table 4.3. (Continued)

Sensor Mission	Organization	Oparation Period	Spatial Resolution (m)	Swath (km)	Spectral coverage (mm)	Number of channels
ASTER (EOS Terra)		1999	15 (VNIR) 30 (SWIR) 90 (TIR)	60	0.52-0.86, 1.60-2.43, 8.125	14
MISR (EOS Terra)		1999	275	360	0.425-0.886	4
ETM+ (Landsat 7)		1999	15 (PAN) 30 (MS) 60 (TIR)	185	0.450-2.35, 10.4-12.5	7
IKONOS		1999	1 (PAN) 4 (NIR)	11	0.45-0.90	4
QuickBird		1999	0,82 (PAN) 3,2 (MS)	6,30	0.45-0.90	4
Hyperion and ALI (EO-1)		2000	10 (PAN) 30 (MS)	185	0.433-2.35	9
MERIS (Envisat-1)		2001	300, 1200	1150	0.39-1.04	Up to 15
ASAR (Envisat-1)		2001	30,0	400	Radar	1
HRG (SPOT 5)		2002	5 (PAN) 10 (VNIR) 20 (SWIR)	60	0.48-1.75 VNIR, SWIR	3
OrbView-3		2003	1 (PAN) 4 (MS)	8	0.45-0.90	4
COSMO Sky/med		Planned	-	-	-	-
SAOCOM		Planned	-	-	-	-
TerraSAR		Planned	-	-	-	-
Pleiades		Planned	-	-	-	-
Radarsat-2		Planned	-	-	-	-
SMOS		Planned	-	-	-	-
METOP		Planned	-	-	-	-
NPOESS		Planned	-	-	-	-
ALOS		Planned	-	-	-	-

4.3.1
Coarse Spatial Resolution Sensors

Even though coarse-resolution image data fall outside of the minimum spatial resolution requirements outlined in Table 4.3, a brief evaluation of the contribution of coarse-scale, large-area sensors to monitoring of forest change is warranted. The National Oceanic and Atmospheric Administration (NOAA) provided a long series of meteorological satellites, including the Advanced Very High Resolution Radiometer (AVHRR), which provides an image every 12 h (actually, every 6 h, because there are always two NOAA satellites working simultaneously with different orbits). Coarse-resolution data have been used for many years to acquire basic land-cover and land-use information over large areas. Stow and Chen (2002) examined the sensitivity of anniversary-date multitemporal AVHRR1 data to map land-cover and land-use change. Recently, Zhan et al. (2002) described the monthly 250-m resolution vegetative cover conversion product generated from Moderate-Resolution Imaging Spectroradiometer (MODIS) data. This product is designed to serve as a global alarm for land-cover change caused by anthropogenic activities and extreme natural events. While these data are too coarse for the purposes of local-level planning and forest management, they could serve as a general "change" product for regional/national agencies especially owing to their great potential for computation of vegetation indices and relative temporal profiles (Ricotta et al. 1998).

4.3.2
Medium and High Spatial Resolution Sensors

Medium-resolution sensors provide information at seemingly adequate scales for a wide variety of applications. The continuation of the Landsat program since 1972 is recognized as a key milestone in remote sensing (Franklin 2001). For 12 years, the Landsat MSS sensor provided image data with a spatial resolution of approximately 80 m in four spectral bands (i.e., visible and near-IR). Although MSS data exhibited significant noise (Schowengerdt 1997), they nevertheless provided a unique opportunity for researchers to investigate and apply remote-sensing data at regional scales. The MSS spatial resolution was also sufficient for general mapping efforts in natural areas. These data are invaluable today for historical change detection studies and form an important component of change detection data bases like the North American Land Characterization (NALC) data set (Yuan and Elvidge 1998).

The launch of the Landsat TM in 1984 produced a new remote-sensing data source that provided higher spectral-, spatial-, and radiometric-resolution data

(Fig. 4.10). Landsat MSS spectral bands and bandwidths were selected by sensor designers for their general utility to map vegetation and geologic features. Landsat TM spectral channels were chosen specifically to map vegetation, soil moisture, and other key landscape features (Jensen 2000). Thus, the TM era has permitted research to be conducted with greater precision over large areas (i.e., the swath width of a TM scene is 185×185 km^2). Landsat TM data have facilitated investigations with thematic issues by an order of magnitude compared with MSS. However, despite these advancements, the planning and land management community still lacked large-area, high-spatial resolution remote-sensing data from space. This situation improved somewhat with the launch of the SPOT-1 satellite in 1986. This sensor provided multispectral data with a slightly higher spatial resolution (20 m) and a panchromatic channel (10 m). The panchromatic data have a high geometric fidelity, so high that they are suitable for photo-interpretion in much the same way as a typical aerial photograph (Jensen 2000). Several projects began to employ image fusion techniques, using panchromatic and multispectral information for improved land-cover monitoring (e.g., Treitz et al. 1992; Pellemans et al. 1993; Muchoney and Haack 1994; Treitz and Howarth 1999). High-spatial-resolution panchromatic information is used effectively as textural information for land-use monitoring (Chen et al. 2001). A 15-m spatial resolution panchromatic band was added to the Landsat Enhanced Thematic Mapper Plus (ETM+). The Landsat series became the most widely used EO data provider all over the world. Overall, the widespread availability of high-spatial-resolution panchromatic data allows for high-order investigation into urban/suburban and natural landscapes (Jensen and Cowen 1999).

In addition to the panchromatic channel, the SPOT sensor was also a major breakthrough in sensor design. The sensor acquires information using a linear array of detectors. This approach is superior since there are no moving parts (i.e., a rotating mirror that scans back and forth across the orbit path). This design provides for a longer "dwell time" or radiance integration period over a target (Schowengerdt 1997) and thereby provides increased sensitivity to radiometric contrasts between surfaces. The overall capability of the SPOT system was enhanced significantly in 1998 by adding a mid-IR channel to the SPOT-4 sensor, a channel that ensures greater utility for land-cover and land-use monitoring purposes (Stroppiana et al. 2002). The SPOT-5 sensor (launched in 2002) collects panchromatic, visible, near-IR, and mid-IR data at 5-, 10- and 20-m spatial resolution, respectively.

The Indian Space Research Organization (ISRO) has also contributed to the suite of medium-resolution sensors. ISRO has launched four linear-array sensors to date (IRS-1A, IRS-1B, IRS-1C, and IRS-1D). In general, the Indian Remote-Sensing Satellite (IRS) sensors offer an intermediate combination of TM/ETM+-like spectral resolution, with SPOT-like-sensor spatial resolution.

IRS-1C and IRS-1D (launched in 1995 and 1997, respectively) offer visible and near-IR bands at 23-m spatial resolution and a mid-IR band at 70-m spatial resolution. Most significantly, these IRS sensors acquire panchromatic information at 5.8-m spatial resolution, which has significant implications for more detailed mapping capabilities.

The contribution of medium-resolution sensors is expected to continue for a long time (Franklin 2001). Indeed, follow-on sensors have already been launched. The Advanced Spaceborne Thermal Emission and Reflection Radiometer (ASTER), an instrument on the Terra platform, acquires visible and near-IR information at 15-m spatial resolution and mid-IR information at 30-m spatial resolution. Further, the Earth Observer (EO-1) platform includes the linear-array Advanced Land Imager (ALI) with ten bands, ranging from the visible to the mid-IR regions of the electromagnetic spectrum at 30-m spatial resolution and a panchromatic band acquired at 10-m spatial resolution (Jensen 2000).

4.3.3 Very High Spatial Resolution Sensors

Remotely sensed data with a spatial resolution of 5 m or better, a resolution often required by forest planners and land managers, became a reality in 1994, when the United States government declassified high-spatial-resolution satellite remote-sensing data (i.e., between 1- and 4-m spatial resolution; Glackin 1998). This groundbreaking decision resulted in several new spaceborne high-resolution sensors providing commercial very high resolution data products.

Technological advances in sensor design allow nearly aerial photographic precision and quality of data from modern high-resolution sensors, precision and quality that permit the investigation of thematic information at the highest order even in natural landscapes that are the most complex in the land-cover classes. The three most advanced sensors are IKONOS-2 (Space Imaging, launched in 1999), Quickbird 2 (DigitalGlobe, launched in 2001), and Orbview-3 (ORBIMAGE). They offer 11-bit visible and near-IR information at 4-m spatial resolution, and panchromatic information at 1-m resolution or better (Jensen 2000; Table 4.3).

Airborne systems have been in operation for many years and are increasingly reliable, cost-effective, and available worldwide (Franklin 2001). The flexibility of airborne platforms means that on-board sensors can acquire data at user-specified times, rather than those of scheduled satellite overpasses. Platform altitude can also be reduced to provide higher-resolution data (as good as or better than stated before). In addition, multispectral sensors and airborne high-fidelity digital frame cameras (capable of acquiring data at 0.2 m across visible and near-IR wavelengths) see wide use in land-cover and land-use applications (Coulter et al. 2000; Jensen 2000; Chen et al. 2002).

4.3.4
Hyperspectral Sensors

Historically, the majority of sensor research, development, and production has been devoted to (1) medium-resolution (i.e., spatial and spectral) large-area image acquisition and (2) high-resolution small-area image acquisition. Today a burgeoning interest in the field of imaging spectrometry for land-cover monitoring is observed (Treitz and Howarth 1999; Köhl and Lautner 2001). Imaging spectrometry is defined as "the simultaneous acquisition of images in many relatively narrow, contiguous . . . spectral bands . . ." (Jensen 2000). The promise is that we may be able to detect and monitor phenomena that could not be detected with a broadband imaging system (Herold et al. 2002). Government agencies and commercial firms have designed a number of linear-array and area-array imaging spectrometers capable of hyperspectral imaging. Franklin (2001) noted a significant increase in the number of airborne multispectral and hyperspectral data providers during the 1990s. The Airborne Visible IR Imaging Spectrometer (AVIRIS) has been operating since 1987 and provides 12-bit data at 20-m spatial resolution across 224 spectral bands. The Compact Airborne Spectrographic Imager 2 (CASI-2), a programmable system (the user can program the sensor to collect a combination of high-spatial-resolution and high-spectral-resolution data) capable of collecting up to 228 spectral channels. Hyperspectral satellite sensors in orbit at the time this book went to print include the MODIS and the EO-1 Hyperion instrument (Table 4.3).

A large number of research studies has examined and developed an understanding of hyperspectral data in natural environments (Treitz and Howarth 1999; Ustin et al. 1999). Yet there is still a significant need for research on quantifying and assessing temporal change in both rural and urban environments (Rashed et al. 2001; Herold et al. 2002).

4.3.5
Microwave Sensors

Active microwave remote-sensing (i.e., radar; Fig. 4.11) technology has been around since 1991 (ERS 1/2). Applied use on the scale of optical remote sensing has so far not materialized. Despite the theoretical expectation of a utility in natural environments, there has been a paucity of applications of active radar to land-cover monitoring (Kasischke et al. 1997). This may be attributed to the lack of readiness and full understanding of the nature of radar data. Reliable and robust methods of analysis remain in short supply. However, a number of SAR systems have been developed for the five SAR systems that

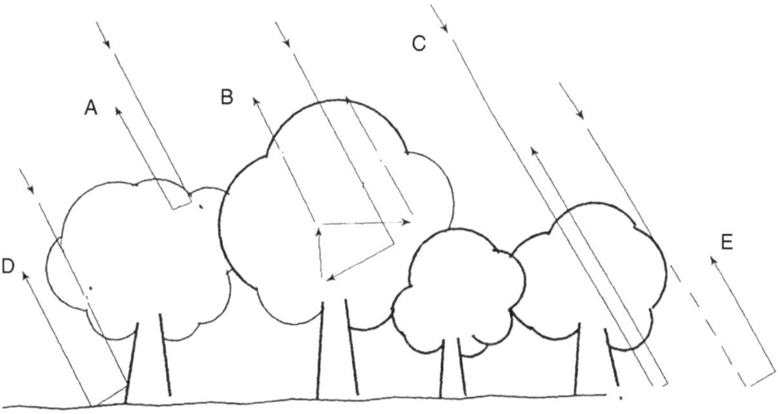

Fig. 4.11. Radar backscatter contributions of a forest scene

have been deployed: SIR-C/X-SAR, ERS-1, ERS-2, JERS-1, and RADARSAT-1. Only RADARSAT-1 (launched by the Canadian government in 1995) and ERS-2 (launched by the European Space Agency in 1995) were operating in 2005. C-band RADARSAT-1 provides a range of spatial resolutions and geographic coverages: data are acquired over 50×50 km^2 at 10-m spatial resolution (fine-beam mode), or, in ScanSAR wide-beam mode, over 500×500 km^2 at 100-m spatial resolution. ERS-2 collects data in C-band wavelengths at 26-m×30-m spatial resolution. C-band data from these sensors have been used effectively in a number of forest mapping and forest change detection studies (Grover et al. 2000; Quegan et al. 2000).

The remote-sensing research community appreciates the potential of active SAR in natural environments, particularly the synergistic application of SAR and optical data (Nezry et al. 1995; Gamba and Houshmand 2001). Several new SAR satellites are planned for launch in the near future, adding polarization diversity (vertical or horizontal) and polarimetry to a range of resolutions and swath widths (e.g., ENVISAT, ALOS, PALSAR, and RADARSAT-2). SARs are capable of operating day and night and given the longer wavelengths, clouds and precipitation are transparent. These characteristics make SARs all-time, all-weather sensors. Current studies can rely on all-time capabilities of SARs and all-weather sensors are at an advantage in frequently cloud covered tropical and boreal biomes. As imaging radars, SARs map that portion of transmitted energy that is reflected back to the sensor's receiver by earth terrain features. SAR "reflection" is typically termed backscatter and earth terrain features scatterers. SAR backscatter at a given wavelength and polarization is known to be determined by structural or geometric properties and dielectric properties. The contributing structural properties of vegetation canopies are:

1. Size distribution of scatterers (for trees, main stem, branches, and foliage) relative to wavelength
2. Orientation of scatterers
3. Number of scattering elements

Volumetric water content, the phase of water (liquid or frozen), and the specific dry density of the scatterers determine the dielectric properties. SAR interaction with larger structural elements of vegetation is in contrast with that of short-wavelength optical sensors, which respond to pigment, structure, and water at the cellular level, and is consistent with the physical principles of the sensor for vegetation properties on the order of the magnitude of the wavelength(s) employed.

To better understand all determinants of SAR backscatter, physics-based models such as the Michigan microwave canopy scattering model (MIMICS; Ulaby et al. 1991) have been developed. Using MIMICS, we can specify sensor and vegetation parameters and model backscatter using radiative transfer theory. Trunks, branches, and needles are modeled as dielectric cylinders and leaves as dielectric disks. Major scattering terms are calculated (Fig. 4.12):

1. Surface and volume scattering from the tree crown (e.g., branches and leaves, needles)
2. Direct ground scattering
3. Direct trunk scattering
4. Ground-to-trunk scattering
5. Ground-to-crown scattering

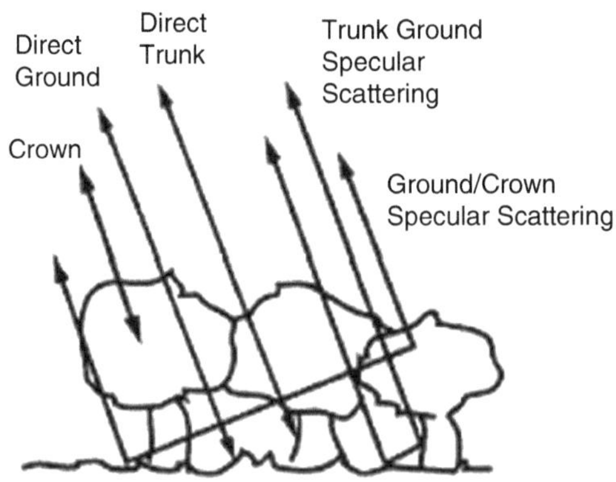

Fig. 4.12. Scattering mechanisms dominating synthetic aperture radar backscatter of forests (after Kathleen et al. 1999)

Analysis of the relative contributions of each of these terms to total SAR backscatter for a given forest type, size, and density is critical in algorithms for land-cover classification and extraction of biophysical parameters. For example, from MIMICS we know that for an array of vertical cylinders (i.e., branchless tree trunks), SAR backscatter is known to be proportional to height squared.

SAR backscatter intensity and interferometric coherence have been used in forest mapping and monitoring (Wegmuller and Werner 1995; Le Toan et al. 1997; Stussi et al. 1997). In particular, tropical forests are known to have a constant backscattering coefficient (σ_0) between –7 and –6 dB in the C-band. The interferometric coherence of the closed vegetated area is typically low compared with that of the clearcuts or sparsely vegetated area.

4.3.6 Laser Sensors

Information about the current state of and recent changes in landscape-creating elements and scenary (such as forests, lakes, and meadows) is important for environmental assessment on different administrative levels. During the last few years, airborne as well as terrestrial laser scanning techniques have developed to the point where they can provide imagelike vertical profiles of vegetation and urban areas. In addition to airborne laser scanning, which has been successfully introduced as a means of collecting information on a regional level, terrestrial laser scanners are also used on a smaller scale to derive higher-precision information. These systems are very flexible and can be used for diverse uses such as topographic mapping, forest height surveying, and digital elevation model (DEM) surveying.

All laser scanners consist of the key items shown in Fig. 4.13. Their performances are determined by the ranging accuracy, the precision of the laser beam deflection mechanism – the opto-mechanical scanner – and the measurement rate. Also, intensity images are becoming of greater interest. Laser scanners

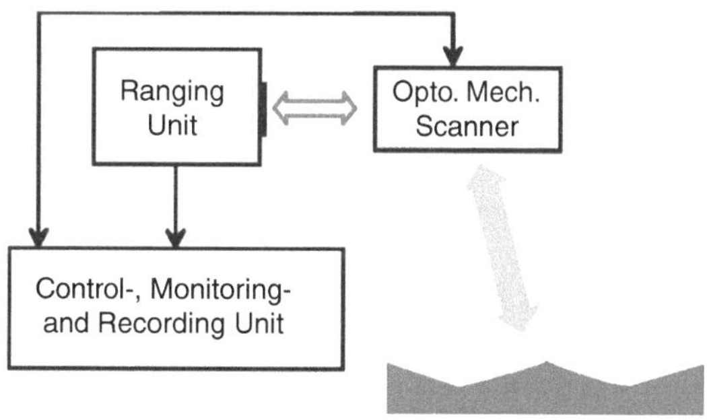

Fig. 4.13. Main components of a generic laser scanner (after Thiel and Wehr 2004)

measure the three-dimensional coordinates by sampling synchronously the slant range and the instantaneous deflection angles of the scanning device.

For airborne laser scanners the positional accuracy of the laser spots depends on the accuracy of the position and orientation system and also the accuracy of the scanner in the slant range. Compared with, for example, multispectral scanners, the depth measurement provided by data from airborne laser sensors is clearly unique (Thiel and Wehr 2004). With the advance of high-precision systems capable of working in most environments, several applications have emerged. In the field of forest surveying, data capturing may be automated. In fact, different sectors along the forest-wood production chain could require precise data on the three-dimensional structure of the environment (Fröhlich and Mettenleiter 2004). In particular, airborne laser scanning has already proven its usefulness for the acquisition of accurate and reliable topographic data of surface characteristics of the canopy (Schnadt and Katzenbeißer 2004).

The most widely used system is lidar, an acronym for light detection and ranging. It transmits laser pulses and detects the backscattered signal. It is an active sensor, very similar to radar (radiowave detection and ranging), where the range to the scattering object is determined from the time delay between transmission and detection. The spectral image data are rectified with help of the surface model to generate true ortho images in true color (red–green–blue, RGB) and CIR. During the last decade, airborne lidar mapping has gained general acceptance as an accurate and rapid method for three-dimensional surveying, despite of its costs and the specialization it entails. It outputs the three-dimensional coordinates of the surface locations hit by the laser pulse (discrete ranges). Most systems are able to distinguish two returns from multiple targets touched by a single laser pulse; some systems provide up to four returns. This information is valuable for forestry applications for deriving several vegetation parameters: tree/vegetation height, vertical canopy expanse, crown cover, height of second, third, and lower levels of vegetation, and the height and density of ground vegetation. Calculation of timber volume, biomass, and other important vegetation descriptors is thus possible with adequate precision (Hug et al. 2004).

4.4 Accuracy Requirements

The accuracy requirement depends on the context in which the measurements are used. In some cases, changes may need to be just detected rather than measured. A major application of this will be when credit is given for preserving an intact forest. Then, obviously, any detectable change due to logging or intentional burns will void the credit for the preservation. The accuracy needed is difficult to know (a sound assessment could be more expensive than the original purpose), however some observations can be made in order to speculate on the approximate needs. If the measurement error, in biomass change, for example, is significantly below the uncertainty level, remote sensing should be considered valid.

The concepts of accuracy, precision, and reliability are essential for the evaluation of the classification of remote-sensing imagery. As there is a considerable amount of confusion in the use of those terms, a short description of the terms as they are used throughout this task is given. *Reliability* is used to describe the closeness of the results obtained (figures, maps, etc.) to the real situation. It is not a statistical term. *Precision* refers to the size of deviations from the estimated mean, $\hat{\mu}$, obtained by repeated application. It is quantified by the standard error or confidence intervals and can be increased by increasing the number of observations. *Accuracy* refers to the size of deviations from the true mean, μ. Increasing the number of observations does not necessarily decrease accuracy. In the remote-sensing literature some different concepts for the terms reliability, precision, and accuracy can be found. Aronoff (1989) defines classification accuracy as "... the probability that the class assigned to location on a map is the class that would be found at that location in the field ...." Story and Congalton (1986) refer to this accuracy as "user's accuracy" or reliability, as being a measure of the value of a map for a potential user. The attribute precision is used by Aronoff (1989) in the context of positional accuracy. Campbell (1996) defines precision as "detail", i.e., the number of classes identified during a classification process.

The objective of using remote sensing for land-cover assessments is to provide the total area or percentage of various classes within the inventory region. If satellite imagery is treated as total tally, the problem of area or percentage is not related to a statistical estimation process and thus does not allow the quantification of the precision of the results. If remote sensing is treated as a sampling procedure, in which land-cover classes are assigned to dimensionless points, statistical estimation procedures – including the estimation of sampling errors and confidence intervals – can be applied. Two possible approaches are (Table 4.4):

Table 4.4. The two alternatives for remote-sensing as a sampling driver and statistical results

Land-cover statistics	Pixels are total tallies of the inventory area	Pixels are systematic point samples of the inventory area
Area of classes	✓	✓
Sampling error of area estimates	–	✓
Proportion of classes	✓	✓
Sampling error of the proportion of classes	–	✓

- Pixels are total tallies of the inventory area. Treating satellite imagery as a total tally of the inventory area could be justified by treating the assigned classes as the total cover of the entire pixel area, i.e., each pixel area covers just one class in reality. This approach is applicable only if the variation of the spatial pattern of classes is much larger than the pixel size, i.e., for small pixel sizes.
- Pixels are systematic sample points of the inventory area. If the class assigned to a pixel is seen as a (random) selection out of the entire classes existing in the pixel area, the individual pixels can be treated as point samples. The pixel area is seen as being covered by a number of discrete classes, each of which occupies a specific proportion of the pixel area. In contrast to treating the pixels as total tallies, this approach allows each pixel to cover more than one class. The classification process selects one of these classes (depending on its proportion and its reflectance characteristics). The selected class is then assigned to a dimensionless point and the pixel becomes the area represented by this point. This approach has been applied by van Deusen (1996) to provide unbiased estimates of class proportions from thematic maps.

4.4.1 Accuracy of Position and of Classification

There is a decisive difference between the accuracy of estimating a statistical key attribute such as area or proportion and estimating the positional accuracy of a classified object. Thus, assessment of accuracy deals with two aspects: positional and thematic accuracy. Positional accuracy of remote-sensing data refers to the accuracy of a geometrically rectified image. Thematic accuracy refers to the nonpositional characteristics of a spatial data entity, i.e., the attributes (Chrisman 1987). Speaking about positional accuracy, in a row–column image the pixel locations are related to a coordinate system of another image data set or to a particular map. The relationship can be established by locating identical ground control points (GCP) in the image within a reference coordinate system. GCP have to be identified well in both the image coordinate system (row and column coordinates i,j) and the reference coordinate system (reference coordinates x,y). These two coordinate systems can be linked by the use of polynomial equations. Thematic accuracy relates to the nonpositional characteristics of attributes. The emphasis is put on the assessment of the classification results (Hord and Brooner 1976). The total classification accuracy can be evaluated with contingency matrix between classification and test sites: the kappa index of agreement should at least be calculated for a synthetic estimate of classifier performance (Congalton 1991).

4.4.2
Testing the Accuracy of Borderlines

A third measure in thematic accuracy validation is to test the accuracy of borderlines. Borderlines are the boundaries between different land-use classes. Skidmore and Turner (1992) used line intersect sampling to estimate the length of borderlines in maps (ground truth). The comparison of two measures of the length of borderlines is regarded to be a rather weak criterion for the assessment of the accuracy of borderline detection. The simple example presented in Fig. 4.14 shows the problem of comparing two length estimates. The locations of the two lines are different; however, they have the same length.

A procedure for the assessment of borderline accuracy has to take into account the location of the borderlines. The transects used for the validation phase are set up in a way that borderlines can easily be detected on the ground. As borderlines have a negligible width, they will hardly coincide with the borderline of two pixels. Therefore, the following rule will be applied to handle borderlines in pixels:

$$\text{Borderline present} \begin{cases} \text{yes, if different land} - \text{use classes are assigned to pixel} \\ n \text{ and pixel n} + 1 \\ \text{no, if identical land} - \text{use classes are assigned to pixel} \\ n \text{ and pixel n} + 1 \end{cases}$$

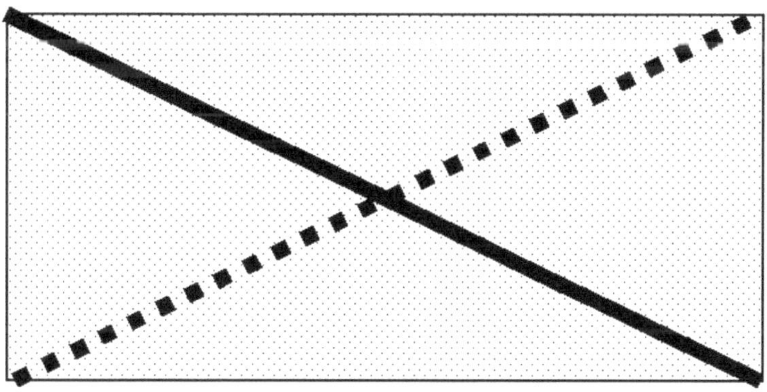

Fig. 4.14. Borderlines detected by two different methods

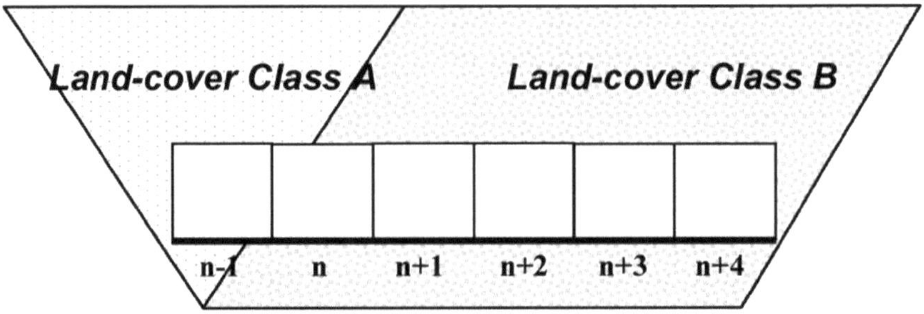

Fig. 4.15. Borderline recognition

Figure 4.15 illuminates this decision rule. Pixel n is located at a boundary line between two classes. Class A is assigned to pixel $n-1$, class B to pixel $n+1$. Depending on the spectral characteristics, one of the two classes A or B will be assigned to pixel n. As the class changes either between pixel $n-1$ and n or between pixel n and pixel $n+1$, a borderline will be recorded. Around the true (mapped) land-cover boundary a buffer will be defined with width equal to the pixel length. Thus, a bivariate attribute can be assigned to each pixel that shows if a borderline is present (yes/no) in the classification and in the reference data.

Geographic and Forest Information Systems 5

5.1 Introduction

The base of any information system is *data.* Data are the recorded values of attributes of objects. They can be raw measurements and assessments according to measurement rules or definitions or can be derived by functions or models. From the perspective of information technology, data are a structured sequence of symbols that follow certain rules.

Data as such do not constitute information of themselves; they are often not of much interest for decision making unless they are transformed or converted into something more meaningful and useful in terms of information needs. *Information* is the linkage of data (syntax) and the associated meaning (semantics). The user needs to know the "correct" rules for interpretation in order to extract information from data. Depending on the vertical level of decision processes, data need to be aggregated and condensed. In information theory, which is a collective name for techniques and methods of data processing and telecommunication, information is understood as a technical measure that is associated with data according to probabilities.

The value and significance of spatial information was recognized independently by numerous cultures throughout the world in over 2,500 years of exploration and development of maps, i.e., analogue representations of spatial order (Dobson 1995). In the 1970s the launch of remote-sensing satellites resulted in a precipitous rise of accessibility and application of spatial information. The availability of spatial information in digital format allowed some of the major deficits in the interpretation of maps to be overcome. The interpretation of spatial relationships by visual examination of a map is often subjective, as the pattern of interpretation may vary from person to person. Maps alone are often not sufficient for analysis of spatial order and spatial associations, as the complexity of spatial relationships may be hidden by generalized maps.

Spatial analyses are dedicated to the spatial order and associations of phenomena or variables. While spatial order describes the organization of geographic entities in the space, spatial association describes the geographical relationships among phenomena. Recent advances in geographic information systems (GIS) link the map-based analysis of spatial patterns with quantitative analytical methods and allow for the objective interpretation of spatial patterns.

The purpose of this chapter is to introduce the quantitative methods offered by GIS-based spatial analysis. For further reading, the books by De Mers (1997), Longley et al. (2001), Chrisman (2001), or Bernhardsen (2002) are recommended.

5.2 Geographic Information Systems

The Environmental Systems Research Institute (1994) defined GIS as "a system for organised collection of computer hardware, software, geographic data, and personnel designed to efficiently capture, store, update, manipulate, analyse, and display all forms of geo-referenced information." Figure 5.1 displays the conceptual framework for GIS.

A GIS according to this framework, has four major components or functionalities:

1. A data input component that collects and preprocesses spatial data from various sources. The input system can handle different types of spatial data and data formats and integrate them in a consistent system.
2. A data-base system component that stores and organizes the spatial data and renders data retrieval, updating, and editing possible.
3. A data manipulation and analysis component that provides operations to select or classify spatial data by certain criteria, to combine them to form new variables, and to alter spatial data for model building. Hypothetical distributions can be generated, and descriptive, explanatory, and predictive models can be derived.
4. A reporting component that displays information in tabular, graphic, or mapped format.

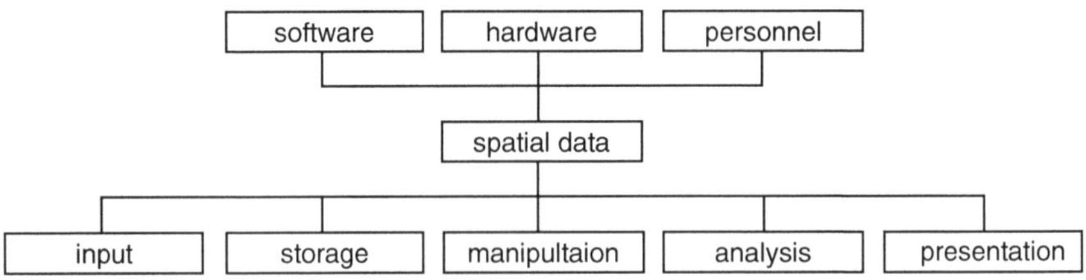

Fig. 5.1. Conceptual framework of geographic information systems

5.2.1 Spatial Data

Quantitative data analysis is either spatial or aspatial. The key difference between these two types of analyses is the inclusions or exclusion of spatial factors. In spatial analyses each object must be characterized by three elements: attribute data, location, and topology. Attribute data provide information about the properties of an object. For example, a line on a map may represent a road, a power line, or an administrative boundary. The exact location of every spatial feature must be available and is expressed by its coordinates on a Cartesian plane. The topology is defined as the spatial relationship between map features. A GIS provides the functionality to handle all three elements of spatial data.

Topology is often misunderstood as a scale model. However, it represents "only" spatial relationships between objects that can be displayed true to scale by taking into account the location of objects. The difference between the two concepts is displayed in Fig. 5.2, where the topology of a public transportation network and a true-to-scale map are presented. The public transportation network is not a true-to-scale representation, but focuses on the presentation of intersections of different bus and tram lines and stops.

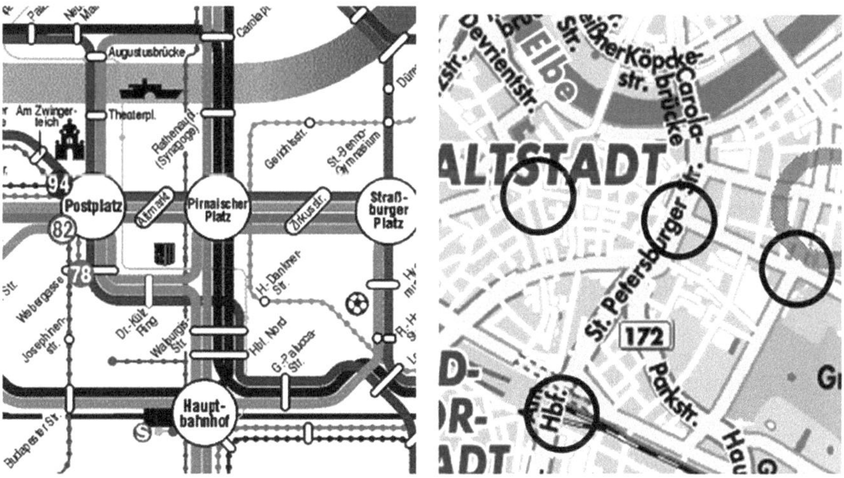

Fig. 5.2. Topology of a public transportation network (*left*) and a true-to-scale map (*right*; *rings* show the location of the four major conjunctions)

Topological features required for spatial analyses include the following:

1. *Adjacency* is required when spatial relationships based on neighborhood information are attempted. For instance, the expected disturbances of a natural forest are often higher than average when they are located close to roads or settlements.
2. *Containment* indicates whether a single feature is contained within the boundaries of another feature. For instance, forest located within a protected area may be subject to specific management regimes.
3. *Connectivity* indicates whether two segments are connected. The concept of connectivity is especially important for transportation and routing analyses.
4. *Intersection* implies that two areas share a common area, which is a subset of both. For instance, the area of a protected forest may be partially located within the boundaries of a private forest, resulting in an area of intersection that is both private property and protected forest.

For efficient data handling and spatial analyses spatial objects need to be represented in a consistent and unambiguous way. The US National Committee for Digital Cartographic Data Standards (NCDCDS) provides definitions for spatial objects. Spatial objects are classified into point, line, and polygon (area) features. Figure 5.3 presents selected cartographic objects defined by the NCDCDS.

A point feature is a zero-dimensional cartographic object. It specifies the geometric location and no other meaningful measurement. The size of point

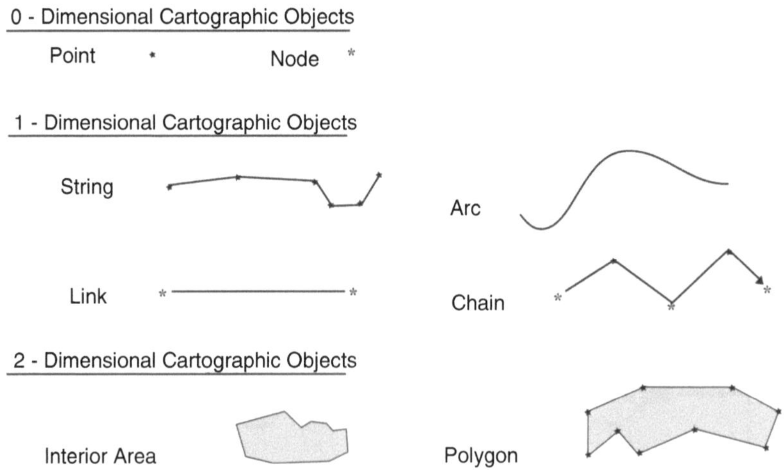

Fig. 5.3. Selected cartographic objects defined by the US National Committee for Digital Cartographic Data Standards

symbols may vary, but the area of those symbols is meaningless. The NCDCDS defines four types of point features:

1. *Entity point.* It is used to describe the location of point features. Here the positional accuracy is of major importance.
2. *Label point.* It is used to identify the position where text associated with map features is presented. Positional accuracy is not an issue, but the label points have to be chosen in a way that a user is not confused with different labels displayed on the map.
3. *Area point.* It gives the position where the attribute information associated with an area is displayed. For example, a point representing the location of the forest service headquarters may be used to display forest area statistics.
4. *Node.* It represents the location of a point that has additional topological characteristics, such as intersections or end points of line features.

In spatial representation, point features are zero-dimensional objects, as the difference in area among different point features (e.g., the area occupied by a utility pole) is considered meaningless. Although line features occupy two-dimensional space on maps they are treated as one-dimensional spatial representations as their width is not considered. Line features present position and direction; length is an important measurement of line features. The NCDCDS definitions for line features are as follows:

1. *Line* is generally defined as a one-dimensional object.
2. *Line segment* is the direct connection between two points.
3. *String* is a sequence of strings without nodes or left and right identifiers.
4. *Arc* is the location of points that are defined by a mathematical function to form a curve.
5. *Link* or *edge* is a connection between two nodes.
6. *Directed link* is a link with a specified direction.
7. *Chain* is a directed sequence of line segments that are nonintersecting or are arcs with nodes at each end.

For directed links and chains, left and right area identifiers are optional. The representation of both area and position leads to area features which are two-dimensional. According to the NCDCDS the following selected area features are defined:

1. *Area* is a two-dimensional, bounded and continuous object.
2. *Interior area* is an area not including its boundary.
3. *Simple polygon* consists of an interior area and an outer ring. The boundary does not intersect itself.

There are two data structures that are adopted by GIS to organize spatial data: the vector and the raster data structure (Fig. 5.4). Some GIS software, especially that developed for handling satellite data, use the raster data structure (e.g., IDRISI by Clark University), while others (e.g., the European Space Research Institute's Arc/Info and Arc-GIS) use vector data structures.

The raster structure is based on picture units (pixels); a map is decomposed into pixels and each pixel is referenced by its row and column position. Each pixel has a given size, and attributes are assigned on a per pixel basis. For a given area for each attribute to be assigned (e.g., ownership, land cover, road networks) an individual raster (data layer) can be constructed. Raster data structures offer a few of advantages:

- They are simple and easy to reference.
- They represent continuous surfaces.
- Combination of different data layers is easy.
- A large amount of spatial information is available in raster format, for example, remote-sensing data, or scanned maps, and data bases can easily be constructed by importing raster data.

However, there are several disadvantages associated with raster data structures:

- In many cases raster data are subject to data redundancy. As a uniform pixel size has to be chosen for each attribute it is likely that a large area is represented by a huge number of pixels all containing the same information.
- The aesthetic appearance of a map depends on the raster size. Linear features are either lost or overrepresented in crude raster formats.
- Topological relationships are hard to capture.
- Spatial data coded in raster format are distorted by transformations. For example, a line feature rotated by a specific angle and then rotated back may be different from its original shape.
- The accuracy in spatial analyses tends to be lower than desired. For instance, the length of a line can be computed exactly if the starting and end points are given, but can only be approximated when the line is represented in raster format (Fig. 5.4)

The vector data structure uses vectors to represent spatial features. A vector is – in mathematical terms – defined by a starting point, direction, and length. The starting point is given by x- and y-coordinates, and the direction by an angle, for example, with reference towards north. A point feature can be treated as a degenerated vector where direction and length are zero. A line feature is represented as a sequence of vectors; as the width of a vector is not meaningful, line features are

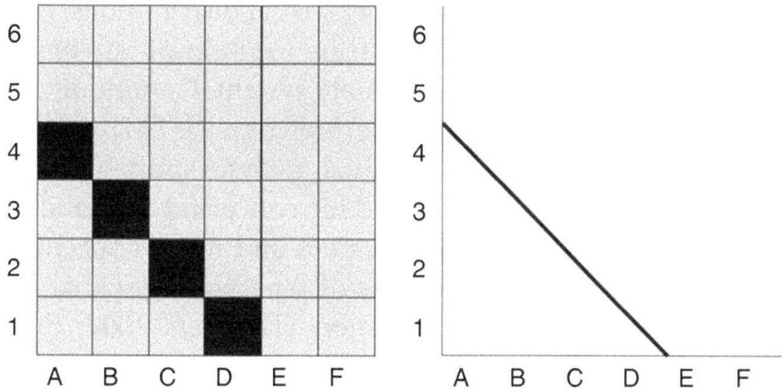

Fig. 5.4. A line organized in a raster structure (*left*) and a vector structure (*right*)

one-dimensional. A polygon is presented in the same way but with a closed set of vectors.

Vector data formats have the following advantages and disadvantages:

- Redundancy. Owing to their organization in a compact format, vector data structures show less redundancy than raster data.
- Topology of spatial features is clearly identified.
- The accuracy in spatial analyses and location is high.
- Map overlay, which is a simple task in raster data formats, becomes a complex task and requires complicated computations.

Most of the commercial programs allow conversions from vector to raster format and vice versa. The data structures determine how information is stored and processed in a GIS and have implications on the performance of spatial analyses.

5.2.2 Spatial Analyses

The analysis subsystem is the heart of any GIS. There exist a vast number of techniques for analyzing georeferenced data, which cover geographic analyses, statistical analyses, systems modeling, and geostatistics. The current chapter is limited to the fundamental methodology of spatial analysis. Interested readers are encouraged to consult textbooks by Foresman (1998), Thill (1999), or Longley et al. (2001) for further reading.

Spatial analyses are based on the three elements of spatial information: location, attribute, and topology. In quantitative spatial analyses the geographic reference allows spatial features to be referenced and mapped, while the analyses of attribute data allow for studying spatial distributions. A key element in spatial analysis is the representation of the location of any object, which is realized by the establishment of a coordinate system. Commonly used coordinate systems are the Universal Transverse Mercator (UTM) or Gauss–Krüger coordinate system. A GIS suitable for spatial analyses must provide means for integrating diverse coordinate systems and for converting spatial data from one to another commonly used coordinate system and to user-specific coordinate systems. Other issues relevant in the scope of representing spatial data are map scale, map orientation, and map projection (De Mers 2000; Burrough and McDonnel 2000).

Once a coordinate system has been established, every point feature can be represented by its *x*- and *y*-coordinates, and basic measurements such as distances or areas of spatial features can be conducted. In most situations not only one but several spatially explicit attributes are to be represented in a GIS. For any attribute (e.g., land use) or attribute classes (e.g., motorways and railroad lines) an independent data layer is generated. The set of individual data layers can be seen as a cascade of information on attributes. With reference to the data layers, spatial analyses can be classified into two major categories: single-layer and multiple-layer operations

5.2.2.1 Single-Layer Operations

Single-layer operations are also called horizontal operations as they consider only one data layer and manipulate data horizontally. There are three groups of single layer operations: feature manipulation, feature selection, and feature extraction. Figure 5.5 presents a summary of single layer operations.

Feature Manipulation

Feature manipulation procedures for single layers include boundary operations and proximity analysis. **Boundary operations** alter the boundaries of

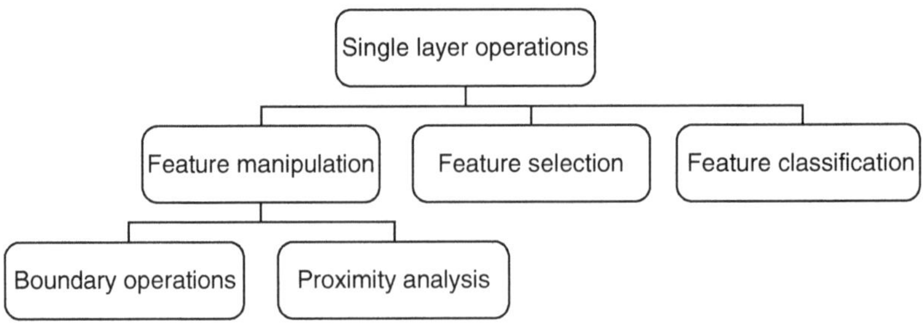

Fig. 5.5. Single-layer operations

objects and define new boundaries. Common boundary operations are clipping, erasing, updating, splitting, dissolving, appending, and elimination.

- *Clip*: A clipping procedure creates a specific set of boundaries and maintains all map features within this boundary (Figs. 5.6, 5.7). The new coverage is a subset of the original map.
- *Erase*: In contrast to the clip procedure, the erase procedure eliminates the portion specified by the erase coverage from the output coverage; the features outside the erase coverage remain on the output coverage (Fig. 5.8).
- *Update*: Update is used to move new or additional information into a coverage and to correct old information (Fig. 5.9). The update coverage contains the information that is to be replaced in the original coverage.
- *Split*: The split procedure is used when a map is to be separated into subdivisions (Fig. 5.10). A map showing the boundaries – or tiles – of the new subdivisions is superimposed on the original cover. New coverages are created according to the tile system of subdivisions.
- *Append*: The append procedure merges separate, adjacent maps and creates a new coverage that contains a single map (Fig. 5.11). Optionally the procedure can rebuild topology. However, the procedure may result in unwanted boundaries and line features do not necessarily coincide.
- *Dissolve/mapjoin*: Dissolve and mapjoin procedures are typically used after adjacent maps have been appended (Fig. 5.12). Dissolve eliminates unwanted new boundaries and links the loose ends of line features. In addition, the topology is rebuilt and nodes between lines with identical attribute values are renamed.
- *Eliminate*: Merging polygons from different sources to a single layer or data errors often produce unwanted lines. These so-called sliver polygons are removed by the eliminate procedure (Fig. 5.13).

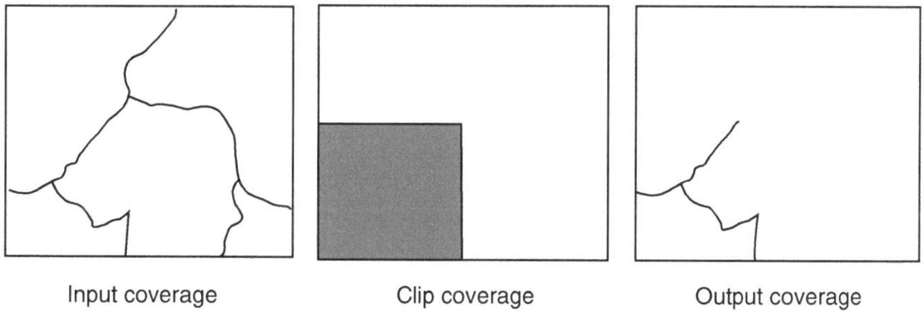

Fig. 5.6. Clip

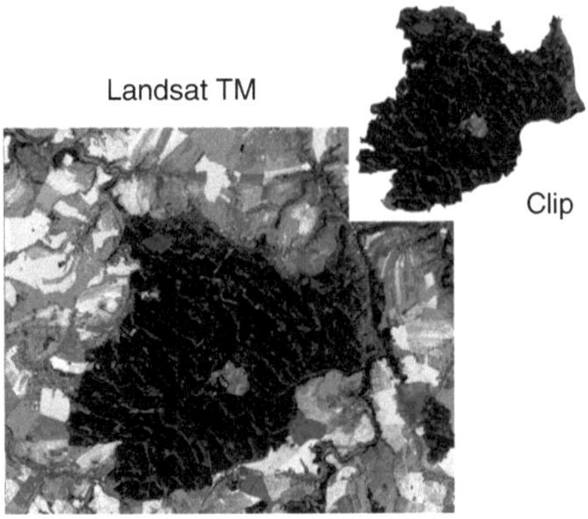

Fig. 5.7. Original coverage (Landsat Thematic Mapper, *TM*) is clipped to extract the forest areas to the output coverage

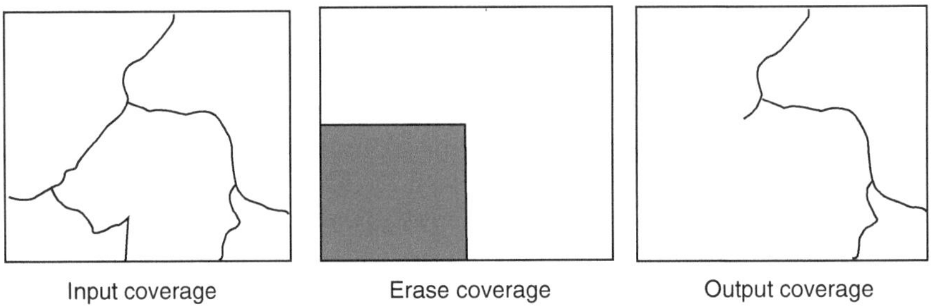

Fig. 5.8. Erase

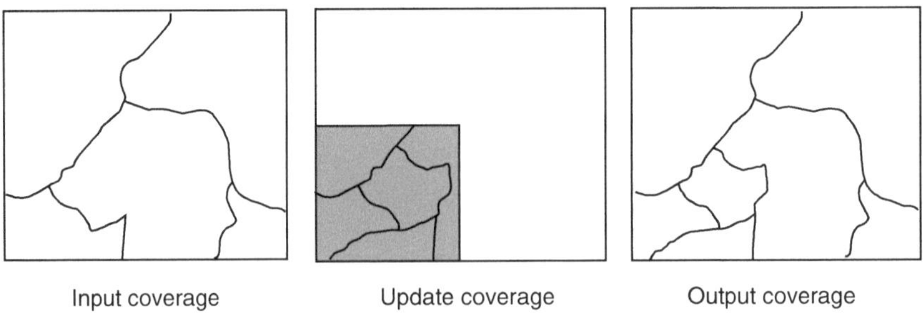

Fig. 5.9. Update

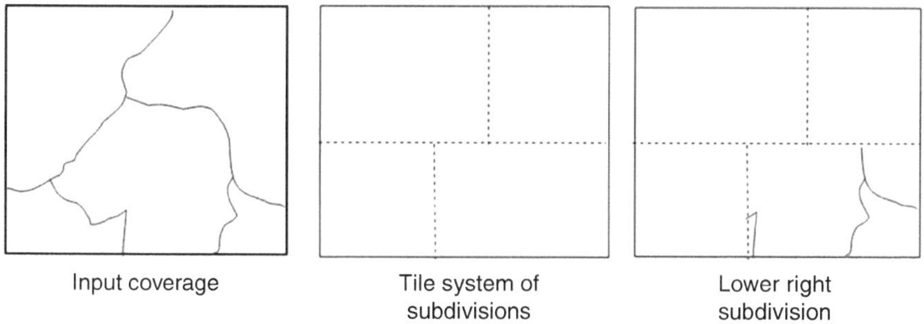

Fig. 5.10. Split

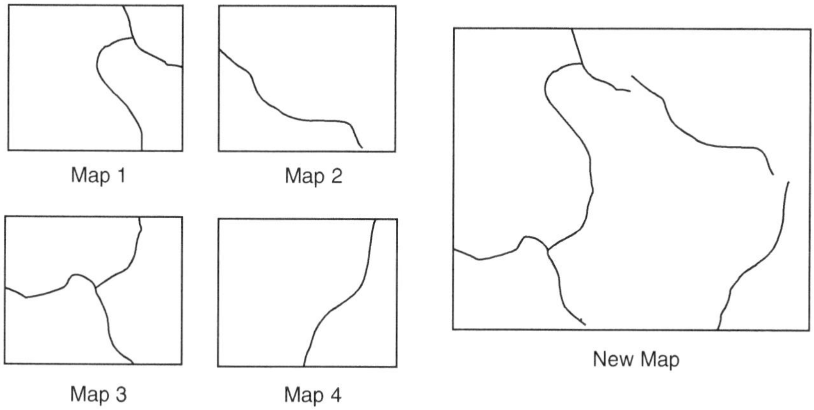

Fig. 5.11. Append

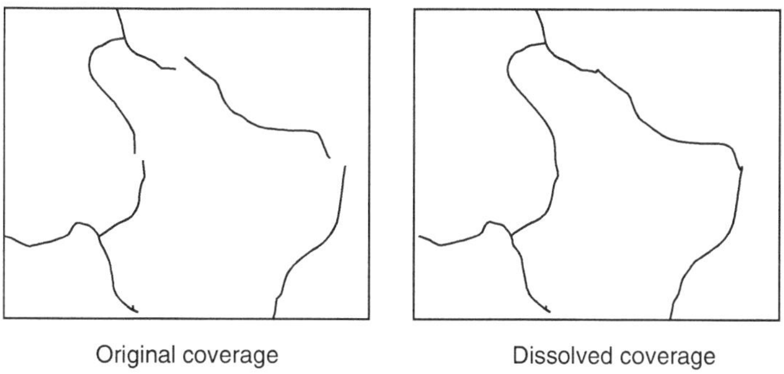

Fig. 5.12. Dissolve/mapjoin

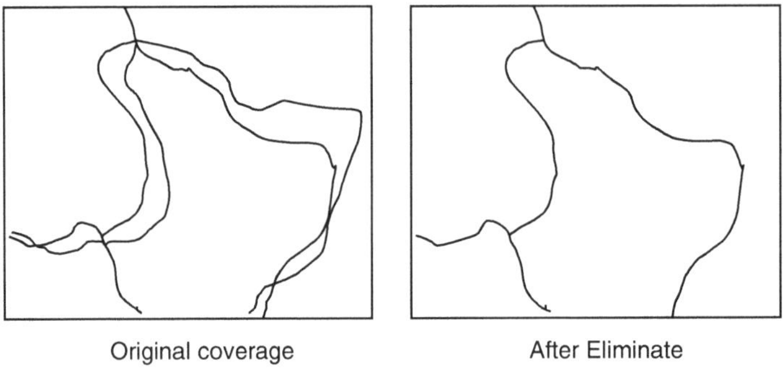

Fig. 5.13. Eliminate

Besides boundary operations, feature manipulation includes **proximity analysis**, which is based on distances from selected features. A common proximity analysis is buffering, where equal-distance zones are drawn around features. A buffer operation can be applied to any spatial feature. Proximity analysis is not limited to single-layer operations. Where different attributes are to be studied multilayer proximity analysis can be applied.

Figure 5.14 presents an application for buffering. In a stand the potential area for natural recovery was estimated by constructing equal-distance buffers around each tree of the species of interest. The analysis can be used for silvicultural planning by identifying areas where the species of interest probably does not regenerate naturally.

Feature Selection

As in aspatial databases GIS offer the possibility to select features from the data space. The selection can be similar to a data-base query by formulating logical

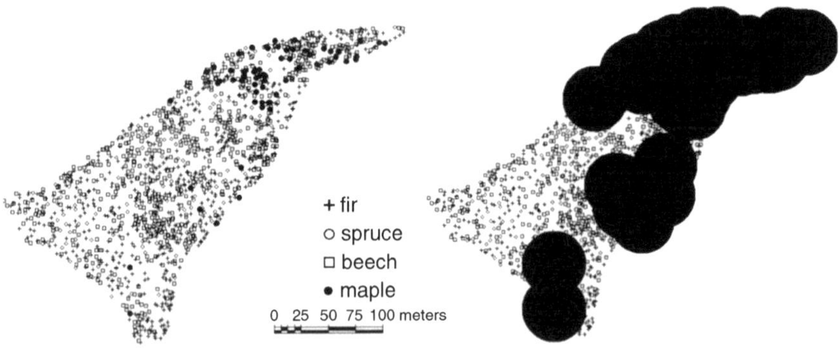

Fig. 5.14. Estimating the potential area of natural recovery of an individual tree species by buffering

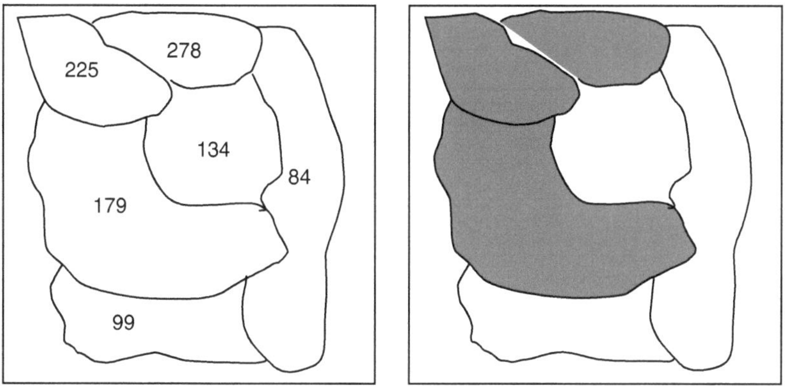

Fig. 5.15. Classification of stand volume (m^3/ha) into two classes (≤150 m^3/ha, >150 m^3/ha)

operations for the selection. For instance, from a forest map all forest with a standing timber volume over 300 m^3/ha could be selected. In addition, GIS allow spatial constraints to be added to the selection procedure. For instance, features can be selected by delineating shapes or specifying distances from a given point; all features that fall inside the defined areas are then selected.

Feature Classification

Where entities show a wide range of values for several attributes, classification might be helpful to extract significant information from the data (Fig. 5.15). The classification can apply constant class intervals. Where the distributions are not uniform a classification with nonconstant width of the class intervals can be advantageous.

5.2.2.2 Multilayer Operations

Multilayer analyses are an extension of single-layer analyses (Fig. 5.16). They operate on multiple data layers and examine the relationships between different layers. Multilayer operation can be used either to combine multiple data layers that cover the same area or to separate data on the same layer into different layers. The basic functions of multilayer operations are:

1. Overlay analysis that uses logical operations to analyze or manipulate several layers
2. Proximity analysis that involves distance measures of features on different layers

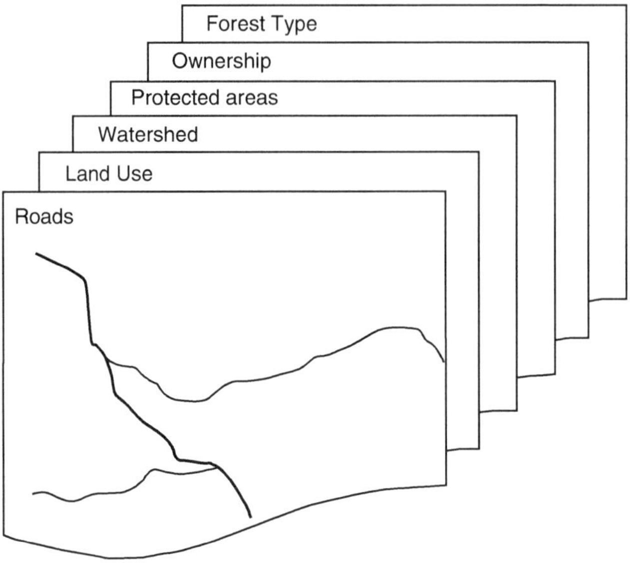

Fig. 5.16. Multiple data layers

Overlay Analysis

In overlay analyses spatial conditions of different layers are studied according to logical conditions. The logical conditions are set up by operators (relationships) and operands (data elements). The relationships specified are mainly based on Boolean operators (AND, OR, XOR, and NOT) and perform a logical check if a condition is true or false. Where ranked or metric data sets are to be studied, operators relating to size or magnitude can be applied (smaller than, greater than, equal to, smaller than or equal to, greater than or equal to). Table 5.1 presents common Boolean operators for the two operands A and B.

When the Boolean operators are transferred into the context of multiple layers the basic operations union and intersection can be identified and can be applied to merge spatial features from separate layers into a new layer.

Table 5.1. True/false conditions for Boolean operators

A	B	A AND B	A OR B	A XOR B	A NOT B
0	0	0	0	0	0
1	0	0	1	1	1
0	1	0	1	1	0
1	1	1	1	0	0

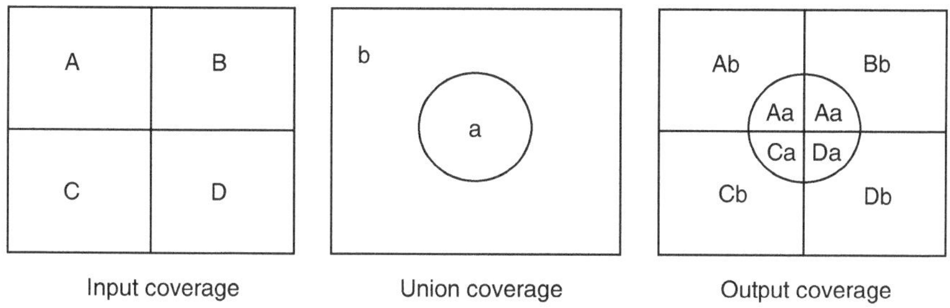

Fig. 5.17. Union operation

Union

The union operation is related to the Boolean OR operation. Two or more original layers are overlaid to create a new output coverage. The procedure is straightforward for point and line features, but may become complex when applied to polygon features. Figure 5.17 presents the union operation for an input layer with four polygons and a union layer with two polygons. The resulting output coverage contains eight polygons. For instance, if the input coverage contains ownerships and the union coverage contains forest types the resulting output coverage will contain the appropriate information on forest type by ownership.

Intersect

An analogue of the Boolean AND operation is the intersect operation (Fig. 5.18). Where tow or more layers are merged, only those portions of the input coverage are maintained that fall in the intersect coverage. While the input coverage can be a point, line, or polygon, the intersect coverage must be a polygon. For instance, the intersect operation may be used to identify private forest land by merging forest areas (input coverage) and private ownership (intersect coverage).

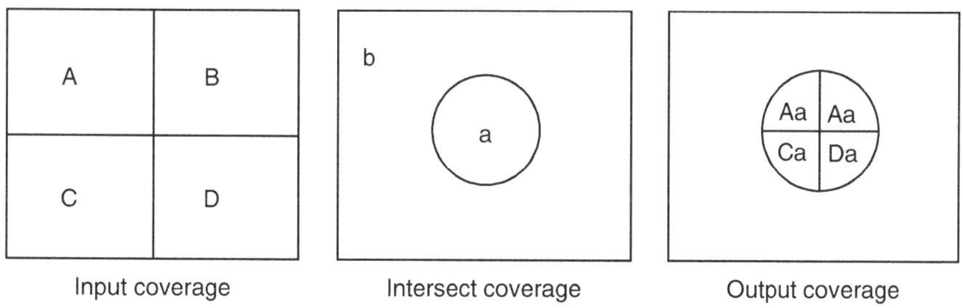

Fig. 5.18. Intersect operation

Overlay analysis can be extended by specifying more complex mathematical or logical operations. Chrisman (2001) described the rules of combination overlay that allow the use of exclusionary rules, weightings, mathematical manipulations, and Boolean logic all at the same time.

Proximity Analysis

Proximity is the quality of being near something. The easiest proximity measure is the distance between two points. The proximity between two polygons can be analyzed by either measuring the shortest separation between polygon perimeters (interseparation) or the distance between the polygon centers (centroid location). Other fundamental concepts of proximity analysis are the identification of the nearest feature in one layer from a given point, line, or polygon in another feature (e.g., from a forest stand the nearest road could be detected), and all the features in one layer within a given distance from a feature in another layer (e.g., all mature forest stands within a distance of 20 km from a sawmill).

5.2.3 Pattern Analysis

The spatial pattern or spatial arrangement is the placement, concentration, connectedness, or dispersion of multiple objects within a defined area. Pattern analysis offers tools to study, describe, and quantify spatial distributions. Among those tools are statistical measures, study of spatial arrangement, and spatial autocorrelation.

5.2.3.1 Statistical Measures

Several statistical parameters are available to describe spatial patterns. One of the first measurements of spatial distributions is frequency, which is the number of features occurring in a specified area. The interpretation of frequency might be misleading, when the size of the area within which the number of features is counted is not specified. The measure of density relates frequency to area (i.e., it is the ratio of frequency to area).

Frequency and density do not provide information on the spatial pattern of point distributions (Fig. 5.19). Besides frequency and density, the spatial features of a point pattern are described by dispersion and by the geometric center. The geometric center of point distributions is characterized by the mean of the x- and y-coordinates, while dispersion is related to the standard deviation of the x- and y-coordinates.

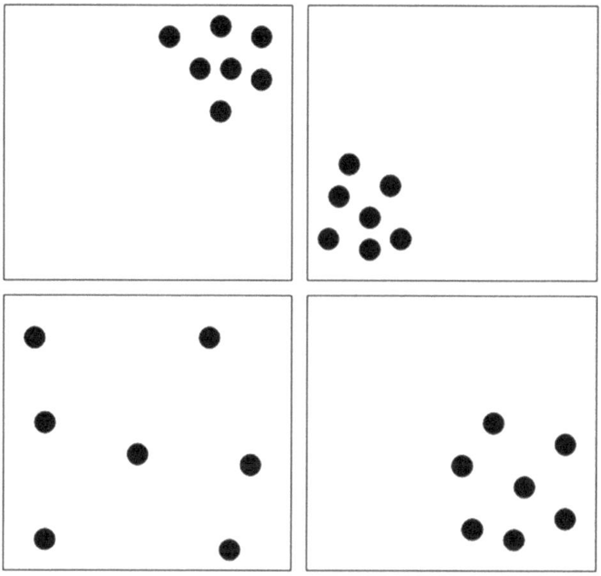

Fig. 5.19. Hypothetical point patterns with the same frequency and density but different dispersions and geometric centers

5.2.3.2 Study of Spatial Arrangement

While frequency, density geometric center, and dispersion provide a quantitative overview of statistical patterns they do not offer information on the spatial arrangement of points. There are three basic types of point patterns (Fig. 5.20):

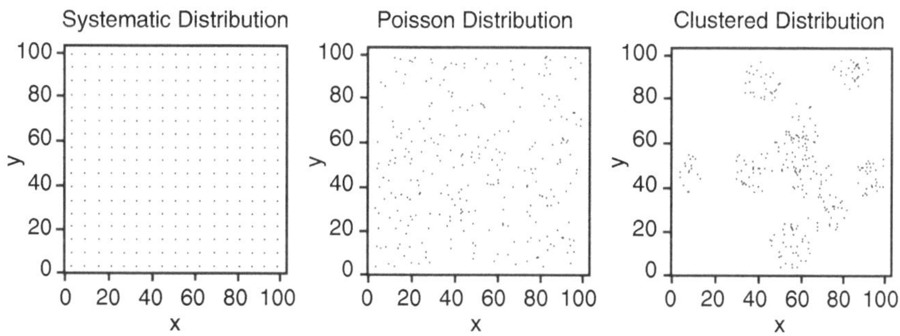

Fig. 5.20. Typical point distribution pattern

1. Systematic or uniform: characterized by regular arrangement of points with a relatively large distance between individual points (e.g., tree locations in plantations)
2. Poisson or random: the points are randomly located within an area (e.g., tree locations in a natural stand)
3. Clustered or aggregated: the point features are concentrated around a few centers and form clusters (e.g., tree locations in a savannah)

The spatial arrangement may have a significant effect on the relationship between features and can be described by a variety of approaches. Pielou (1970, 1977) and Ripley (1981) provided methods for the analysis of point pattern analyses. A simple and straightforward method is the nearest-neighbor index, which is the ratio of the average nearest-neighbor distance to the expected value of the nearest-neighbor distance. The average nearest-neighbor distance, $\bar{d}$, for a set of n points and utilizing the distance d_i, from a point i to its neighbor is

$$\bar{d} = \frac{\sum_{i=1}^{n} d_i}{n}.$$

For a random point pattern within a given area, A, the expected value of the nearest-neighbor distance, d', is

$$d' = \frac{1}{2}\sqrt{\frac{A}{n}}.$$

The nearest-neighbor index is

$$\mathrm{NNI} = \frac{\bar{d}}{d'}.$$

For a random spatial pattern, the nearest-neighbor index is 1. With increasing clustering effects, the nearest-neighbor index becomes smaller than 1; nearest-neighbor index values larger than 1 indicate a systematic pattern.

5.2.3.3 Spatial Autocorrelation

The concept of spatial autocorrelation combines the similarities in the location of spatial features and their attributes. A pattern is characterized by positive spatial autocorrelation when features that are located close together are also similar in their attributes. Where features which are close together tend to be more dissimilar in attributes than features which are further apart, a pattern is said to exhibit negative spatial autocorrelation. Zero autocorrelation is found in situations where attributes are independent of location. Figure 5.21 presents a situation with extremely negative spatial autocorrelation.

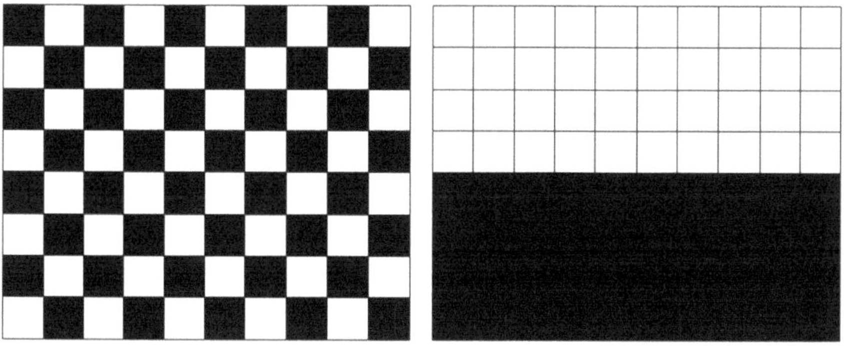

Fig. 5.21. Arrangement of cells with extreme negative (−1) spatial autocorrelation (*left*) and extreme positive (+1) spatial autocorrelation (*right*)

5.2.4 Network Analysis

In the GIS community a network is defined as a system of connected lines. Each line has a starting point and an end point and intermediate points (called vertices) that define the shape of a line. Vertices are defined by their location and simply delineate a line but do not carry information on topology. Nodes do carry information on both location and topology. Thus, starting and end points of lines as well as all points defining locations where two or more lines are connected are nodes.

In addition to nodes and vertices to each segment, an impedance factor and additional attributes may be assigned. The impedance factor is related to the size of a segment and can be its length or the time needed to travel from one end to the other. Attributes may be information on one-way roads, load capacities, or speed limits.

The structure of a network is characterized by its relative complexity and connectivity. A straightforward approach for analyzing network structures is the γ index, which is the ratio of the actual number of links in a network to the maximum possible number of links. The possible number of links in a planar network with n nodes is $3(n–2)$. The γ index is

$$\gamma = \frac{l}{l_{\max}} = \frac{l}{3(n-2)},$$

where l is the number of links in the network

The value of the γ index ranges from 0 to 1, where a value closer to 1 indicates a network with a complex structure with many links. Figure 5.22 presents two hypothetical networks with a simple and a complex structure. Both networks have eight nodes (n=8). While the simple network has only five links the complex network has 11 links, resulting in γ indices of 5/18=2.8 and 11/18=6.1, respectively.

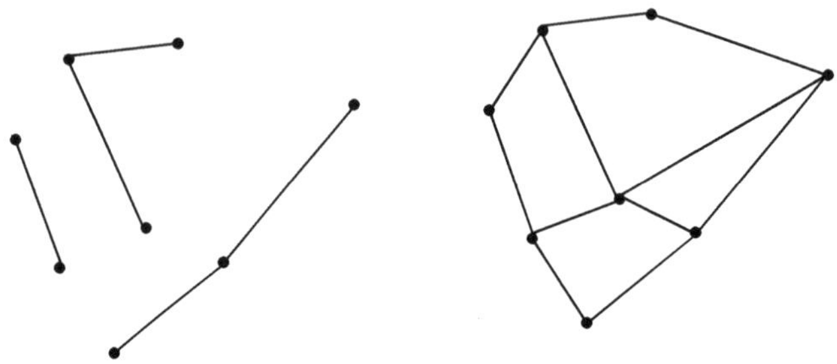

Fig. 5.22. Two networks with simple (*left*) and complex (*right*) structures

Other measures for describing networks are the network diameter, network connectivity, and network accessibility. Network diameter defines the maximum number of steps needed to move from any node to any other node through the shortest possible route. In the complex network just described, the network diameter would be 3. Network connectivity is based on the maximum number of links directly connected to any node (four in the complex network). The total number of unique ways through which one can move directly or indirectly from one node to another through the network is quantified by network accessibility.

Besides studying the complexity and structure of networks, network analysis is widely used to solve routing or transportation problems. The most common applications include:

1. Routing problems: building the shortest pass between two specified locations
2. Traveling salesman scenario: visiting a set of stops (locations) in the most efficient way
3. Shipment problem: minimizing total transportation costs within a transportation network
4. Location allocation: finding the location that minimizes total distances or traveling times to a number of points

5.2.5 Surface Analysis

Where a third dimension variable is available in addition to *x*- and *y*-coordinates the variation of the surface with respect the third dimension can be analyzed. The third dimension variable – often called the *z* variable – can be any quantifiable attribute, for instance, physical attributes such as elevation, aspect, slope, precipitation, or temperature or economic variables such as income, value of the standing crop, potential yield, or stand volume.

Surface analysis covers two major issues: (1) the organization of information about a surface, and (2) spatial interpolation. The latter allows the construction of a continuous surface from a set of discrete data points.

5.2.5.1 Organization of Surface Information

The organization of surface information is driven by the nature of the data source. In a vector-based GIS the *z* attribute characterizing a surface can be represented by:

- A set of irregularly spaced point features (Fig. 5.23a)
- A set of regularly spaced point features that form a grid (also called a lattice) (Fig. 5.23b)
- A set of digital contour lines (Fig. 5.23c)
- A set of triangulated networks (Fig. 5.23d)

5.2.5.2 Spatial Interpolation

Despite the fact that surface information is represented by discrete point information in digital data, surface analysis requires that data values can be estimated at any location. Spatial interpolation techniques provide the means to convert

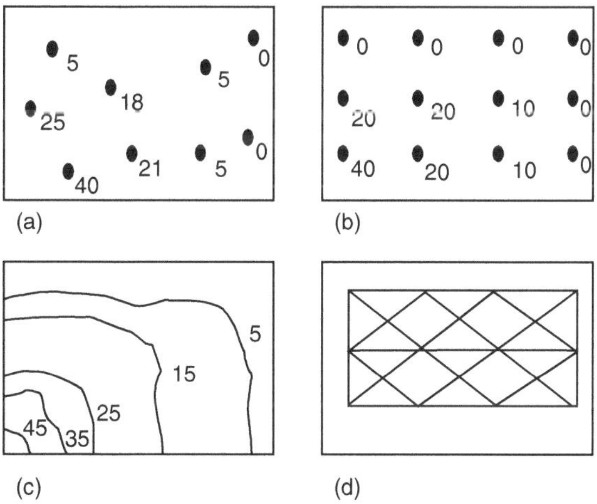

Fig. 5.23. Spatial organization of surface information

discrete point information into continuous surface information. An implicit assumption in spatial interpolation is that the *z* variable is spatially dependent. This assumption is used to estimated from *z* values of a given set of points *z* values for any point on a surface.

Several techniques exist to derive regularly spaced systems of *z* values. Local estimation derives the *z* value for a specific point on a surface based upon a limited number of known *z* values in its neighborhood. As an example, for local estimation the various-radius search method is presented in Fig. 5.24. A radius of given size is specified for the point for which a *z* value has to be estimated. If this radius does not contain the required number of known points, the radius is enlarged until the specified number of known points is included. The local estimate is then derived based on the selected points. Local estimation approaches reflect the surrounding information, but need decisions on the extent of the neighborhood and the number of known points to be used.

Global approximations use the entire set of known *z* values. The estimation procedure relates to the entire known surface points and does not need any constraints on the neighborhood to be included. An example for global estimation is kriging, which is described in Chap. 3.

Widely used applications of surface analysis are perspective diagrams which visualize the distribution of *z* values on a surface in a three-dimensional representation. Commonly used perspective diagrams display altitudes on a surface (Fig. 5.25).

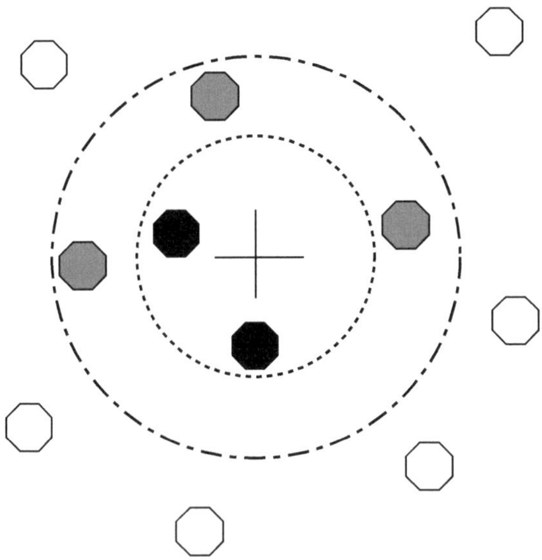

Fig. 5.24. Various-radius search method

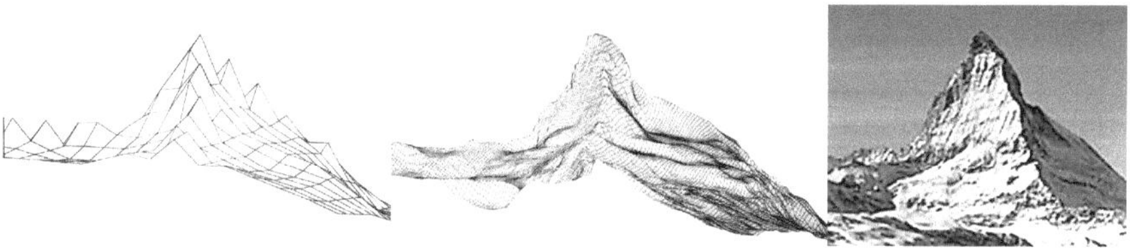

Fig. 5.25. Perspective diagram of a surface in two different spatial resolutions and the real situation (Matterhorn, Switzerland)

5.2.6 Grid Analysis

In contrast to surface analysis, which uses vector data, grid analysis is based on raster data. The regular arrangement of geographic units in raster data does not render any spatial interpolation necessary. Grids are common data formats for many data sources of spatial information, such as satellite imagery or scanned aerial photographs or maps. The identical size and shape as well as the spacing of geographic units make multilayer operations easy and efficient and allow for organizing different features in the same layer. However, grid formats have some disadvantages. Redundancy occurs where grid cells are smaller than the spatial variation. Grid cells which are too large to resolve spatial detail problems result in difficulties in assigning grid values.

Values can be assigned to cells by different methods (Fig. 5.26), among which are:

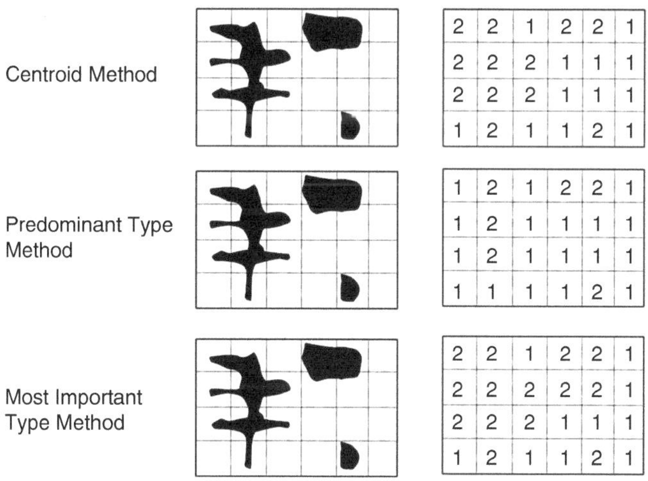

Fig. 5.26. Data value assignment

- The centroid method, which assigns a value to a cell according to its centroid location
- The predominant type methods, where grid values are assigned according to the type of surface value with the largest area in the cell
- The most important type method, where surface values are ranked according to their importance and grid values are assigned according to the most important type within a cell

Grid-based spatial analysis can be performed on local, focal, zonal, or global levels. The operations performed are based on mathematical functions or Boolean operators. Local functions are performed on individual cell values and are executed on a cell-by cell basis. Focal functions include the values of neighboring cells. The neighborhood to be considered is often specified by an $n \times n$ kernel, where n is typically an odd number so that the cell for which the value is to be computed is in the center of the kernel. In applying zonal functions, two different grids have to be available: (1) a grid that contains the cell numbers on which operations are to be performed, and (2) a grid that defines the shape, size, and location of zones. The entire grid is considered by global functions, for instance, when the distance of a cell to the nearest cell having the same value is searched for. Most grid analysis are some combination of local, focal, zonal, and global analyses.

5.2.7 Geostatistical Methods

Geostatistical methods utilize point information for spatial interpolation. They are a direct extension of methods developed for analyzing time series (Ripley 1981). The basic methods of geostatistics were developed by two mining experts. Matheron (1965), working in France, developed the theoretical formulation of spatial statistics, generally known as the theory of regionalized variables. Krige (1951, 1966), working in South Africa, developed geostatistical methods empirically and applied them to the location of gold deposits. Since then, several comprehensive studies have been published (e.g., David 1977; Journel and Huijbregts 1978; Clark 1979; Akin and Siemes 1988; Webster and Oliver 1990; Goovaerts 1999; Deutsch 2002).

A set of random variables, $Z(\mathbf{x}_i)$, determined from each point $\mathbf{x}_i$ on a plot or area, can be regarded as a realization of a sample from a random function $Z(\mathbf{x})$. The random variables are correlated according to the distance between any two given points and their orientation. One observation, $Z(\mathbf{x}_i)$ at point $\mathbf{x}_i$, can thus be regarded as the realization of a random variable, and a set of observations, $z(\mathbf{x}_1)$, $z(\mathbf{x}_2)$, $z(\mathbf{x}_3)$,..., as the realization of a random vector $\mathbf{Z}(\mathbf{x})$. Through this approach, locally dependent (regionalized) variables characterized by randomness and spatial dependence can be addressed.

The development of a formula for this relationship can be illustrated with a simple example of two sites of known distance from each other. The random variables $Z(\mathbf{x}_1)$ and $Z(\mathbf{x}_2)$ have value z_1 at the first location, and value z_2 at the second location. The relationship between the two values can be described by the difference $(z_1 - z_2)$ and by the variance:

$$s^2 = (z_1 - \bar{z})^2 + (z_2 - \bar{z})^2 = \frac{1}{2}(z_1 - z_2)^2,$$

where

$$\bar{z} = \frac{z_1 + z_2}{2}.$$

This equation can be extended for all possible pairs of locations by denoting the two locations as **x** and **x+h**, where **x** represents the position of one point, and **h**, termed lag, is a vector giving the distance and the orientation of the other point.

It follows that

$$s^2 = \frac{1}{2}[z(\mathrm{x}) - z(\mathrm{x} + \mathrm{h})]^2.$$

If the observation extends over m points, all separated by the vector **h**, it follows that we can estimate the mean $s^2(\mathbf{h})$ by

$$\bar{s}^2 = \frac{1}{2m}\sum_{i=1}^{m}\left[z(\mathrm{x}_i) - z(\mathrm{x}_i + \mathrm{h})\right]^2.$$

In sample surveys, the practical application of geostatistical methods presents a similar problem. The observation x must be measured, where x is a realization of X. Similarly, where a random function $Z(\mathbf{x})$ is considered, only the one observation of the random variable $z(\mathbf{x})$ is known. In order to draw statistical inferences from $z(\mathbf{x})$, assumptions regarding the stochastic model must be made.

The first homogeneity assumption is that the expectation of the random function is independent of the location **x** and is constant:

$$E[Z(\mathrm{x})] = \mu.$$

The second isotropy assumption concerns $\frac{1}{2}s^2(h)$, the expected squared difference between the values for the location separated by the lag **h**,

$$E\left\{\left[Z(\mathrm{x}) - Z(\mathrm{x} + \mathrm{h})\right]^2\right\} = 2\gamma(h)$$

This indicates that the variance of differences depends only on **h**. The semivariance $\gamma(\mathbf{h})$ is the expected value of the variance s^2 for the lag **h**.

In linear geostatistics, only the first two moments are considered, so it suffices to limit the definition of stationarity to these two moments. What is termed the stationarity of order 2 is given when:

- The statistical expectation $E[Z(\mathbf{x})]$ exists and depends on the support point
- The covariance for each pair of random variables $\{Z(x), Z(x + h)\}$ is present and dependent on the vector h.

Where the spatial properties of $Z(\mathbf{x})$ are invariant under translations, the random function displays strict stationarity. This indicates that the k-component vectorial random functions are identical for the two random functions $\{Z(\mathbf{x}_1),....,Z(\mathbf{x}_k)\}$ and $\{Z(x_1 + h),..., Z(\mathbf{x}_k + \mathbf{h})\}$ and are independent of the size of the vector h.

Thus, the presence of a covariance needed to formulate a hypothesis for stationarity of the second order demands a finite variance, var $\{Z(\mathbf{x})\} = \gamma(0)$. According to Journel and Huijbregts (1978), the two assumptions on which $E[Z(\mathbf{x})]$ and $E[Z(\mathbf{x}) - Z(\mathbf{x} + \mathbf{h})^2]$ are based express the intrinsic hypothesis, although Matheron (1965) formulated it rather more broadly. The consequences resulting from failure to meet this hypothesis are discussed in detail by Akin and Siemes (1988).

Where the intrinsic hypothesis is valid, the measurements from several pairs of samples, n_h, with the distance of h can be used to estimate $\hat{\gamma}$. The semivariance $\hat{\gamma}(h)$ can be computed for several lag vectors to produce a set of semivariances $\hat{\gamma}(1), \hat{\gamma}(2), \hat{\gamma}(h), ...$ The basic equation for $\hat{\gamma}(h)$ is

$$\hat{\gamma}(h) = \frac{1}{2n_h} \sum_{i=1}^{n_h} \left[z(i) - z(i+h)\right]^2.$$

This very general representation is only valid for one-dimensional problems, for example, soil samples along a given line. For this discussion we must restrict ourselves to the previous comments and refer those interested to the literature, as even one-dimensional problems illustrate the value of the variogram – the most valuable tool for describing a regional variable.

The semivariogram is the plot of estimates $\hat{\gamma}(h)$ against lag $\mathbf{h}$ (sometimes termed a variogram). Figure 5.27 shows one possible version of such a variogram, aimed at illustrating general characteristics. The variance increases with increasing lag until it reaches a maximum, at which point the curve flattens out. This maximum point is termed the sill. The lag at which the sill is reached is termed the range and it denotes and shows the limits of the spatial dependence.

Another aspect is reflected by the nugget effect. Although the variance for lags tending towards zero also tends towards zero, there is often a considerable residual variance. This residual variance, or nugget variance, may stem from different causes, for instance, measurement errors. Frequently, the structures concerned are small in comparison with the lag and can only be resolved through additional, less widely spaced samples. This must be taken into account when interpreting variograms.

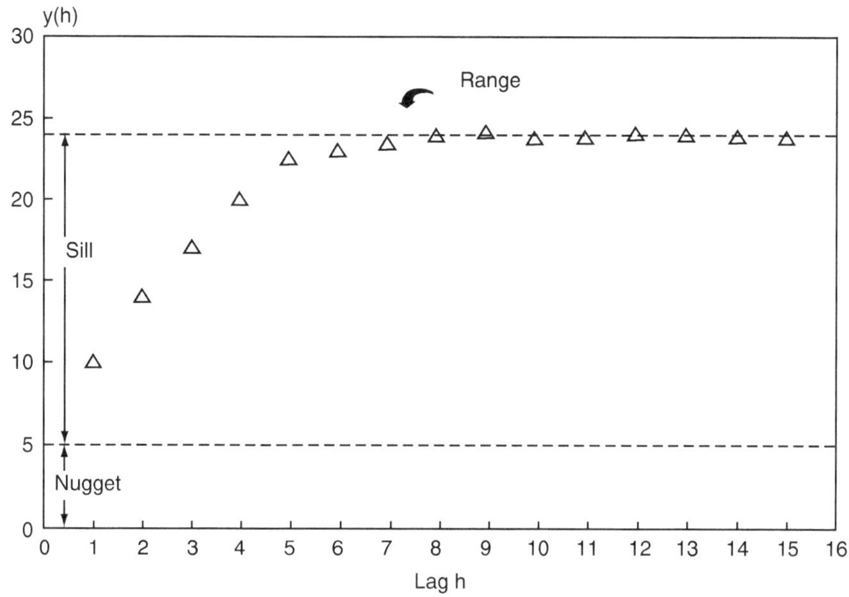

Fig. 5.27. Variogram

Naturally there are other variograms with different, more complex forms than that shown in Fig. 5.27; however, many of them, even those computed in this study, conform to that pattern.

Where variograms at one level are computed on the basis of the one-dimensional case, as discussed before, they can be constructed for various directions. Ideally, they are elaborated in four directions: the two directions parallel to the ordinate and abscissa and the two diagonals. If the variograms for the different directions match, it can be assumed that the variation is independent of direction. This circumstance is termed isotropy. Where variograms for the different directions exhibit the same sill but different ranges, the term geometric anisotropy is used. In a third case, termed zonal anisotropy, both the sill and the range differ.

A more advanced step in the application of geostatistical methods is to seek models to fit the sample values in a variogram. The elaboration of models for fitting and their parameters is essential for local estimation. Models are commonly fitted on the basis of least squares. There are a whole range of functions for fitting, for example, exponential, Gaussian, linear, and spherical models. These functions are iteratively fitted to the points in the variogram, in a procedure analogous to that of "fit-by-eye." The ideal fitting of the functions is still the subject of debate. Webster and Oliver (1990, pp. 239–240 ff.) weighed the various approaches and arguments against each other.

Estimates for point or small unit areas (blocks) at one level are possible through the use of linear kriging, which can be regarded as an estimation procedure, in which a given number of samples, *n*, within a neighborhood are taken into account through the method of weighted means. The estimate for a block, *B*, is computed as follows:

$$\hat{z}(B) = \sum_{i=1}^{n} \lambda_i z(\mathrm{x}_i).$$

Since $\sum \lambda_i = 1$ is valid, the estimate is unbiased. The variance of $\hat{z}(B)$, the krige variance, is given by

$$\hat{\sigma}^2(B) = \sum_{i=1}^{n} \lambda_i \bar{\gamma}(\mathrm{x}_i, B) + \psi - \bar{\gamma}(B, B).$$

Here, $\bar{\gamma}(B, B)$ is the within-block variance, ψ is the lag range, and $\bar{\gamma}(\mathrm{x}_i, B)$ is the mean semivariance between the block and the *i*th sample point. The resulting estimates are unbiased and exhibit minimum variance. The semivariances and λ_i are computed via the fitting functions of the variogram, which demonstrates the importance of the variograms for the krige procedure.

Though the theory of regionalized variables has been widely applied in mineralogy, prospecting, and soil science, it has so far been used comparatively little in forestry. Webster and Oliver (1985) employed geostatistical methods in soil mapping in the Wyre Forest, England. They showed that the greatest variation occurred between points 66 m apart. The variables correlated through this method corresponded with the occurrence of different soil types.

Payn and Clough (1988) presented an approach based on geostatistical methods for the mapping of plantations subjected to different fertilizer treatments. The spatial variability of heavy metals with a stand covering 1 ha was described by computing variograms according to Wopereis et al. (1988).

Palmer (1988) employed a combination of ordination and geostatistical methods to describe the spatial pattern of plant associations. Using the results as a basis, he deduced various square plot sizes and distances between plots for phytosociological studies.

Ramirez-Maldonado (1988) discussed the theory and method of geostatistics in the context of forestry. Sample plots which are close to each other are often correlated; it is usually assumed that observations of the sample plots are independent in space. He then fitted the variogram of basal area measurement, constructed by means of point sampling, to a spherical model, and found that small gauge constants are to be recommended as being more efficient than large ones. He also used geostatistical methods for the analysis of the ten-point

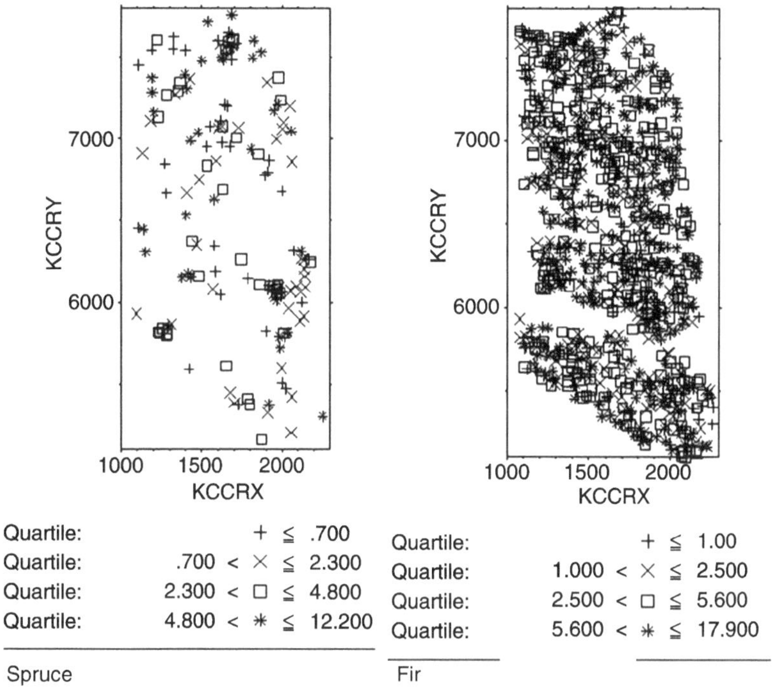

Fig. 5.28. Raw data (shown are tree locations and diameter at breast height growth between 1976 and 1987)

cluster employed by the US Forest Service and for two-phase sampling. He presented a new theoretical approach to classic point sampling in which point sampling is regarded as a multinomial process.

Using growth in height of *Melia azederach* in Karnal, India, as an example, Samra et al. (1989) investigated the potentials of spatial interpolation by means of krige procedures for the mapping of forest inventory results. Within the study data, tree height varied in relation to both the lag **h** and the direction within a given stand (anisotropy). Over 70% of the heterogeneity of tree height in the northwest and southeast directions could be explained through age.

Working in a forest near Zurich, Mandallaz (1991) compared various estimation procedures based on geostatistical methods and reported on their suitability for stand and enterprise inventories. A new approach termed double-kriging permits the use of auxiliary information such as aerial photographs or maps (Mandallaz 1994).

Jost (1993) applied geostatistical methods for the estimation of standing volume. For her study she used 28 data sets from Germany and North America ranging from 5 to 5,000 ha in size. In 60% of the 28 data sets she found autocorrelative structures. The spherical and exponential models were sufficient to model semivariograms. The lags ranged between 100 and 150 m for stand data and between 500 and 700 m for larger forested areas. Anisotropy was found only in rare cases. Jost applied geostatistical methods in order to calculate the sampling error in systematic sampling.

The potential of geostatistical methods for the analysis of forest decline surveys was investigated Köhl and Gertner (1997). They analyzed a data set from the Swiss National Forest Health Monitoring Program and could show distinct spatial patterns of changes of forest condition. Köhl et al. (2000) applied geostatistical methods for the investigation of the spatial pattern of tree growth. For a mixed-species, uneven-aged forest stand in the Bernese Pre-Alps in Switzerland (Fig. 5.28) the pronounced spatial pattern of growth rates of individual trees could be shown (Fig. 5.29).

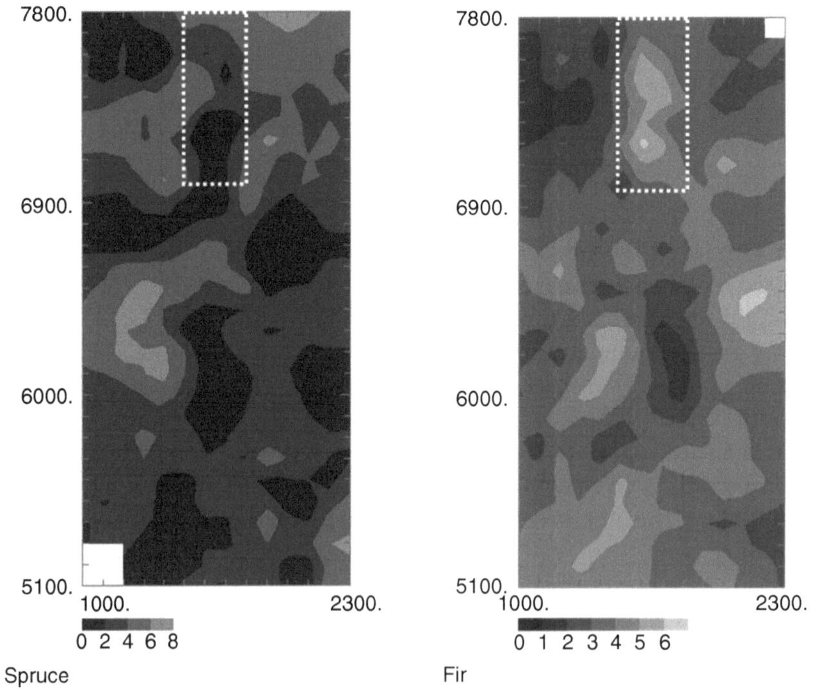

Fig. 5.29. Results of kriging for diameter at breast height growth between 1976 and 1987

5.3 Forest Information Systems

The sustainable management of forest ecosystems requires decisions that have to incorporate a diversity of specialized knowledge and are subject to long-term impacts. Forest information systems (FIS) are a modern tool that provides decision makers with comprehensive information and results in better management decisions. The increasing amount of available data, the need for spatial reasoning, the desire to share distributed information, and the impressive potential of modern computer technology need to be made available for the thorough management of forests. FIS offer the potential for informed decisions and support the maintenance and enhancement of the multiple forest functions. However, problems on the semantic level are much harder to be solved than the technical problems.

FIS should provide a comprehensive set of information to the decision maker, support the implementation of more timely decisions, and improve the quality of decisions (Dykstra 1997). As many decisions incorporate a spatial domain, GIS have become a widely used component of FIS to facilitate handling of spatial data (McCloy 1995). Because a FIS is a management tool rather than a playground for computer enthusiasts, it is essential that the system interface to the user is easy to learn, use, and understand. The current trend in applying computer graphical interfaces improves the convenience by simultaneously requiring minimum training. A FIS is often based on sophisticated technology and is implemented by experts with sophisticated technological skills. However, a user has generally minor interest in technical solutions, but demands a system that is efficient, easy to use, flexible, and reliable.

A FIS will comprise at least five components:

1. Input
2. Analysis
3. Estimation and prediction
4. Decision support
5. Presentation and visualization

FIS structured in those components deal with technologies such as computer sciences, remote sensing, GIS, image processing, modeling and simulation, or internet and communication protocols. Depending on the technical realization of the system, the data utilized are either stored in a central data base or are retrieved from distributed databases. Merging data from various sources and sectors is not straightforward as different spatial, temporal, and thematic resolution as well as data formats render harmonization necessary. Metadata standards such as the Dublin Core Metadata Initiative (DCMI)[1] offer a possibility to join different data

[1] http://dublincore.org

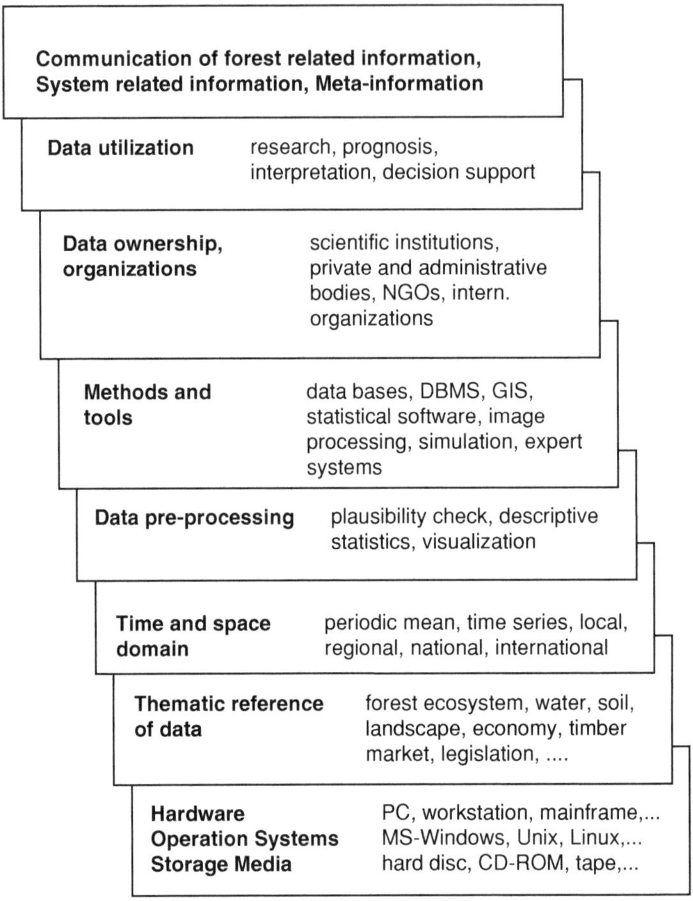

Fig. 5.30. Dimensions of forest information systems

sources. Where data sources are to be assessed via the Internet, tools for information retrieval and resource allocation are required. Figure 5.30 presents the dimensions of a FIS.

Decision support provides the interface with the decision maker. Here the information retrieval capacities of the system and the information needs of the user must be brought together. The method of information use affects the form in which the information is presented, for example, as statistics, digital data, graphs, or maps. It is advisable to supplement any output of the system by information on its reliability.

FIS recognize three types of information needs (FAO/ECE/ILO 1992):

1. Information for strategic planning and analysis
2. Information for tactical (medium and short-term) planning
3. Information for operational management and control

Information management is associated with cost. Data on phenomena have to be assessed, the data have to be analyzed in order to extract information, the information is used for decisions, and the decisions have to be translated into actions. Costs originate from assessments, analyses, and decisions. A utility function contrasts costs and benefit (Fig. 5.31).

Implementation of a FIS does not lead to a benefit per se. In order to optimize the utility function it is advisable to customize FIS to the specific purposes and conditions. However, considerable savings can be realized by modular designs that assemble tried and tested components and reduce the development of new components to a minimum (Günther and Humbolt 1998).

Information systems can be assigned to one of the following categories. The classification criterion is the achieved analysis level of information.

1. *Monitoring and control systems.* These systems can be assigned to the classic task fields of "measuring, controlling and regulating." They are used either for environmental monitoring (pollution of air, water, soil, or vegetation as well as noise and radiation) or the control of technical processes in the scope of computer-assisted process control.

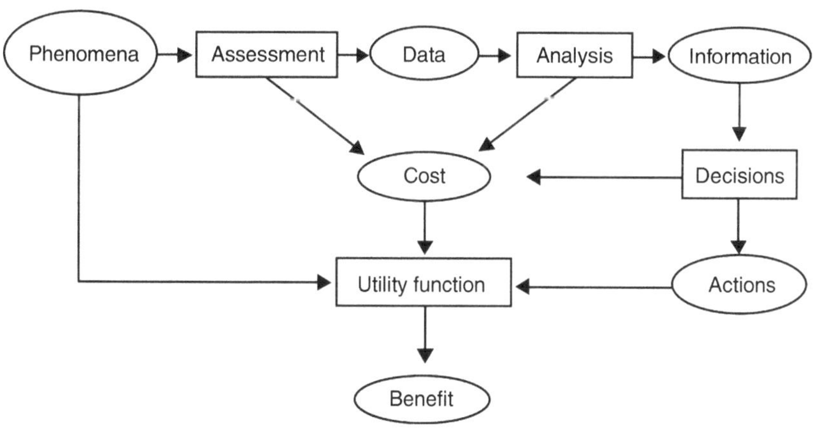

Fig. 5.31. Cost and benefits of forest information systems

2. *Conventional information systems.* This class groups systems for the storage, analysis, organization, integration, and presentation of different data sources and data formats, such as
- Measurements from monitoring systems
- Formatted data such as areas, timber volumes, growth figures, mortality, or potential cut
- Unformatted documents such as laws and regulations, or relevant literature

Often specific requirements in relation to time and space are to be considered. Concerning the contents little more than a targeted analysis and compilation of data can be achieved.

3. *Evaluation and analysis systems.* Evaluation and analysis systems allow for the processing of data by means of complex mathematical and statistical methods and models. Associated with this area are dispersal and prognosis models, image analysis methods, and simulations. The results of the analyses are information about the potential impact of alternative management plans. Examples of evaluation and analysis systems are the prognosis of potential timber supply under different silvicultural treatments or the prognosis of changes of habitat suitability by introducing new landscape management regimes.
4. *Decision support systems.* Decision support systems offer a decision maker direct support during the decision process by providing assistance for the evaluation of alternatives or for the substantiation of decisions. In contrast to evaluation and analysis systems, decision support systems include explicit valuation and rating methods. In addition, inference methods can be realized; here especially knowledge-based systems originating from the domain of artificial intelligence are demanded.
5. *Integrated information systems.* Currently information systems are not only implemented in the forestry sector, but are becoming a widely used tool in all environmental fields of activity. This includes both different thematic alignments as well as different hierarchical levels (e.g., enterprises, communes, federal states, or international organizations). Many of the systems implemented cannot be assigned to one of the specific categories already mentioned, as they combine different components. Such systems are called integrated information systems. It is to be expected that those systems will gain importance in the future and will be implemented as distributed systems. The integration of different information technology concepts, data formats, and system classes (e.g., simulation systems, knowledge-based systems) is a particular challenge for applied computer sciences.

5.4 Methodical Components of Information Systems

When developing and implementing information systems, the entire spectrum of computer sciences and computer technology goes into action; however, some concepts and methods are more relevant than others. Table 5.2 presents the relevance of concepts and methods for the different categories of information systems.

1. *Data base systems and geographic information systems.* Data bases are undoubtedly the most important tools for information systems. Owing to the spatial context many FIS can be considered to be extended GIS, which contain in addition to georeferenced data and relevant methods, additional

Table 5.2. Methodical components of information system categories. The table combines the methodical components with the different categories of information systems. Data-base and geographic information systems, computer graphics and visualization, user interfaces and software ergonomics, as well as integration are relevant for all information system categories. (After Page and Hilty 1995)

	Monitoring and control systems	Conventional information systems	Evaluation and analysis systems	Decision support systems	Integrated information systems
Data base systems and geographic information systems	✓	✓✓	✓✓	✓	✓✓
Modeling and simulation	✓	×	✓✓	✓✓	✓
Knowledge-based systems	✓✓	×	✓	✓✓	✓
Computer graphics and visualization	✓✓	✓	✓✓	✓	✓
User interfaces and software ergonomics	✓✓	✓✓	✓	✓✓	✓✓
Artificial neural network and fuzzy logic	✓	×	✓✓	×	✓
Integration	✓✓	✓✓	✓	✓	✓✓

✓✓ of particular relevance, ✓ relevant, × not relevant

thematic and nonformatted data with a temporal relation (Bettinger and Wing 2003). Specific problems arise when the complex structured objects of the forest and environmental sector need to be implemented into the data models of conventional data-base systems. Another challenge is the integration of different heterogeneous data sources in distributed systems, which are a prerequisite for genuine exhaustive information systems for the strategic managerial level. Orientation and user guidance within extensive information systems with numerous heterogeneous data bases requires the creation of metadata bases.

2. *Knowledge-based systems.* Increasingly attempts are undertaken to utilize knowledge-based systems, especially expert systems, for information management (Schmoldt and Rauscher 1996). The increasing number of projects in the environmental context cannot belie that the proportion of systems applied in everyday business is relatively small. Within the environmental sector most of the expert system developments are in the category of diagnosis and interpretation. The knowledge-based support of the utilization of data bases, the processing of information request, and monitoring tasks as well as training and education are promising fields of application.
3. *Modeling and simulation.* Modeling and simulation methods have a long tradition in the forestry sector; their practical applicability is proven. Modeling and simulation are essential for the analysis of complex, dynamic systems, as they are typical for the forestry and environmental sector. Modeling and simulation components will increase the number of problems that can be treated by information systems. While simulation and modeling techniques were up to now to a large degree developed as standalone systems, the demand for their integration into information systems will increase in the near future.
4. *User interfaces and software ergonomics.* A prerequisite for the wide application of FIS is the assignment of modern, mainly graphical concepts for the design of user interfaces. From experience especially within the group of "occasional users" the supply level can be multiplied by offering user-friendly interfaces. No task specific language should be used, but general (natural) languages should be sufficient for queries.
5. *Computer graphics and visualization.* The application of computer graphics as an instrument for sophisticated visualization is of vital importance in any decision process. As the human eye can catch complex relationships best, data visualization is an essential component of explorative data analyses. Many tasks in forest ecosystem management indicate the comparison and combination of numerous parameters referring to different geometric and geographic objects or different time scales. Computer graphics and

visualization facilitate data evaluation and decision processes by presenting complex problems in an intuitively clear way.

6. *Artificial neural networks and fuzzy logic.* In the context of ecology and natural resources management, uncertain information and variables that do not permit simple yes/no categorizations are rather common and allow us to use more than just true or false. Artificial neural networks and fuzzy logic can be used to make decisions where uncertainty occurs. Neural networks are systems inspired by biological nervous systems that process information by passing data between many simple processing elements. Computing systems mimic the brain through a network of highly interconnected, processing elements, which give them learning capabilities and enable them to recognize, and to understand, subtle or complex patterns.[2] Fuzzy logic provides an approach to approximate reasoning in which the rules of inference are approximate rather than exact. Fuzzy logic is useful in manipulating information that is incomplete, imprecise, or unreliable. Also called fuzzy set theory, fuzzy logic extends the simple Boolean operators, can express implication, and is used extensively in artificial intelligence programs. Fuzzy logic allows computers to work more easily with phrases such as "fairly," "rarely," or "almost."[3] In the context of information systems, neural networks and fuzzy logic are important as they support structuring and analyses of large data sets that are subject to uncertainty.
7. *Integration.* Integration is an essential task in environmental reasoning. Besides system-relevant problems of integration (such as the integration of data bases, or simulation systems) the problems of integrating different players and fields of expertise are to be solved. The additional integration of isolated solutions as well as the integration of methods and data require innovative approaches.
8. *Other relevant methods.* Among the other relevant methods used for the design and implementation of information systems are remote sensing and image analysis, which provide georeferenced data and facilitate environmental monitoring and planning. The availability of distributed data bases, data warehouses, and method bases as well as their accessibility via the Internet makes computer networking a substantial component and challenge for information systems.

[2] http://www.japaninc.net;; http://www-agecon.ag.ohio-state.edu/class/AEDE601/glossary/glossa.htm
[3] http://stott.feis.herts. ac.uk/support/technical_glossary.htm

Multiresource Forest Inventory 6

6.1 Introduction

All over the world, the idea of sustainable, close-to-nature, and multifunctional forestry has progressively replaced the unbalanced perception of forests as a source for timber (Kohm and Franklin 1997; Corona and Zeide 1999; von Gadow 2000). The sustainable management of the multiple functions of forest resources requires a substantial amount and sensitivity of information for decision-making processes. Information needs originating from the consideration of ecological, environmental, or socioeconomic aspects are hardly met by adding some "new" attributes to existing lists of attributes of traditional and established forest inventory approaches (Avery and Burkhart 2001). Forests are dynamically connected to their surrounding areas, and the spatial and structural composition of border zones as well as the interconnection of forests to other land-cover classes are driving factors for ecological processes at the landscape level (Forman and Godron 1986). Information needs originating from the consideration of ecological, environmental, or socioeconomic aspects render sampling designs necessary that widen the scope from timber production to the diverse functions and services provided by forests.

This chapter describes different facets of multi-resource forest inventories. As forest are complex (eco-)systems that show substantial variation with respect to climate, management regime, stand history, disturbance, and needs of local communities, it is clear that the need to quantify spatially explicit functions of the services and needs of the forest will also vary considerably (Bachmann et al. 1998, MCPFE 2003. The full range of potential forests functions is so vast that we can afford to give only a few examples here. They are restricted to the assessment of forest utilization, nonwood goods and services (NWGS), landscape analysis, and forest fires.

6.2 Forest Production

6.2.1 From Tree Volume to Utilized Timber Volume

Sustainability in tropical forests, when pursued, mainly applies to timber production rather than to ecosystem conservation and to the range of goods and services it offers (ITTO 2004). The *sustainable timber management* is usually based on simple quantitative parameters such as (Marchetti et al. 2005) an *annual allowable cut* for each *forest management unit:* it is very important to remember that commercial tropical species grow, on average, around 1 m^3/ha/year, with a possible strong decline in the second and subsequent rotations. At present, logging impacts are mittigated (*reduced impact logging vs. conventional logging)* by focusing on spatial patterns (Asner et al. 2004) and prescribing a *minimum diameter cutting limit* (30–45 cm) and the maximum number of exploitable trees (2–14/ha depending on the species).

Volume estimates from published forest inventories refer almost exclusively to the gross volume of standing (live) trees. Only a part of this volume is extracted during a harvest operation, and is called a drain. The volume

Fig. 6.1. Logging losses in a pine stand (*Pinus radiata*) in New Zealand

fractions remaining in the forest are called logging losses (Fig. 6.1). The fraction of the gross volume actually recovered from a certain site depends on many factors. For instance, harvest technique, access to the harvest site, market conditions, as well as quality, size, and species requirements set by consumers and buyers. Consequently, information on the usable or net volume recovered in the process of harvest and transport is needed for both consumer information and economic and planning purposes.

The net volume is often computed through rough deductions from the gross volume. Deductions are often subjective and determined, in part, by different interests. To gain objective deductions it is useful to conduct a special recovery study, otherwise known as a harvesting study or forest utilization study. Here, representative net volumes are computed from samples with known total (gross) volume. Such studies are particularly important in regions and for forest estates where the volume to be cut/harvested by logging contractors is to be contractually fixed in terms of the net volume of timber to be recovered from logging sites. Results from harvest studies allow forest managers a priori calculation of extractable volume and possibly the number of stems to be harvested. Those figures are essential for determining a sustainable level of logging in both an economic and an ecological context.

Recovery, or harvesting, studies are carried out within the framework of specially planned harvesting operations or, where this is not feasible, in logging operations for which the plans have already been finalized. Their objective is to obtain objective volume reduction factors for particular trees, tree sizes, species, or tree species groups. There is little point in computing a single global reduction factor since specific recovery factors vary greatly both in magnitude and in impact. Impact depends on the harvest volume applicable to a specific reduction factor.

Because of the continually changing patterns of demand and supply in the timber market, and the constant introduction of new technologies, the results of a recovery study may only be valid for a limited period of time and for limited areas.

Recovery studies are oriented towards typical harvesting techniques. They involve three common steps:

1. Determination of the number of trees not to be felled within the area of harvest. Reasons for not felling a particular tree are besides poor growth form, poor quality, or obvious defects, the desire to leave some mature trees for regeneration, maintenance of gene pools, aesthetics, or wildlife.
2. Determination of the proportion of trees felled but not utilized. Reasons for nonusage may be splitting, breaking, internal defects not detected before felling, or poor logging practices.
3. Determination of the actual usable volume of the felled trees, classified according to quality grades and species.

Fig. 6.2. Overexploited forest area in Zimbabwe

On the basis of these three steps, the reduction factors are calculated. Recovery studies are essential for planning and checking harvesting operations and for identifying wasteful and ecologically unsustainable practices, including illegal logging and overexploitation (Fig. 6.2). Various aspects of recovery studies are discussed by Hutchinson (1985) and Torelli (1987).

Especially when harvesting is done by concessionaires it is advisable to monitor the logging operations in order to assure ecological compliance. Bhandari and Hussin (2003) have explored the potential of Landsat 7 Enhanced Thematic Mapper Plus (ETM+) imagery for the detection and quantification of forest logging activities in a concession area in East Borneo. Multitemporal imagery and field data, which were collected from recently logged sites and sites representing the conditions that prevailed before logging, were combined and enabled reliable monitoring of the logging activities. The development of impact indicators and guidance to improve implementation of ecocertification principles and indicators will increase with the awareness of sustainable forest management: *national forest programs* and *community forestry* are continuously improving access to certification for small and low-intensity managed forests.

6.2.2 Access Studies

Accessibility to a forest areas is an important aspect to be considered for all utilitarian aspects of forest production and functions. Human access to forests has become important for the debate on human-induced disturbances or irreversible destruction of forest ecosystems. Issues of access to a forest area can be a major factor for the cost of field assessments carried out during the normal process of a forest inventory. Hence, accessibility has a major implication on the inventory design.

From the viewpoint of economics and management of the forest enterprise and its resources, the interpretation of inventory data must be done in the context of accessibility of the various resources. Consequently, accessibility studies should be conducted for all surveys, especially in areas where the access to forest resources is a limiting factor. These studies are already a standard part of preliminary investigations for national inventories in regions with a developed infrastructure of transport corridors (Zinggeler et al. 1999).

Access to a resource can be gauged from maps and aerial photographs, but a final assessment should be made on the basis of a field survey. For this reason, accessibility studies are best conducted within the framework of a forest inventory, not as an independent survey. A combined survey of access and resources allows the analyst to estimate the available resources by access classes that indicate the ease, constraints, or simply the cost of access.

In order to estimate accessibility, two types of data must be compiled:

1. Density of the existing transportation networks by type (roads, trails, railways, water channels)
2. Quality of transportation networks

The density of a forest road network is the total length of the network divided by the area under consideration. Matérn (1964) showed that the overall length of a transportation network can be determined efficiently from a line survey. Further details are given by von Segebaden (1964) in his description of the application of line surveys in the Swedish NFI. His concept is based on the surveying of parallel, equidistant lines, with distance b between lines, oriented in a randomly chosen direction. The total length of a network (L) can be computed from the number of points (n) at which these lines intersect the network. The estimator is

$$L=(\pi/2)bn.$$

Where the transportation network runs parallel to the survey lines, a considerable bias in the estimate of L is to be expected. However, the procedure can make use of two perpendicular line patterns to balance out the bias. The line pattern(s) can be traced on maps or aerial photographs or applied in field surveys, for example, on the lines connecting the plots in a tract. Detailed discussions are given by Matérn (1964), von Segebaden (1964), Stierlin (1979), and Stierlin and Zinggeler (2001), while an analysis of a line survey may be found in EAFV (1988). Jung et al. (2001) used the line-intersection method for forest road route selection.

Where traffic routes are available in a georeferenced format, geographic information systems (GIS) can be used for studying accessibility. Network diameter (the distance between two nodes in a network) and network connectivity (the number of nodes between any two nodes of the network) are essential aspects in the spatial analysis of road network structures (Chou 1997; Haining et al. 2003).

Table 6.1. Terrain classification in accessibility studies

Terrain parameters
Terrain pattern (e.g., rock outcrops, drainage pattern)
Width and depth of rivers
Ground roughness (occurrence of obstacles, friction)
Slopes (gradient, length of slope, aspect)
Area inundated by floods (duration, periodicity, frequency)
Soil parameters
Bearing capacity
Potential of erosion
Suitability for road construction
Climatic parameters
Temperature
Moisture
Rainfall (maximum, average, periodicity)
Parameters relevant for felling operations
Undergrowth, vegetation
Branchiness of trees
Slope
Climate
Microtopography

Other decisive factors for the accessibility of forests include the terrain, the climate, the vegetation, and the soil type. The International Union of Forest Research Organizations (IUFRO) Working Party on Terrain Classification (von Segebaden et al. 1967) developed a terrain classification. Table 6.1 shows some of the pertinent points to be considered in estimating accessibility.

6.3 Nonwood Goods and Services

6.3.1 Historical Perspective

Temperate and boreal forests have a long history of providing a large suite of goods and services to local, regional, and larger communities. The mix of goods and services has seen dramatic changes over time in response to demands and needs of dependent societies. Wood extraction for cooking, housing, ships, tools, and fencing went along with extraction of numerous other products, notably resin, tannin, fodder, litter, medicinal plants, fruits, nuts, roots, mushrooms, seeds, honey, ornamentals, and exudates. The use of the forest was generally only regulated by virtue of ownership privileges or traditional rights. The ability of the forest to sustain the production of goods and services was rarely taken explicitly into account. Locally, demands could outstrip what could be produced sustainably, often with long-lasting negative impact on the forest and local communities. A well-known example was the overuse of many central European forests for grazing and extraction of litter. Forest soils became deficient in essential nutrients, which triggered a long-lasting cycle of progressive degradation of the forest community.

Over time, as the industrialized society took hold, the utilization of nontimber products became marginalized as many forests became part of a wood-producing business enterprise, a trend that simply mirrors the fact that the economic benefits of wood production by far exceeded the economic potential of NWGS. The shift happened through a complex interaction of several different processes happening simultaneously: an increasing disconnect and less dependence of people on their local surroundings, an increasing disregard of subsistence use of forests, the decline and even disappearance of small-scale rural industries, technological substitution of wood products, intensification of agricultural production, and a general increase in prosperity.

Today we seem to have shifted towards a reevaluation of the opportunities for forests to produce goods and services beyond timber. Conservationists, foresters, forest owners, social development advisors, or indigenous people are showing increasing interest in nontimber goods. From a national economy point of view, greater importance is attached to the multitude of services provided by forests, for example, their role in the global carbon cycle, water supply, protection, recreation, spiritual values, and well-being. It is widely accepted that in distinct areas the value attributed to nontimber functions of forests exceeds the value of timber production (Köhl et al. 1995). The sustainable use of nontimber forest resources is a well-accepted basic premise.

This paradigm shift has been accompanied by a proliferation of studies on nontimber functions of forests. Many of these studies were triggered by local circumstances and the peculiarities of the resource under study. Studies focusing on a holistic concept for the assessment of nontimber forest functions for larger areas are, however, rare. In the following, definitions and classification systems of NWGS as well as approaches for the assessment of NWGS are presented.

6.3.2 Definition of Nonwood Goods and Services

The socioeconomic contribution of forests to livelihood and the impact of their use on the environment are essential components of modern concepts for sustainable forest management (MCPFE 2003). A great deal of new information on nontimber forest resources has been compiled. However, there is no unique definition of nontimber resources. In contrast, a proliferation of new terms is being introduced such as nonwood goods and benefits, NWGS, other forest products, secondary forest products, special forest products, wild crafted products, biodiversity products, natural products, minor forest products, nontimber forest products, nonwood forest products, forest products other than wood/timber, and tree crops (Vantomme 2003). One of the most versatile yet precise definitions was given by De Beer and McDermott (1989):

"The term 'Non-Timber Forest Products' (NTFPs) encompasses all biological materials other than timber, which are extracted from forests for human use."

According to FAO (2004) nonwood forest product data collections fall into one of 16 main categories. (Table 6.2).

In *nonwood goods and services*, the word *services* is often interpreted in a broad sense to include aspects of the environment (e.g., biodiversity, provision of habitats), socioeconomics (e.g., income), recreation (e.g., scenic beauty), protection (e.g., erosion, water supply, avalanches, rockslides) and heritage values, even though services in the strict sense are quantifiable products or services actually produced and used (e.g., managed grazing). Chandrasekharan (1992) presented the following definitions:

Timber Wood in forms suitable for heavy construction; sawn wood exceeding a specified width and thickness; not included is fuel wood, wood for carving, pulp wood, small-dimensional wood.

Goods Things, articles, objects worth attaining; movable properties; merchandise; wares; services of value. Economic goods are defined as any physical object, natural or manmade, or service rendered, which could command a price in a market.

Services Provision of assistance; act of serving; work done to meet expressed or assumed needs; intangible, nontransferable economic goods distinct from physical commodities.

Products Things/substances/articles produced by a process; output of goods and services resulting from the input of resources or factors of production used to produce them.

Benefits Advantages; favorable effects; output; profit. In forestry they include products and nonmonetary favorable influences attributed to forests

Table 6.2. Nonwood forest product data categories (from FAO 2001)[a]

Plant products		Animals and animal products	
Category	Description	Category	Description
Food	Vegetable foodstuffs and beverages provided by fruits, nuts, seeds, roots, mushrooms, etc.	Living animals	Mainly vertebrates such as mammals, birds, and reptiles kept/bought as pets
Fodder	Animal and bee fodder provided by leaves, fruits, etc.	Honey, beeswax	Products provided by bees
Medicines	Medicinal plants (e.g., leaves, bark, roots) used in traditional medicine and/or for pharmaceutical companies	Bushmeat	Meat provided by vertebrates, mainly mammals
Perfumes and cosmetics	Aromatic plants providing essential (volatile) oils and other products used for cosmetic purposes	Other edible animal products	Mainly edible invertebrates such as insects (e.g., caterpillars) and other «secondary» products of animals (e.g., eggs, nests)
Dying and tanning	Plant material (bark and leaves) providing tannins and other plant parts (especially leaves and fruits) used as colorants	Hides, skins for trophies	Hide and skin of animals used for various purposes
Utensils, handicrafts and construction materials	Heterogeneous group of products including thatch, bamboo, rattan, wrapping leaves, and and fibers	Medicine	Entire animals or parts of animals such as various organs used for medicinal purposes
Ornamentals	Entire plants (e.g., orchids) and parts of the plants (e.g., pots made from roots) used for ornamental purposes	Colorants	Entire animals or parts of animals such as various organs used as colorants
Exudates	Substances such as gums (water soluble), resins (water insoluble) and latex (milky or clear juice), released from plants by exudation	Other nonedible animal products	Bones used as tools

[a]Source: FAO; http://www.fao.org/documents/show_cdr.asp?url_file=/DOCREP/004/41997E/y1997cof.htm

The global *Forest resources assessment 2000* (FAO 2001) provides a summary of the most important nonwood forest products utilized in Africa, Asia, South America, Central America, the Caribbean, Europe, and North America.

Major services provided by forests are (Gottle and Sène, 1997; MCPFE 2003):

- Protection of water resources, soil, and other ecosystem functions
- Protection of infrastructure and other managed natural resources
- Influence on the local climate and reduction of the impact of gas emission.
- Conservation of the natural habitat and biological diversity
- Recreational and other social functions of forests
- Protection of cultural and spiritual values

In the international context, the term "nontimber forest products" is widely used, for example, by the Food and Agriculture Organization (FAO)[1] and the Center for International Forestry Research (CIFOR) (Cheng Tan et al. 1996).

6.3.3 Classification Systems for Nonwood Goods and Services

Various classification systems have been provided for assessment of NWGS. The approaches include internal trade reporting, biodiversity inventories, ethnobotanic studies, as well as assessments in the scope of forestry, wildlife ecology, and land resource management. In the following, some of the more common classification systems are presented. They demonstrate both the diversity and inevitable shortcomings of any system that tries to classify NWGS; shortcomings that reflect the fact that each system serves a different purpose.

6.3.3.1 National Nonwood Forest Product Accounting

Chandrasekharan (1995) presented a typology for national nonwood forest products accounting that uses four classes: live plants and part of plants (A), animals and animal products (B), prepared and manufactured products (C), and services (D). Table 6.3 summarizes the subcategories under the four main classes. Chandrasekharan's typology is based on the product type and requires reporting in different measurement units, including weight, volume, and market value. The approach focuses on quantities that are marketed, but could easily be expanded to include potential harvests or changes of harvesting levels over time. The approach is aimed at national-level statistics with no provision to monitor local sustainability.

[1] http://www.fao.org/docrep/004/y1457e/y1457e09.htm

Table 6.3. National nonwood forest product accounting

A. Live plants and parts of plants	Live plants
	Parts of plants (fresh, cut, dried. or crushed), collected for specific uses
	Specific parts of plants with multiple uses, not included under the previous group
	Vegetable materials not elsewhere classified
	Raw exudates and similar natural products
B. Animal and animal products	Live animals
	Animal products
C. Prepared/manufactured products	Prepared (provisionally preserved) edible products
	Prepared beverages
	Prepared animal feed/fodder
	Vegetable oils/fats
	Animal fats/oils
	Prepared waxes of animal or vegetable origin
	Dying and coloring extracts of plant or animal origin
	Phytopharmaceutical/medicinal extracts, galenicals, medicaments
	Essential oils and their concentrates
	Resin and resin derivatives
	Processed gums and latex
	Fuels and alcohols
	Other basic organic/phytochemicals
	Prepared bark products
	Plaited products
	Products of natural fiber
	Tanned leather, fur, and products of taxidermy
	Miscellaneous products, manufactured from nonwood forest raw materials
	Other nonwood plant and animal products
D. Services	Forest-based services

6.3.3.2
End Use and Plant Use Classifications

Wyatt (1991) presented a classification system that has been applied in Ghana. It has a focus on the end use of extracted material but does not include services. Categories defined by Wyatt (1991) are as follows: sponges, chewing sticks, tooth cleaners including fibers, jute, cloth, wool, cloth, pestles, aphrodisiacs, and basketry, foodstuffs, including wild fruit, edible leaves, sweeteners, intoxicants, beverages, water, medicinal plants, latex, rubbers, gums and resins, decorative beads, and seeds. Ethno-botanists such as Prance et al. (1987), Boom (1989), Edwards (1991), Salick et al. (1995), and Valkenburg (1997) have designed plant-use classification systems mainly for application in tropical forests. They classify NWGS according to local end uses and traditions. Ethno-botanic classification systems include categories such as edible, food, fuel, construction, intoxicants, remedies, religion, exudates, oils, poison, resins, animal fodder, and medicinal products.

6.3.3.3
Classification Based on Life Forms and Plant Parts

Classification systems based on life forms and plant parts are often developed in the context of forest resource or wildlife assessments. As such they are easily incorporated in multiresource forest inventories. McCormack (1998) presented a simple classification that first separates animal- and plant-derived NWGS. While no further subdivision is proposed for the animal-derived NWGS class, the plant NWGS are further split into subcategories. This system can be seen as a classification tree with classes (1) perennial species and products [trees with subclasses wood and bark, nontrees with subclasses climbers (lianas, rattans) and nonclimbers (palms, bamboo, shrubs, epiphytes)], (2) ephemeral products from perennial species (e.g., nuts, seeds, and fruit), and (3) ephemeral species (e.g., herbs, wild honey, and mushrooms). A classification system aimed at applications in forest inventories was developed by Kleinn et al. (1996). They defined (1) nonwood parts of trees such as fruits, leaves and twigs that, for modeling and estimation purposes, can be related to the size of the tree trunk, (2) products from "treelike" plants, such as bamboo or rattan, with easily measurable size-related attributes, and (3) herbs and other plants. Dunn et al. (1994) and Wong (1998) have given us similar classification systems.

6.3.3.4
Classification According to Management Characteristics

Classification systems based on aspects and characteristics of resource management have a natural tendency to focus on issues related to access, extraction, utilization, processing, and commercialization. Wiersum (1999) presented a classification of NWGS according to management characteristics (Table 6.4).

6.3.4
The Assessment of Nonwood Goods and Services

The need to include an assessment of NWGS in a local or regional forest survey is evident in many parts of the world. Whenever NWGS play or could play

Table 6.4. Classification by resource management issues (after Wiersum 1999)

Supply characteristics	1. Production characteristics
	Degree of ecological sustainability of extraction
	Ease of vegetative or regenerative propagation
	Ease of cultivation under different environmental conditions
	Ease of stimulating production by technological means
	2. Organization of production
	Access to nontimber forest product resources
	Gender division of production responsibilities
Demand characteristics	1. Opportunistically collected products for subsistence consumption not related to main household needs (e.g., snack foods)
	2. Occasionally collected products purposively collected in times of emergency (e.g., medicinal products, emergency foods during droughts)
	Products for regular household consumption
	Easy to substitute with products of other species (e.g., various food products, fodder, fuel wood)
	3. Difficult to substitute with products of other species (e.g., preferred forest foods)
	4. Products for sale at various market types (local, regional/national, international)
	High degree of competition with substitutes
	Low degree of competition with substitutes
	5. Products demanded in manufactured form, and which can be produced locally, giving them added value (e.g., palm sugar, liquors)

an important role for poverty reduction, a source of income, recreation, traditional use, and spiritual values to local communities, they need to be addressed within a holistic framework for sustainable forest resource management (Jones 2004; Sim et al. 2004). NWGS are now commonly included in NFIs within the European Union. Traditions in northern and eastern Europe are reflected in a strong emphasis on the assessment of berries, mushrooms, resins, and medicinal plants. Assessments of animals, small herbs, or epiphytes are rare in Europe. In the tropics an assessment of export products such as bamboo or rattan is now a core component of many multiresource inventories (Wong 1998; Serna 1990).

A seamless integration of an assessment of NWGS into a large-scale forest resource survey may be difficult. A large suite of NWGS do not have the attributes of a physical commodity and many can only be defined unambiguously in the context of time, space, culture, and behavior. Furthermore, their importance may vary over short distances in time and space. Pelz (1995) listed several other problems associated with the assessment of NWGS:

- Seasonality: Some of the functions, variables or attributes can only be observed during specific seasons (e.g., fruits, mushrooms).
- Area reference: Some NWGS functions can be related to a single or a group of sample plots, others only to a specific, usually small area, and some only to a large area (e.g., wind protection extending beyond the forest).
- Assessment is only partially possible during an inventory. Recreational use, for example, can only be assessed by indicators of use and not actual use per se.
- Attributes are qualitative (e.g., conservation).
- Potential vs. actual use of the NWGS function (e.g., recreation – a forest might be suitable for recreation, but the actual use depends on the proximity to settlements and access).
- The assessment is often very costly (e.g. measuring herbaceous vegetation or faunal diversity) compared with more traditional forest resource inventory attributes.

The current lack of a unified methodology and approach to quantify NWGS has limited the use, analysis, and comparison of NWGS data. Only a widely accepted classification scheme can remove the major obstacle to the wider use and dissemination of NWGS data.

NWGS can be described (1) in quality terms, (2) as indices, (3) in quantity (physical) terms, or (4) in value (monetary) terms (Linddal 1995). Lund (1998) identified four types of assessments that are needed for reporting on NWGS:

1. Biodiversity inventories (list of species)
2. Cultural studies
3. User, market, or product surveys
4. Resource inventories

Biodiversity inventories provide basically "*a list of biological entities from a particular site or area*" (Stork and Davies 1996). They can provide useful information on the presence of taxa, but rarely note which taxa are potential nonwood forest products and too often lack information (estimates) on abundance.

Cultural studies employ social science techniques to gain local knowledge and are based on participatory approaches. Participators approaches include those of rapid rural assessment (RRA), participators rural assessment (PRA), participators learning and action (PLA), gender analysis, objective-oriented project planning (ZOPP), appreciation–influence–control (AIC), and social assessment (Davis-Case 1990; Nichols 1991; Inglis 1991). Key features of these techniques are the selection of methods and tools depends on the informants, a multidisciplinary approach towards involving researchers and integration of methods from different fields, and an adaptive approach to planning. No method is formalized into a single rigid assessment protocol; rather, in view of the overall objective to collect local knowledge, the best in situ approach is chosen. Cultural studies rarely report on statistics related to quantitative aspects of the NWGS resource(s).

Anthropological approaches view NWGS from a cultural perspective and study the interaction between humans and their environment. They apply quantitative methodologies and use quantifiable measures which lend themselves to statistical analyses. Zent (1996) provided the following examples: spatial distribution analysis for the study of spatial relationships between human and resource communities; resource accounting with records of resources procured or utilized by a community in a given period; human activity studies with time spent in various resource-related behavior; and input–output analysis with descriptive statistics on the interactive relationships between humans and resources.

User, market, and product surveys assess the income generated by NWGS and their contribution to markets. They use a variety of econometrics tools and methods that are often different from those applied in natural resource inventories. Most common are market surveys and market research, supply-and-demand studies, investigation of trade networks, and studies quantifying the share of NWGS utilization in local incomes (Godoy et al. 1993; Chopra and Kumar 2004; Svarrer and Olsen 2005).

Resource inventories are based on quantitative approaches and provide statistical estimates for a well-defined population. They consider the availability and abundance of quantifiable NWGS and, when repeated over time, indicators of the sustainability of current and past patterns of NWGS use.

All four types of assessments require rules to assign numbers to objects in such a way as to represent quantities of attributes. The measurement rules for the assessment of nonwood products depend on the life form of the product concerned. Wong (2000) proposed a suite of methods for different life forms (Table 6.5). In practical applications difficulties may occur when objects are hidden (e.g., truffles), short-lived, or have the capacity to avoid or escape detection.

In defining the nomenclature for NWGS it is essential to decide whether the assessment should include all goods available in a forest or it should be limited to those that have actually been either extracted or used. A similar decision must be made for services (Wullschläger 1982).

The quantification of nonwood services is far more complicated than the assessment of nontimber products. A forest may have the potential to provide a specific service, but the service may not be utilized. Further complicating the

Table 6.5. Assessment methods for different life forms (after Wong 2000)

Method	Life form	Description
Tally	Any – sessile	Counts of target individuals in sampling unit
Presence/absence	Any	Record occurrence of target in sampling unit
Size/age measurement	Larger plants and animals	Measure size of all individuals in a sampling unit (e.g., leaf width, stem diameter, height, life stage – juvenile/adult, etc.)
Cover	Plants	Record percentage of sampling unit covered by target species
Relative abundance	Any	Score density of target in sampling unit into classes (e.g., low, medium, high, or Braun–Blanquet scales for plants)
Trapping	Mobile – animals and fruit/seeds of trees	Capture individuals for counting and measurement (e.g., mist netting, Sherman traps, seed traps)
Partial trapping out	Small animals (where loss from population is not critical)	Capture individuals and remove them from the sampled population; repeat over a period of time and use an exponential model of decreasing capture rates to estimate by extrapolation the size of the initial population
Mark–recapture	Animals	Capture individuals, mark (toe clipping, tags, paint, etc.), release and recapture, use numbers recaught to estimate total population.
Distance sampling	Animals	Record distance from observation point to a target and via (for example) a Fourier analysis estimate the size of the target population
Response to playback	Birds	Play recordings of bird calls and count the number of responses to a particular call
Indirect/index methods	Any	Record hair, dung, nests, or any other easily observable sign of presence; use regression methods to estimate the size of the target population

assessment are the difficulties associated with measuring a service and defining unambiguous reporting units when natural units are missing. Information on nonwood services is usually presented in the form of models that combine field measurements, auxiliary data, subject knowledge, and expectations based on historical data (Brassel 1995; Pelz 1995). Brändli (2001) describes models for the estimations of protective functions of a forest. The model was developed for Switzerland with data from the second Swiss National Forest Inventory (NFI). Protection demands were defined as a function of hazard and damage potentials, the presumed threats to human lives and property. Areas and objects exposed to elevated natural hazards were identified by the application of a generalized gradient method, with slope, flow direction, and flow height, as well as a generalized gradient for endangered objects as explanatory variables. Hazard rates were calculated for each pixel in a digital terrain model (DTM). The model allows for a prediction of potential starting zones and trajectories for rock-falls, avalanches, and landslides. Forests located in starting zones and trajectory areas are viewed as providing a protective function.

Lesslie and Maslen (1995) presented an approach to quantify and identify "wilderness." The relative variation in remoteness from human activity across a landscape plays a key role in their approach. The remoteness and absence of human and technological influence enters into the Australian wilderness index (Lesslie and Maslen 1995; Husby 1995). Specifically they use four indicators:

1. Remoteness from settlement (remoteness from places of permanent habitation)
2. Remoteness from access (remoteness from established access routes)
3. Apparent naturalness (the degree to which the landscape is free from the presence of permanent structures associated with modern technological society)
4. Biophysical naturalness (the degree to which the natural environment is free from biophysical disturbance caused by the influence of modern technological society).

Volk and Schirmer (2004) provided guidelines for mapping of forest functions. For a set of services they gave attributes that are associated with the specific service (Table 6.6). Those attributes can be assessed in the field or can be derived from spatial data and can be used as input variables for models quantifying NWGS.

6.4 Forest Ecosystems and Biological Diversity[2]

The 1997 the International Union for the Conservation of Nature and Natural Resources (IUCN)[3] Red List of Threatened Plants indicates that nearly 34,000 plant species, or 12.5% of the world's vascular flora, are threatened with extinction. Even more alarming is the 1996 IUCN Red List of Threatened Animals, which revealed that 11% of all birds and 25% of all known mammal species are threatened. As forests are the repository of much of the world's biodiversity (Kapos and Iremonger 1998), the forest society has to share responsibility for the maintenance and conservation of forest biodiversity.

[2]The terms biodiversity and biological diversity are used here as synonyms.
[3]http://www.iucn.org

Table 6.6. Attributes describing forest services (after Volk and Schirmer 2004)

Service	Attributes
Protection of water resources, soil, and other ecosystem functions	Soil, humus layer, precipitation, climate, management regimes, silvicultural system, slope, slope form, vegetation, topography, stand age, stand density, crown closure, application of pesticides, erosion, regulation of access and traffic
Protection of infrastructure and managed natural resources	Slope, slope length, slope from, soil, geology, exposition, precipitation, climate, management regimes, vegetation cover, stand age, slope distance to objects to be protected
Influence on local climate and reductions in impacts of gas emissions	Vertical structure of stands, growth rate, biomass, melioration, exposition, slope, species composition, stand age, management regimes, wind direction, average and maximum wind speed, distance to potential emissions (industries, power plants, disposal sites, mining sites), distance to residential areas, recreational areas, or protected biotopes
Conservation of natural habitats and biological diversity	Patch size, structure and fragmentation of forest margin, species composition, species mixture, stand age, vertical and horizontal structure of stands, soil, diameter at breast height distribution
Recreational and other social functions of forests	Road density, accessibility, distance to next settlement, parking spaces, length of hiking trails, number of visitors per hectare and day, natural conditions (climate, relief, forest structure), recreation infrastructure (e.g., camp sites, picnic sites, cross-country ski runs), limiting factors (e.g., local emissions, landfills, traffic noise), scenic values, historic sites, evidence of traditional use (e.g., forest grazing, coppice), and religious significance

One of the key agreements adopted at the 1992 Earth Summit in Rio de Janeiro was the Convention on Biological Diversity (CBD) as a comprehensive component for sustainable development. The Convention establishes three main goals: the conservation of biological diversity, the sustainable use of its components, and the fair and equitable sharing of the benefits from the use of genetic resources.

There exists no single definition for biological diversity (Kaennel 1998). The term "biological diversity" was introduced by Williams (Fisher et al. 1943) and is unilaterally related to the number of species or species richness. The inadequacy of this definition becomes obvious when two populations with the same number of species but different species abundance are studied. Thus, diversity is nowadays defined as a function of the number of species within a defined area and the distribution of individuals within the species. According to Pielou (1975), diversity is equivalent to variance. While variance characterizes the variety of quantitative measures, diversity is related to the variety of qualitative measures. Reid and Miller (1989) provided a commonly used definition of biological diversity which refers to the "variety and variability amongst living organisms and the ecological complexes in which they occur." Commonly biodiversity is divided into genetic, species, and ecosystem diversity. The European Nature Information System (EUNIS) habitat classification glossary[4] gives the following definitions:

Genetic diversity	The variation between individuals and between populations within a species
Species diversity	The different types of plants, animals, and other life forms within a region
Community or ecosystem diversity	The variety of habitats found within an area

The complex concept of diversity renders the consideration of several dimensions necessary (Vanclay 1998), which may be characterized in part by:

- Units, ranging from genes, species, structural aspects (e.g., morphological groups, size classes) to landscape components
- Grain, ranging from individual samples and microhabitats, habitats, regions to large biogeographic areas
- Patterns comprising richness, evenness, contagion, and fractal dimension (Olsen et al. 1993; Sect. 6.5.)

Forest biodiversity cannot solely be described in the context of quantity. A large area may offer the potential of biological diversity, but it is not necessarily biologically diverse and valuable. Planted with exotic, nonnative species, managed by silvicultural systems that aim at even-aged, single-species stands, polluted or fragmented into small patches, even large forest areas may not contribute significantly to biological diversity (Fig. 6.3). An indication of forest biodiversity is only valuable when quantity and quality are linked. Biodiversity indicators have been developed that combine the aspects of quality and quantity of biological diversity. For instance, in the boreal region the assessment of dead wood has been established as one indication of biological diversity (Sect. 6.4.6).

[4]http://eunis.eea.eu.int/habitats.jsp

(a)

(b)

Fig. 6.3. Forest biodiversity: low biodiversity in an even-aged spruce stand **a** and high biological diversity in a mixed-species mountain forest in Switzerland **b** and in a tropical forest in Malaysia **c**

(c)

Fig. 6.3. (Continued)

In order to implement biodiversity maintenance and conservation, policy makers need sound information on (Puumalainen et al. 2002):

- The current situation of biodiversity (current state)
- Changes over time (time series)
- Cause–effect relationships

Maintaining forest biodiversity can be achieved by two complimentary approaches: (1) biologically and geographically representative networks of protected forest areas and (2) the conservation of forest biodiversity outside protected areas by, for example, sustainable forest management. As a consequence, forest monitoring concepts need to consider if a combined inventory concept are approached, or if separate monitoring systems are needed.

Besides the assessment of biological diversity inside forests, multiresource inventory systems have to address the interaction between forest ecosystems and other land-use regimes. Information on forests in the landscape context (Sect. 6.5) and information on the transition zones between forests and other land-cover types are corequisite (Sect. 6.4.2). The interactions of humans and forest ecosystems require the assessment of forest management practices and potential threats to the naturalness of forests.

6.4.1 Biodiversity Indicators

Biodiversity indicators are information tools, summarizing data on complex environmental issues to indicate the overall status and trends of biodiversity (Ott 1978). They need to address the structural, compositional, and functional aspects of biological diversity. Besides providing information on the current state and changes of biological diversity in managed and unmanaged forests, they can also be used to assess national performance and to signal key issues to be addressed through policy interventions and other actions. The development of indicators is, therefore, important for monitoring the status and trends of biological diversity and, in turn, feeding back information on ways to continually improve the effectiveness of biodiversity management regimes.

For the quantification of diversity a suite of methods have been described (Pielou 1975; Magurran 1988). The objective of all methods is to depict multidimensional variability such as relative frequencies, spatial patterns, and development over time in a one-dimensional ranked order.

The diversity of a population can be described by two components: (1) the number of species or species abundance and (2) the relative frequency of the individual species. Pielou (1975) presented an approach for describing diversity by theoretical distributions, which are fit to the frequencies of species. The parameters of the fitted distributions can serve as a measure for diversity. However, this approach is rarely applied in practical applications, as it is based on complex and hardly realistic assumptions (Sai und Mishra 1986). An alternative approach is based on descriptive statistics and derives diversity indices, which can be applied for any desired population without reference to the underlying theoretical frequency distributions. Peet (1974) gave a comprehensive overview on diversity indices. Several of the most frequently applied diversity indices are species richness, species count, and Shannon's and Simpson's indices (Shannon und Weaver 1949; Simpson 1949; Gove et al. 1994):

Species richness:

$$\Delta_{SR} = \sum_{i=1}^{s} \left(\frac{1}{\pi_i}\right) \pi_i = s,$$

Species count:

$$\Delta_{SC} = \sum_{i=1}^{s} \left(\frac{1}{\pi_i} - 1\right) \pi_i = s - 1,$$

Shannon:

$$\Delta_{Sh} = \sum_{i=1}^{s} (-\log \pi_i)\pi_i = -\sum_{i=1}^{s} \pi_i \log \pi_i,$$

Simpson:

$$\Delta_{Si} = \sum_{i=1}^{s} (1 - \pi_i)\pi_i = 1 - \sum_{i=1}^{s} \pi_i^2,$$

where s is the species richness for a community and π_i is the relative abundance of the ith species, so that $\sum_{i=1}^{s} \pi_i = 1$.

Peet (1974) and Swindel et al. (1984) showed that Shannon's index is sensitive to changes in the importance of the rarest class, while Simpson's index is largely influenced by the most abundant class. Köhl and Zingg (1995) applied the four diversity indices, species richness, species count, and Shannon and Simpson indices, to monitor changes of diversity over time and showed that Simpson's and Shannon's indices may fail in describing changes in species composition over time.

The development of biodiversity indicators is driven by the permanently increasing understanding of ecological processes and the links between ecological, social, and economic development. Their application depends on the specific local conditions. Peet (1978) pointed out a major problem in the application of biodiversity indicators: "To understand patterns of diversity a necessary first step is to place these patterns in an appropriate perspective. Diversity values are exceptionally difficult to interpret when taken out of context, and little justification exists for their publication unless the perspective is provided."

6.4.2 Assessment of the Forest Edge

According to Forman and Godron (1986), the primary significance of shape in determining the ecological significance of patches in a landscape appears to be related to the edge effect. An edge is the outer band of a patch, where the environment is significantly different from the interior of the patch. The edge effect is related to the different species composition and abundance found at the edge. Forest edges can be described by various attributes, such as length, width, shape, vertical and horizontal structure, density, and interior-to-edge ratio. Brändli et al. (1995) described an assessment procedure for the forest edge that was operationally applied in the second Swiss NFI (Fig. 6.4). In the Swiss NFI, sample plots are located in a systematic grid. Whenever a forest has margins within 25 m of the center of a field plot, an assessment of the forest margin is mandatory. A taxation line of 50-m length forms the basis for the forest edge assessment.

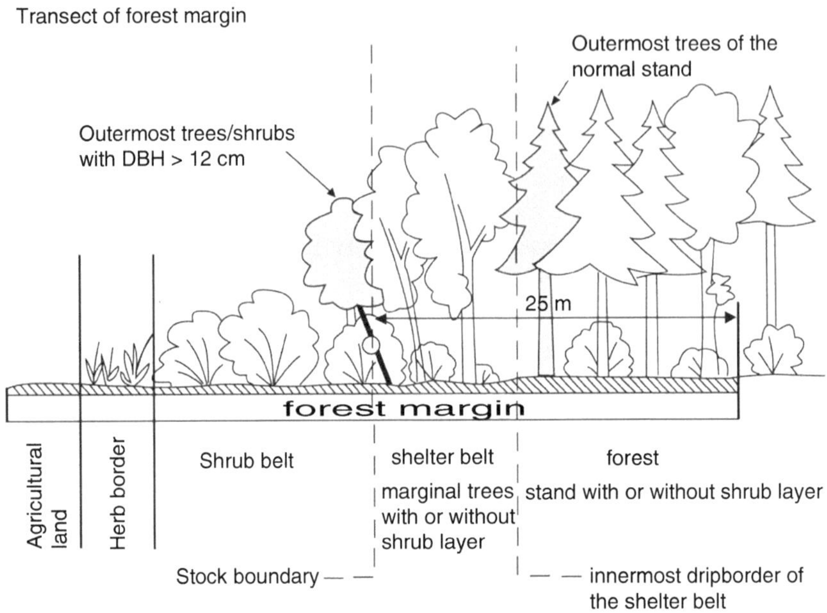

Fig. 6.4. Transect of forest edge. *DBH* diameter at breast height. (Source Brändli et al. 1995)

Determination of botanical diversity, description of important habitats for various animals, especially birds and insects, and judgment of the aesthetic value for recreational purposes are important objectives of the edge assessment. Results of the forest edge survey are given by Brändli and Ulmer (1999).

6.4.3 Sampling Diversity

Sampling for biodiversity can incorporate different scales and aspects. Alpha, beta, and gamma diversity are common concepts to approach the different scales of biodiversity monitoring (Table 6.7). Alpha diversity refers to the species abundance within a community. Beta diversity is related to the degree of changes of species abundance between different communities or environmental gradients. Gamma diversity describes diversity at the landscape level.

In many extensive inventories, biodiversity is assessed by lists of key factors (Tomppo 1997). The FAO *Forest resources assessment 2000* (FAO 2001) used the IUCN[5] categories for protected areas as an indicator for biodiversity. Larsson

[5]http://www.iucn.org

Table 6.7. Scales of diversity assessment (after Bischoff and Dröschmeier 2000)

	α-diversity	β-diversity	γ-diversity
Definition	Diversity within a habitat	Diversity within a mosaic of habitats, including borderline effects	Diversity within biogeographical region or country
Pressure	Nutrients Structure Access techniques Management	Heterogeneity Length of border Size of areas of a defined area type	Area shift Species formation Species extinction
Major protection strategy	Develop/optimize access techniques	Biotope protection Compensatory areas Biotope networking	Species protection Reintegration Large corridors Possibly isolation
Sensitive species	Common widespread species	Widespread, uncommon species	Rare species
Suitable size of unit	Units of a defined area type	Regions Altitude bands	Biogeographical regions

et al. (2001) presented a preliminary list of key factors of European forest biodiversity at the national and regional levels which address structural, compositional, and functional diversity (Table 6.8).

Where the presence and the relative abundance of species within well-defined patches are to be assessed, the definition of patches (points, habitats, regions, scale) and the corresponding sample size are critical (Vanclay 1998). Odum (1968) presented the number of organisms found ranging from 1,021 per square meter for soil bacteria to 5–10 per square meter for deer. No general

Table 6.8. List of key factors of European forest biodiversity (after Larsson et al. 2001)

Scale	Structural key factors	Compositional key factors	Functional key factors
National or regional	Total area with respect to Forest types Legal status/utilization or protection Forest ownership Tree species and age classes Old growth or forest left for tree development Afforestation (deforestation)	Native species Nonnative or not "site-original" tree species Forest types	Natural disturbance: fires, wind and snow, biological disturbances Human influence: forestry, agriculture and grazing, other land use, pollution

advice for sample sizes and sampling units can be given as they depend on the interdependency between organisms and patch size.

Another problem in sampling for biodiversity is the basic assumption of randomness in sampling theory. Vanclay (1998) points out that "many factors (predators, competition, habitat, etc.) lead organisms to aggregate, making it unlikely that individuals will be randomly sampled." Adaptive cluster sampling and plant density estimation may help to implement sound sampling concepts for real-world applications.

6.4.4 Assessment of Rare Species by Adaptive Cluster Sampling

Monitoring species diversity is an important objective for forest ecosystems that we wish to maintain in a state considered as "close to nature." Random or systematic allocation of sampling units involves the nontrivial risk of missing rare species (Chap. 3.7). Thompson (1990) describes the adaptive cluster sampling (Chap. 3.7.1) as an efficient method for the assessment of spatially clustered species with low abundance. Many rare species fall into this category. Low abundance and spatially clustered means that there are only a few locations where the species can be found. Consequently, they are easily missed in a conventional survey. Adaptive cluster sampling allocates sampling units in two steps: (1) an initially fixed number of sample plots is randomly or systematically distributed over the sampling area; (2) in each plot, where the rare species under concern is found, the neighboring plots are also surveyed. For every additional plot containing the species of interest a new set of neighboring plots is established. The procedure is continued until no newly added plot contains the species of interest. Despite the fact that a priori sample size cannot be calculated, the procedure can be very efficient (Roesch 1993). In a simulation study utilizing computer generated spatial patterns, Smith et al. (2003) confirmed the efficiency of adaptive cluster sampling relative to that of sampling with randomly and systematically distributed plots. He found that adaptive cluster sampling was superior for spatially clustered species occupying less than 5% of the inventory domain.

6.4.5 Plant Density Estimation

The problem of estimating plant densities commonly arises in forest surveys when nontree species are considered. In vegetation surveys quadrat count methods are a widespread tool, but according to Clayton and Cox (1986) "in many times it is extremely time-consuming or impractical to map all the events or carry out quadrat sampling." Accordingly, distance methods are often preferred to quadrat count methods. Distance-based estimators involve measuring the distance from randomly selected points to a defined number of neighbors. As distance methods often fail to find the analytic form for the distance distribution when processes are nonrandom, Upton and Fingleton (1985) and Ord (1990) provided estimators which are robust to departures from spatial randomness. However, the practical applicability of these proposals seems to be limited owing to the prohibitive number of measurements involved. Patil and Taillie (1979) established a relation between the plant density and the probability density function of squared point-to-plant distances under fairly general conditions. On the basis of this relation they proposed a plant density estimator which just requires point-to-plant distances and whose results are consistent and asymptotically normal. Barabesi and Fattorini (1995) suggested collecting point-to-plant distances by the use of ranked-set sampling. Patil et al. (1994) give an overview of ranked set sampling. The method involves the selection of m random samples with m units in each sample from an infinite population. In the next step, the units in each sample are ranked by visual assessment or any other procedure that does not require exact measurements of plant distances. The unit with the smallest rank is selected from the first sample, the unit with the second smallest rank from the second sample, and this procedure is repeated until the unit with the highest rank is selected from the mth sample. The procedure requires the quantification of the m units out of the m^2 units originally selected. The procedure can be repeated r times in order to obtain enough quantification for inference, resulting in $n=mr$ quantified units out of m^2r selected units. A considerable potential for efficiency can be obtained whenever the ranking is reasonably correct. Barabesi and Fattorini (1995) showed for various spatial plant patterns the improvement on simple random sampling that can be achieved by estimating plant density by a ranked set sampling of point-to-plant distances.

6.4.6 Assessment of Deadwood by Transect Relaskop Sampling and Guided Transect Sampling

The assessment of coarse woody debris has become an important part of forest surveys. Dead decaying wood is important for the survival and presence of many species depending on forested ecosystems. Ståhl and Lämås (1998) and Travaglini et al. (2005) discussed the performance of circular plot sampling, strip surveying, line intersect sampling, and transect Relaskop sampling for the assessment of downed coarse woody debris. While the first three methods are well known, the transect Relaskop method was introduced only recently by

Ståhl (1998). It can be considered as a combination of Relaskop sampling and line intersect sampling. Along survey lines, a wide-angle Relaskop is used and – as in Relaskop sampling – all downed logs that appear to be larger than the Relaskop angle are tallied. The procedure allows an estimation of the length of downed logs by simply counting the logs included in the sample.

Ringvall and Ståhl (1999) described an innovative method called guided transect sampling that exploits auxiliary information as a guide for a transect with a starting point determined by conventional statistical design criteria. If the inventory area is subdivided in quadrates or pixels, a surveyor has the choice of moving to one of the three connected pixels in front of him or her. Auxiliary information is used to assign probabilities to the pixels to be entered next. Information from remote sensing can be utilized to assign probabilities according to some probability proportional to size (PPS) rule to individual pixels and thus to determine the route (see also Chap 3.7.3). In sampling for scarce objects the probabilities should be selected in a way that favors the detection of a large number of the rare objects.

6.5 Landscape Analysis

In recent years landscape ecology has intensively studied quantitative methods for describing spatial structures and fragmentation of spatial landscape elements. Turner and Gardner (1991) highlighted the significant need for research in this field by saying that "... at landscape-level research requires new methods to quantify landscape patterns, compare landscapes, identify significant differences, and determine relationships of functional processes to landscape patterns ...."

Any landscape is characterized by a unique structure. Several statistical measures, so-called landscape indices, have been described that allow us to quantify the structure of a specific landscape (O'Neill et al. 1988; McGarigal and Marks 1994; Walz 1999). Individual landscape indices describe different structural characteristics and can be summarized in eight groups:

1. Area metrics
2. Patch density metrics
3. Edge metrics
4. Shape metrics
5. Core area metrics
6. Nearest-neighbor metrics
7. Diversity metrics
8. Contagion and dispersion metrics

For each of these eight groups, several indices are available. The terminology of several indices creates problems; some indices are defined slightly differently by different authors, while seemingly different indices may lead to similar results (Haines Young and Chopping 1996; Giles and Trani 1999). According to Herzog et al. (1999) "half a dozen indices is sufficient to describe the main structural characteristics of a landscape."

The calculation of a single index or a set of indices does not necessarily allow a sufficient description of landscape patterns, as the variability of a landscape depends on the scale selected and is not quantified consistently for varying scales (He et al. 2002; Leimgruber et al. 2002). Landscape indices may perform above the small-scale variability and may smooth differences in structure; hence, calculating a single measure for an entire landscape may not reveal complete information, as the variability of a landscape is not necessarily captured. This dilemma can be resolved by presenting landscape indices in mapped format, which permits us to present the spatial dispersion of indices. Maps are constructed by selecting subareas from the region under study and the subareas are studied independently. One approach for selection of subareas is the so-called windows concept.

6.5.1 The Theory of Windows

Windows, also known as kernels, masks, or filters, are used for digital image analysis to characterize spatial information of neighborhoods (Gonzales and Woods 1992). The traditional kind of window, the geometric window (Ricotta et al. 1998, 2003b), is a rectangle with dimensions of *n* (number of rows) by *m* (number of columns), where *n* and *m* should be odd numbers to allow a true center pixel. The value calculated for the $n \times m$ pixel of the window will be projected to the center pixel.

Raster data are often analyzed by the "moving-window technique" (Fig. 6.5). The window moves stepwise, pixel by pixel, through the entire image. At each position the value for the neighboring pixels within the window is calculated

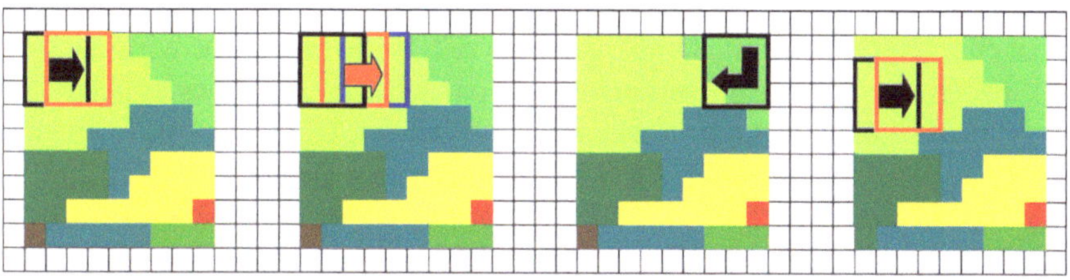

Fig. 6.5. The moving-window technique

and assigned to the center pixel. This process is repeated until the last pixel of the raster data set is reached. If the values of the center pixels are color-coded, output images can be produced that visualize the spatial distribution of the attribute of interest.

An example for a simple algorithm is the projection of the mean of the $n\times m$ values calculated for the pixels within the window to the center pixel. The moving window technique computes moving averages for the entire input image. In the case of large-scale landscape analyses, landscape indices can be calculated inside the window and projected to the center pixel, thus allowing the display of the spatial distribution of the given index.

In general, the size of the window should be chosen to provide an "optimal" adaptation to the calculated landscape index within the context of the study. Therefore, the extension of the window depends on the purpose and the object of the study as Chavez and Bauer (1982) pointed out: "... there is no constant rectangle ... size that provides the best results for every image because the optimum size is dependent on the 'busyness' of the individual image."

Geometric windows generate artefacts along edges of the domain of interest, because calculations close to borders of the input domain can lead to incorrect results owing to a less than complete set of neighbouring pixels inside the domain of interest. This results in output images having a border area with a number of rows (n_r) and columns (n_c) that are not analyzed. Depen-ding on the $n\times m$ window, we can calculate them using the following equations:

$$n_r = \frac{(n-1)}{2}$$

and

$$n_c = \frac{(m-1)}{2}.$$

Usually the border of the output image is padded with zeros in order to maintain constant image dimensions between input and output image dimensions.

Another problem using the geometric window method for landscape analyses is the use of a single size and shape of window. That means that the current local part of the landscape is forced into a rectangular neighborhood of pixels defined by the window size. This results in an influence of the window size on the local spatial characteristics of the landscape analyzed. If the size of the window is very small relative to the mean dimensions of patches, the windows situated completely inside a patch get a uniform value. Deviations from this "uniform value" are only to be expected when the window includes a patch boundary and thus covers two or more patches. This effect is demonstrated in Fig. 6.6 for the example of a 3×3 geometric window. Therefore, geometric windows can be used for edge detection, but they do not satisfactorily reflect the spatial variability of

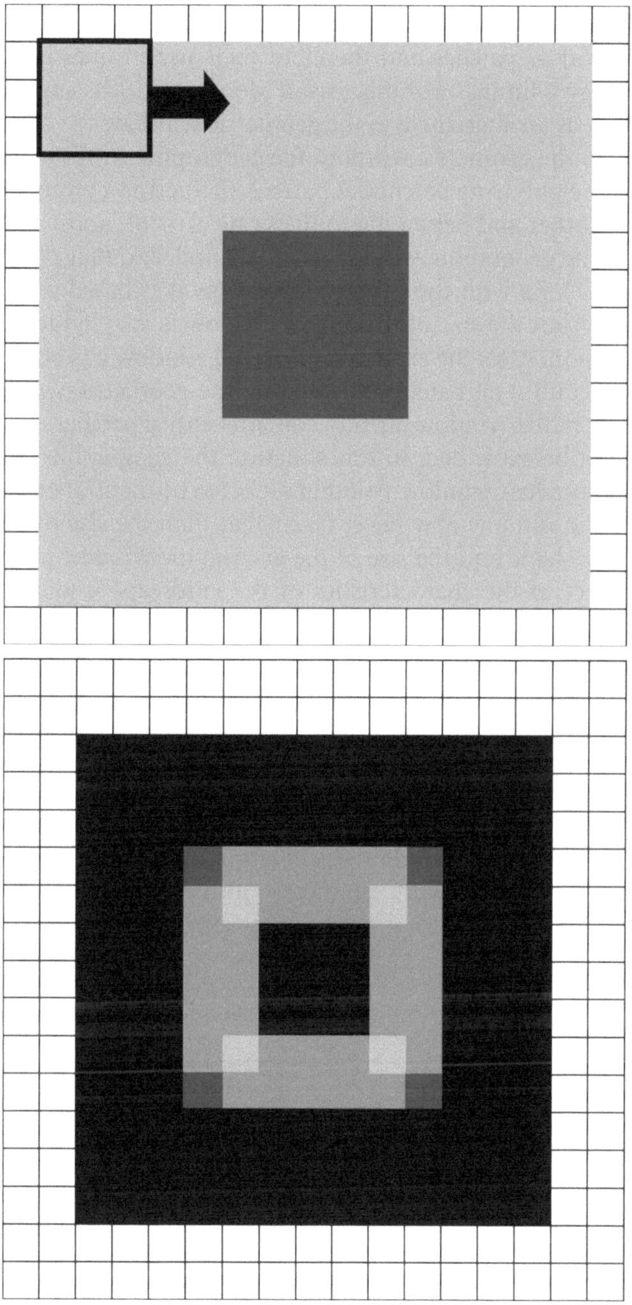

Fig. 6.6. Assessment of a landscape with a 3×3 geometric window. The upper figure shows a landscape segment and the starting point and direction for 3 × 3 geometric window, the lower figure presents the result

landscapes. Geometric windows analyze the neighborhoods of pixels instead of the neighborhood of patches and therefore their use in questions concerning landscape ecology is limited. For this reason Merchant (1984) suggested the geographic window as an alternative to the geometric window.

In contrast to the geometric window, the geographic window method operates on a set of neighboring patches. A patch is defined as a group of pixels bordering on each other and belonging to the same class of landscape.

The idea of the geographic window is quite similar to that of the geometric window (Fig. 6.7). As with the geometric window it is based on a rectangular "initial window," but the size of this initial window is adapted to the objectives under investigation. After the extent of the initial window has been defined, the window expands until all patches touched by the geometric window are fully included. If the initial window already contains entire patches, the geographic window will not be expanded. In this situation the geographic window corresponds to the geometric window. Another aspect to be mentioned is that the size of the geographic window may never be smaller than the size of the geometric one. In sum, the shape and the size of the geographic window change dynamically with respect to the characteristics of the landscape under investigation. Because of this adaptation, patches are never truncated in the rectangle borders of a window; they are always completely included in calculations.

Figure 6.7 presents a geometric and a geographic window applied to the same subset of a landscape consisting of four different patches. Figure 6.7a demonstrates a 5×5 pixel geometric window with the center x. Figure 6.7b demonstrates how the geographic window expands to include the entire area of patches.

The definition of the optimal size of a window is not straightforward. Oehmichen (2001) studied window size in relation to the overall patch size and showed that calculations with geometric and geographic windows yield approximately the same values if the window size is large compared with the patch size (Fig. 6.8).

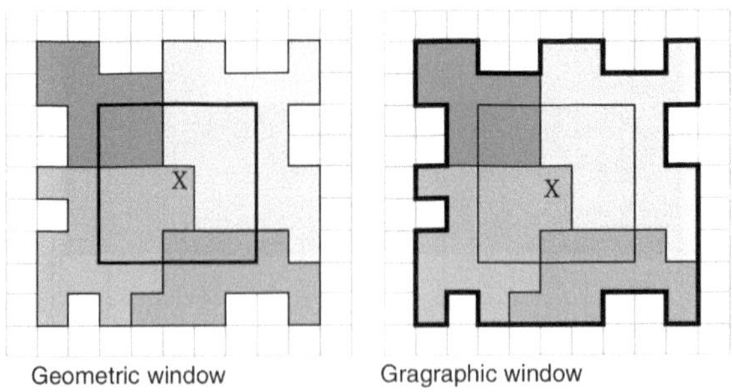

Fig. 6.7. Selection of patches by the geometric window **a** and the geographic window **b**

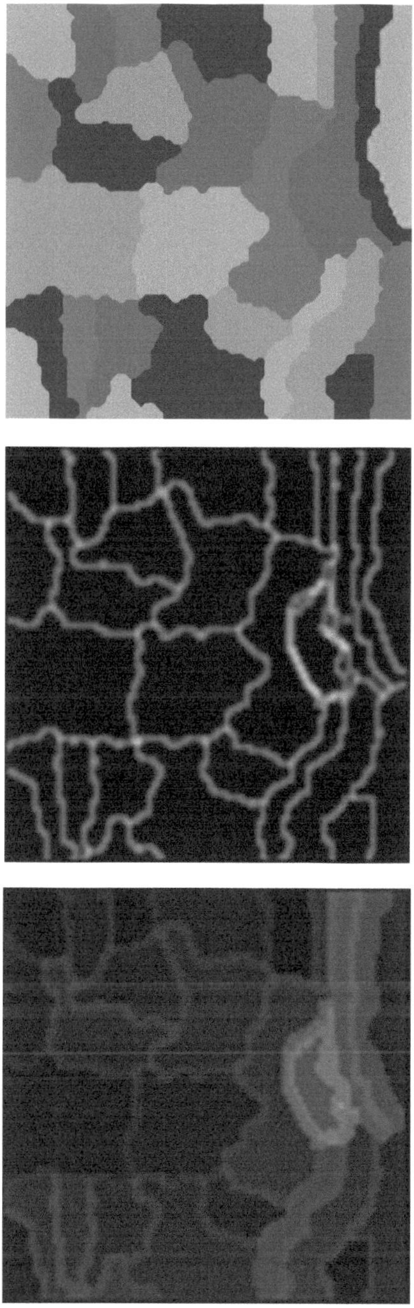

Fig. 6.8. Influence of window size

Fig. 6.8. (Continued)

Oehmichen (2001) and Köhl and Oehmichen (2003) studied the influence of the patch shape on the results of both window concepts to spot the impact of the complexity of landscapes. They used computer-generated landscapes to alter the shape and the edges of the patches from regular to irregular appearance. The more complex a landscape becomes, i.e., the more irregular the shape and the edges of the patches are, the more pronounced are the differences in the index values between geometric and geographic windows (Fig. 6.9).

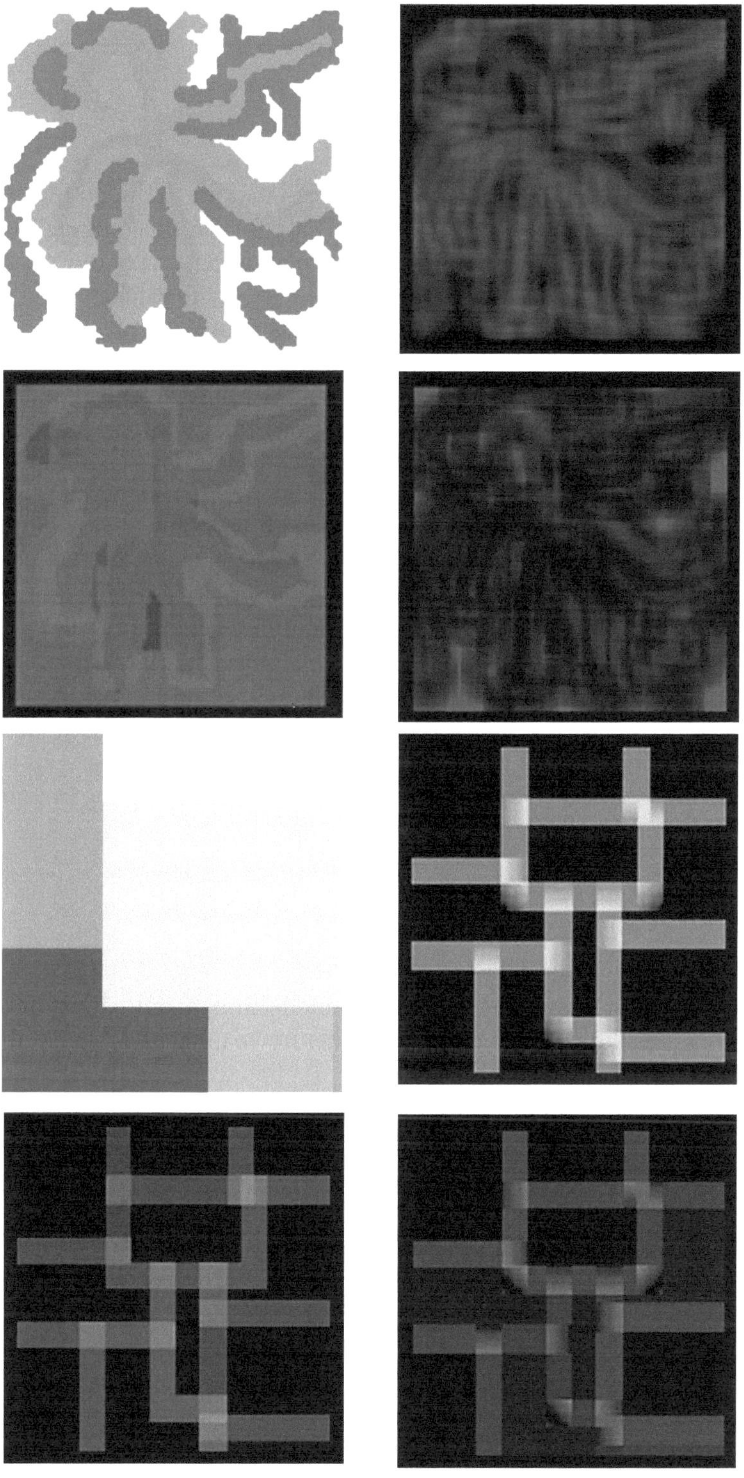

Fig. 6.9. Influence of landscape complexity

6.5.2 Trees Outside Forests

Tree resources outside the forest comprise trees that are excluded from definitions of forest and other wooded lands. They are formations ranging from single trees to systematically managed trees in plantations, trees in line, such as windbreaks or hedgerows, and trees grouped in small patches (Koukal and Schneider 2003). They are located mostly on farmlands, but are also located in built-up areas. Many of them are planted, others are relics of former extensive forests: the fragmentation and degradation of the forest cover is mainly due to agricultural conversion and firewood extraction, particularly in regions where the population pressure on natural resources is heavy, and where the natural ecosystems and those influenced by humans are highly vulnerable. The importance of nonforest trees is increasing with continuing destruction and fragmentation of closed forests.

They provide wood and nonwood products, are part of habitats for animals and plants, contribute to carbon storage and protection of soil and water resources, and have much influence on the appearance of the landscape. Currently these tree resources are not included in common forest inventories. Methods for mapping and monitoring tree resources outside forests have been developed mainly as a combination of very high resolution and aerial photographs remote-sensing methods and field sampling methods, in general adopting line-intersect sampling procedures (FAO, 2001).

6.6 Forest Fires

Wildfires, which are uncontrolled, unwanted human-caused or lightning-caused fires, constitute one of the biggest phenomena that affect renewable natural resources on local and global scales. In the scope of the International Geosphere and Biosphere Program (IGBP) the presence of one or more fires was recorded for a total area of 6.5×10^6 km^2 in the period between April 1992 and March 1993. Seventy percent of that area was located in the tropical zone, with a pronounced focus in Africa, leading NASA[6] to define Africa as the "earth fire belt." At present, the operational program of fire monitoring from the National Oceanic and Atmospheric Administration Advanced Very High Resolution Radiometer (NOAA-AVHRR) data exists for tropical forests, such as the Amazon Basin, controlled monthly by the Brazilian federal government (http://www.cnpm. embrapa.br;http://www.mma.gov.br).

[6]http://earthobservatory.nasa. gov/Observatory/Datasets/fires.trmm.html

At the UN World Conference on Natural Disaster Reduction held in Yokohama in 1994[7] the following averages of areas burned annually were provided:

- 10–15×10^6 ha in the boreal and temperate forests
- 20–40×10^6 ha in the tropical rainforests
- 500–1,000×10^6 ha in the savannahs

Multiresource inventories and remote sensing facilitate forest fire assessment by providing information on fire *risk*, *danger*, and *hazard*. *Risk* is related to the causal agents of the event, *hazard* refers mainly to the fuel situation, and *danger* takes into consideration the meteorological condition. Within this context, satellite images provide input to specifically designed GIS, to map the spatial distribution of fire occurrences risk, hazard and danger. The diversity of factors that affect the start and the spreading of a fire dictate the use of an integrated approach to assessment. Topics to be considered in assessments are:

- Detection of fire spots
- Cartography of burned areas, to locate and estimate the extent of such areas.
- Estimation of gas emissions
- Assessment of fire effects on vegetation and the damages suffered by forest stands
- Checking the ability of the ecosystem to naturally recover after the fire and the assessment of the dynamics (pattern and speed) of natural recovery
- Management of postfire assessment of large burned areas
- Checking the outcome of any eventual restoration intervention
- Monitoring forest fuel conditions (vegetation moisture stress)
- Development of risk models, by the integration of field assessments, remote sensing, and GIS.

6.6.1 Assessment and Modeling of Wildfire Risks

The amount and condition of forest fuel is a critical variable for quantifying a fire hazard (Burgan and Rothermel 1984; van Wagner et al. 1987). Field measurements to quantify forest fire hazard concentrate mainly on the estimation of fuel wood and litter. In the assessment of wildland fire risk, a key variable for communities and ecosystems is the fire regime condition class (FRCC). It

[7]http://www.reliefweb. int/ocha_ol/programs/idndr/yokohama/

describes increasing risk levels from wildfire behavior and effects (Hann and Bunnell 2001; Hardy et al. 2001; Schmidt et al. 2002), according to the relative risk of losing key components that define an ecosystem. Condition classes are generally equivalent to low, moderate, and high departure from the natural or historical range of variability, considered a baseline for coarse-filter assessment of risks to ecosystems, habitats, and social values (Morgan et al. 1994; Hann et al. 1998; Landres et al. 1999). Hardy et al. (2001) and Schmidt et al. (2002) proposed mapping the FRCC and associated information to support national-scale strategic fire and fuels planning in the USA using estimates for 1-km^2 data units derived from a combination of fire regime, cover type, forest density, potential natural vegetation group, classification of biophysical environments, land stratification based on site and disturbance conditions that support a regime of natural vegetation (Küchler 1975; Hann et al. 1998; Hann and Bunnell 2001). They defined and mapped five historical natural fire regimes using a rule set approach to unique combinations of biophysical data. The fire regimes, based on Heinselman (1981), Brown (1995), and Morgan et al. (1996), define fire frequency or interval as "the average number of years between fires" and severity as the "effect of the fire on the dominant overstory vegetation" (Table 6.9). The relative risk is determined by comparing the current with the historical or natural baseline based on combination of fire regime, cover type, and forest density or nonforest for each potential natural vegetation group. A simplified description of the FRCC and associated potential risks is given in in Table 6.10.

The classification (Table 6.10) is based on changes to one (or more) of the following ecological components: vegetation characteristics (species composition, structural stages, stand age, canopy closure, and mosaic pattern); fuel composition; fire frequency, severity, and pattern; and other associated disturbances (e.g., insect and diseased mortality, grazing, and drought). Earth observation (EO) data could greatly help to provide information, especially at large scales.

Coarse woody debris and dead fuel are very difficult to sense remotely. In spite of the low resolution of Landsat Multispectral Scanner (MSS) images, significant correlations between downed fuel classes and classification results were estimated in the past (Rabii 1979). Most of these studies used topographic data to help in the identification of some fuel categories (Shasby et al. 1981; Agee and Pickford 1985). Other authors are not as confident in the application of EO, because of the difficulty in identifying the understory component of forest stands (Salazar 1982).

The physical principles underlying fire direct remote observation are based on the specific reflectance patterns of burning and burned areas. The sun, the main energy source, is directly responsible for the sensing in the visible and near-IR bands (0.4–1.3 μm). Longer wavelengths are more related to the object's emittance, which is proportional to its temperature. This spectral region is called the thermal IR, located between 8 and 14 μm. Between near-IR

Table 6.9. Fire regime condition classes for modeling forest fires

Fire regime class	Frequency (mean fire return interval)	Severity	Modeling assumptions
I	0 – 35+ years, Frequent	Surface and mixed	Open forest, woodland, and savannah structures maintained by frequent fire; also includes frequent mixed severity fires that create a mosaic of different age post-fire open forest, woodland, shrub, or herb patches that make a mosaic of structural stages. Mean fire interval can be greater than 35 in systems with high temporal variation.
II	0 – 35+ years, Frequent	Replacement	Shrub or grasslands maintained or cycled by frequent fire: fires kill non-sprouting shrubs which typically regenerate and become dominant within 10-15 years; fires remove tops of sprouting shrubs which typically resprout and dominate within 5 years; fires typically remove most tree regeneration.
III	35 – 100+ years, Infrequent	Mixed and surface	Mosaic of different age post-fire open forest, early to mid-seral forest structural stages, and shrub or herb dominated patches generally <40 hectares, maintained or cycled by infrequent fire. Interval can range up to 200 years.
IV	35 – 100+ years, Infrequent	Replacement	Large patches generally >40 hectares, of similar age post-fire shrub or herb dominated structures, or early to mid-seral forest cycled by infrequent fire. Interval can range up to 200 years.
V	200+ years	Replacement, mixed, and surface	Variable size patches of shrub or herb dominated structures, or early to mid to late seral forest depending on the type of biophysical environment. Cycled by rare fire or other disturbance events. Often have complex structures influenced by small gap disturbances and understory regeneration.

and thermal IR wavelengths, there is a transition band called the mid-IR (1.3–8 μm) where reflectance and emittance processes are mixed together. The sun, with a radiative temperature of 6,000 K, has its peak in the visible band, around 0.48 μm. The Earth's surface emits strongly in the thermal band, around 9.66 μm, corresponding to an average temperature of 300 K. One of the main effects

Table 6.10. Condition classes vs. fire regimes

			Examples of Key Ecosystem Component Susceptibility to Changing Fire Regime Condition Classes			
Condition Class	Fire Regime	Example Management Options	Species composition and structure	Invasion by non-native species	Smoke production hydrology, and Soils	Insects and disease
Condition Class 1	Fire regimes are within the natural (historical) range, and the risk of losing key ecosystem components is low. Vegetation attributes (species composition, structure, and pattern) are intact and functioning within the natural (historical) range.	Where appropriate, these areas can be maintained within the natural (historical) fire regime by treatments such as fire use.	Species composition and structure are functioning (historical) range at both patch and landscape scales.	Non-native species are currently not present or present in limited extent. Through time or following disturbance sites are potential vulnerable to invasion by non-native species.	Functioning within their natural (historical) range.	Insect and disease populations functioning within their natural (historical) range.

Condition Class 2	Fire regimes have been moderately altered from their natural (historical) range. Risk of losing key ecosystem components is moderate. Fire frequencies have departed from natural frequencies by one or more return intervals (either increased or decreased). This result in moderate changes to one of more of the following: fire size, intensity and severity, and landscape patterns. Vegetation and fuel attributes have been moderately altered from their natural (historical) range.	Where appropriate, these areas may need moderate levels of restoration treatments, such as fire use and hand or mechanical treatments, to be restored to the natural fire regime.	Species composition and structure have been moderately altered from their historical range at patch and landscape scales. For example: Grasslands – Moderate encroachment of shrubs and trees and/or invasive exotic species. Shrublands – Moderate encroachment of trees, increased shrubs, or invasive exotic species. Forestland/Woodland – Moderate increases in density, encroachment of shade tolerant tree species, or moderate loss of shade intolerant tree species caused by fire exclusion, logging, or exotic insects or disease. Replacement of surface shrub/grass with woody fuels and litter.	Populations of non-native invasive species may have increased, thereby increasing the potential risk for these populations to expend following disturbances, such as wildfires.	Have been moderately altered from their natural (historical) range. Water flow typically less. Smoke and soil erosion following fire typically greater.	Insect and disease population have been moderately altered from their natural (historical) range.

(*Continued*)

Table 6.10. (Continued)

Condition Class	Fire Regime	Example Management Options	Examples of Key Ecosystem Component Susceptibility to Changing Fire Regime Condition Classes			
			Species composition and structure	Invasion by non-native species	Smoke production hydrology, and Soils	Insects and disease
Condition Class 3	Fire regimes have been substantially altered from their natural (historical) range. The risk of losing key ecosystem components is high. Fire frequencies have departed from natural frequencies by multiple return intervals. Dramatic changes occur to one or more of the following: fire size, intensity, severity, and landscape patterns. Vegetation attributes have been substantially altered from their natural (historical) range.	Where appropriate, these areas may need high levels of restoration treatments, such as hand or mechanical treatments, before fire can be used to restore the natural fire regime.	Species composition and structure have been substantially altered from their historical range at patch and landscape scales. For example: Grasslands – High encroachment and establishment of shrubs, trees, or invasive exotic species. Shrublands – High encroachment and establishment of trees, increased shrubs, or invasive exotic species. Forestland/Woodland – High increases in density, encroachment of shade intolerant tree species, or high loss of shade intolerant tree species caused by fire exclusion, logging, or exotic insects or disease.	Invasive species may be common and in some cases the dominant species on the landscape. Any disturbance will likely increase both the dominance and geographic extent of these invasive species.	Have been substantially altered from their historical range.	Insect and disease population have been substantially altered from their natural (historical) range. Typically higher mortality or defoliation.

of fire on the environment is the overheating of the surface, caused by reduced transpiration and by the increment of the solar radiation absorbed by soil. Lambin and Erlich (1996) observed that ash and charcoal form a dry layer that does not allow cooling processes, increasing by 7–8 K the superficial temperature during the day in recently burned areas. Forest fires have flame temperatures of between 548 and 693 K (Robinson 1991), so their band of maximum emittance is located in the range 4.18–5.29 μm (mid-IR). In consequence, this can be the most adequate spectral interval for fire detection. The limit in fire spot detection is the temporal resolution. To avoid the unmanageable spread of a fire, the attack time must be around 20–30 min. The thermal band is more sensitive to lower temperatures, so it can be utilized for recently burned areas.

The destructive effect of forest fires has initiated efforts for understanding and modeling the mechanisms of fire propagation in different land-cover types. It presents a complexity which involves several parameters, such as physical (combustion process), biological (biomass characteristics), and environmental (meteorology, topography) ones. Most of the studies put emphasis on fire behavior experienced with different combinations of fuel, weather, and topography. Perry (1998) reported two main aims in fire modeling: the fire growth prediction, spread rate quantification, and the fire area prediction of zones that can be affected. Fire simulation models are used by several forest services in different countries and almost all rely on the Rothermel equation of fire spreading rate (Andrews and Queen 2001). In the last two decades, several studies have been carried out on technical improvement of fire modeling (Richards 1990, 1995; Hirsch 2000; Fall and Fall 2001; Nelson 2002; Reed and McKelvey 2002) and on the development of spatio-temporal GIS fire modeling environment (Yuan 1997; De Vasconcelos et al. 2002). Fire simulation models have been mainly developed for temperate ecosystems and have to be calibrated for tropical and equatorial environments.

Critical parameters in wildfire modeling that need to be assessed are:

- Topography of the area
- Vegetation fuel layout (type and density)
- Meteorological conditions
- Transportation networks
- Public services and facilities, etc.

6.6.2 Detecting Fires and Emissions

The mid-IR is the optimum spectral band for monitoring hot spots. This band is less affected by smoke, so images obtained with this band can show active fires more easily than visual observation. The first experiments using sensors with information from this band were made in 1970, when an aerial scanner was used to detect forest fires in Montana (Wilson et al. 1971). During 51 h of observation, over 800 hot targets were detected and delineated. More than 200 of them were wildfires, and 45 were observed by the sensor before they were visually observed from the 59 lookout towers in the area. The scanner also provided real-time information useful to fire fighters on the ground. Some years later, in 1988, a similar scanner was used to help extinguish the big Yellowstone fire (Ambrosia and Brass 1989). Thermal-band observation has also been used for fire fighting using scanners on board helicopters (Bosiak 1986).

Besides detecting hot spots it is possible to remotely sense particulate and gas emissions. Most of the work related to gas release is based on the determination of optical thickness, particle size, and single scattering albedo using shortwave bands. Kaufman et al. (1990) in Canada, Cheasepeake Bay, developed a self-consistent algorithm for simultaneous determination of smoke emissions and characteristics. This method is based on visible and near-IR bands. The algorithm derives the aerosol characteristics from the difference in upward radiances between two images of the same area, one acquired on a clear day and one during hazy conditions. A close relation was found between the aerosol estimation from AVHRR data and ground measurements. A more qualitative approach was adopted by other authors to estimate the area affected by smoke plumes in large-scale forest fires (Chung and Le 1984; Jijia et al. 1989). From the analysis of space photographs taken during the Skylab and Space Shuttle missions between 1973 and 1986, a tenfold increase in smoke pall surfaces was estimated (Herfert and Lulla 1990). Setzer and Pereira (1991) also estimated the amount of gases released to the atmosphere from the 1987 Brazilian Amazonian fires. Based on a simple method which relates the total dry matter burned with fire emissions, total amounts of 1,700 Tg of CO_2, 520 Tg of C, and 94 Tg of CO were calculated. More detailed projects to measure gas emissions from tropical forest have been carried out since then, using satellite, airborne, and field sensors (Kaufman et al. 1990, 1992).

In spite of its interest, the use of AVHRR data for fire detection is complicated by several problems (Chuvieco and Martin 1994): the limited spatial resolution of the sensor, the difficulty in geolocating image pixels, and – most important – the limited spectral sensitivity. Originally designed for sea surface temperature estimation, channel 3 is saturated at 320 K. Therefore, confusion between active fires and other hot surfaces (i.e., bare soil during the summer season) may easily arise. On the other hand, the strong radiometric contrast between a fire and the surroundings may cause a pixel to appear as being burned even if only a small portion of it is actually occupied by the fire (Kaufman et al. 1990). Fire observation requires the presence of active fires at the time of satellite overpass. Most environmental satellites have a coarse cycle (one image every 2 or 3 weeks), which makes it difficult to use them as a tool for real-time fire detection. An alternative is the use of meteorological satellites. The drawback in using images from these satellites is their low spatial resolution, 1.1 km at nadir. Therefore, their use is more suitable for large fires, although this limit is balanced by the possibility of detecting fires in remote lands where no visual detection is available. But even with high temporal resolution sensors, such as AVHRR, fires may be missed because of its brevity or because of cloud cover.

6.6.3 Mapping Burned Areas

Evaluation of forest fire effects includes two aspects: (1) the identification and mapping of burned areas and (2) the evaluation of the damage in terms of surface and affected vegetation and soil. Fires have two principal effects on the forest areas (Robinson 1991): char production and amassing and fire scars. Char represents an unmistakable fire action signal, but it tends to be almost completely removed by wind and rain a few weeks after the fire; thus, forest structure alteration is more an operational indicator. However, structure cannot be associated with fire in an unambiguous manner: partial or total forest canopy destruction can be the consequence of different actions such as cutting, grazing, or pathogenic agents. This distinction is important to keep in mind remote-sensing techniques to determine the location and extent of burned areas. To recognize the spectral proprieties of burned areas, it is important to distinguish white ash from black ash, when applying (Chandler et al. 1983). Ash consists of light-colored mineral residues, produced by the complete combustion of vegetal materials and caused by an intense fire (Riggan et al. 1994); in contrast, charcoal indicates incomplete biomass combustion, caused by less disastrous fires (Chandler et al. 1983; Ambrosia and Brass 1988). These two combustion products are often described as ash, confounding spectral information and interpretation of burned areas.

One of the first experiences in forest fire satellite mapping came from Colorado (Hitchcock and Hoffer 1974). Two images after a fire were digitally processed and two areas of different intensities were delineated on the map. Sometimes, problems in identifying different damage levels occur (Minick and Shain 1981), because of the range of species affected differentially by the fire. A 58% agreement between burning intensities collected on the ground and detected by satellite may be regarded as satisfactory accuracy considering the high costs of field survey.

A very important aspect in fire evaluation is timeliness. Remote sensing provides a quick tool to monitor the area affected by a fire and this makes it possible to mitigate some of its consequences (unprotected soil may be eroded). In order to apply rehabilitation measures, the burned land surface should be quickly evaluated. This was the objective of a pilot study conducted in Oregon by Isaacson et al. (1982). An image before the fire was used to identify burned areas for potential reforestation. In the previously forested areas, annual grasses were planted, while in areas of no previous forest, perennial grasses were used instead. The former do not prevent tree growth, while the latter do. Thanks to this quick evaluation, a big storm 2 months later did not create any mud-slide or serious soil erosion. Remote-sensing studies of fires are usually carried out using optical/IR sensors to provide some of the essential information, such as the NOAA-AVHRR, Landsat Thematic Mapper (TM) (Pereira and Setzer 1993) and SPOT-1–3 (Liew et al. 1998). The AVHRR sensors are able to detect the presence of fire hot spots using channel 3 (3.8 μm) and provide images showing the spatial distribution and temporal evolution of fire hot spots (Matson et al. 1987; Robinson 1991). Channels 1 and 2 could also provide information about aerosol characteristics and distribution of the smoke haze (Ferrare et al. 1990). However, owing to the coarse resolution (1.1 km) of the AVHRR sensors, the exact locations of fires and the types of land cover on fire cannot be determined. Measurements from hot spot images generally overestimate the area burned but underestimate the total fire count (Kaufman et al. 1990; Malingreau 1990). Burned areas can be mapped with high resolution, using the spectral characteristics of the fire scars. Individual smoke plumes can be observed in the images and hence the precise locations of the active fire areas can be determined. Land cover can also be determined from the spectral and contextual features of the fire areas.

Digital processing can be less than satisfactory if it is only based on spectral data. Often, burned lands do not have a singular spectral signature but rather a wide range of reflectance values, causing some difficulties:

1. Confusion between slightly burned and nonaffected vegetation (Benson and Briggs 1978).
2. Difficulties in separating burned vegetation from other cover classes, especially from urban land. Confusion is greatest for areas where the previous vegetation was sparse (Tanaka et al. 1983).
3. Confusion between burned areas and shade and water (Tanaka et al. 1983).

To identify a suitable spectral range for burned areas, it is necessary to study characteristics of the neighboring vegetation not affected by fires. It is has been demonstrated for different ecosystems that the near-IR spectrum is most suitable to study burned areas (Hall et al. 1980; Tanaka et al. 1983; Richards 1984; Langaas and Kane 1991; López Garcia and Caselles 1991; Pereira and Setzer 1993; Caetano et al. 1994; Marchetti et al. 1995; Razafimpanilo et al. 1995; Pereira 1999). In the near-IR spectral range sound vegetation shows a high reflectance and can clearly be differentiated from surfaces with burned material, which have lower averages values of reflectance than any other land-cover type (except water). In the visible region, burned areas appear similar to agricultural or sparsely vegetated lands.

Chuvieco and Congalton (1988) indicated a strong reduction of near-IR values as a consequence of a fire: particularly they noted a decrease in the Landsat TM4 band (0.76–0.90 μm). This result agrees with Koutsias and Karteris (2000), who reported in addition a lower and nonmeaningful decrease in the TM5 band (1.55–1.75 μm). Jakubauskas et al. (1990) found a reduction of reflectance in the near-IR region that is proportional to fire intensity; this relationship proved to be suitable for classifying severity classes from Landsat TM. Pereira (1999) combined AVHRR near-IR and mid–IR bands in small-scale burned area monitoring and considered Landsat TM4 and TM5 as the most adequate to identify burned areas. López Garcia and Caselles (1991) showed that Landsat TM4 and TM7 (2.08–2.35 μm) results were of most operational interest.

Pattern and texture variations after fire can be observed for a variable period of time depending on land-cover type. In general, in forest ecosystems they persist for 1 year, while in prairie they are detectable only for 2 months. A reason could be the dependency of the size of coal particles on the size of original material. In woods, flammable material has different sizes (leaves, branches, and stems) and can remain on the ground as coal. In prairie, fire transforms burned herbs almost exclusively into ash that can easily be dispersed by wind and rain.

Mapping of burned areas should be based on a multitemporal approach, as it avoids confusion with land-cover types that show behavior similar to that of a burned area. A widespread critique against multitemporal analysis pertains to radiometric and geometric corrections that would guarantee the highest precision and accuracy in the overlay phase. Inaccuracies concerning the size and radiometric properties of images can produce random errors and lead to a biased estimate of burned areas. In addition, the magnitude of the errors depends also on other parameters, such as the size and fragmentation of burned areas. Where only images from the postfire situation are available, maximum-likelihood and minimum-distance classifiers have been widely used. Those classifiers are more difficult to apply in multitemporal analysis of burned areas (Siljestrom and Moreno 1996; Marchetti et al. 1995; Chirici et al. 2002). French et al. (1995) and Pozo et al. (1997) applied maximum value composites of the original images for mapping the postfire situation. Other techniques for multitemporal burned area monitoring are:

- Visual classification of the postfire mutitemporal images. A color composite (Fig. 6.10) that includes the green, red, and near-IR is used to identify burned scars. Tentative burned areas on the postfire images are confirmed as burned areas when they are identified as not burned on the previous image.

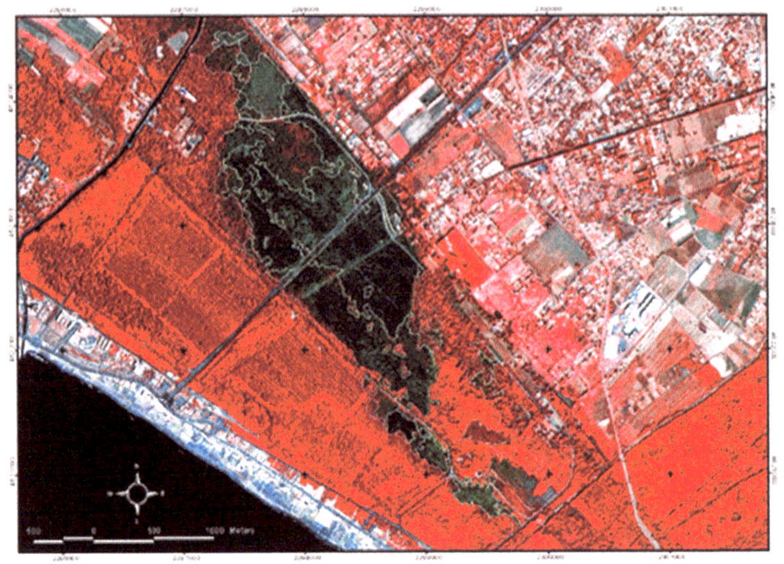

Fig. 6.10. Ikonos composite 124 image used for mapping burned pine forest (*P. pinea*) close to Rome after a fire event in 2000

- The identification of vegetated areas using thresholding of the normalized-difference vegetation index (NDVI) on the prefire image. A mask of pixels with positive NDVI values is created. This mask retains areas that are vegetated on the prefire image and are susceptible to suffer forest fires. In the next step the a "burn index" (orthogonal transformation of three spectral bands in the range of the green, red, and near-IR) is applied to the postfire image. The burn index identifies clearly burned areas. However, it may also confuse some urban features, which should be removed when applying the NDVI mask from the prefire image. The intersections of the mask obtained on the prefire image with the burned areas identified on the postfire image isolate vegetated areas that have been burned in the time lapse of the acquisition of the two images.
- The use of a nonparametric and nonsupervised classification algorithm. Fukunaga (1990) used a gradient modified function that searches for local modes on bivariate histograms. The algorithm expands from the modes identified to the border of the burned area, which is recognized as belonging to a different spectral class. This approach was developed using prefire and postfire imagery and was then applied to postfire imagery. Frequently many of the modes detected in the bivariate histograms may not correspond to modes of spectral classes, which may be a consequence of a simple modification of the image (e.g., normalization process).

6.6.4
Vegetation Indices and Forest Fires

Several linear transformations of reflectance values or biomass or vegetation indices (VI) have been developed. The use of VIs, such as the NDVI, is widely used for global-change studies (Goward et al. 1991). Forest fire assessment through remote sensing may focus on different objectives, including mapping of burned areas, evaluating the intensity of damage, or monitoring postfire regeneration. VI are used frequently (Jakubauskas et al. 1990; Bovio 1990). They have improved classification accuracy and have significantly contributed to the identification of burning intensities (Milne 1986) mainly by enhancing the separation of vegetation and soil. Utilizing AVHRR, a simple difference between NDVI values of images acquired before and after the fire gives a reliable estimation of burned lands for the large-area events (Chuvieco and Martin 1993; Kasischke et al. 1995). Principal component analysis was used to discriminate among burned areas using multitemporal sets of images by Richards (1984), who detected changes in vegetation cover by comparing modified spectral information. VI captures relations between spectral bands in parts of the spectrum where major differences between vegetation and other land-cover types occur (Tucker 1979). VI is commonly correlated to some physiologic parameters of vegetation, such as leaf area index, biomass, photosynthetic activity, and productiveness (Sellers 1987; Baret and Guyot 1991; Asrar et al. 1992). Owing to their ecological value, VI are often used for mapping fires and monitoring regeneration processes. They can be "intrinsic" (simple red/IR or NDVI) or "line soil related" (perpendicular VI or weighted-difference VI) (Gomarasca 2004). Figure 6.11 gives an example of Landsat bands TM3 (0.63–0.69 μm), TM4 (0.76–0.90 μm), and TM5 (1.5–1.75 μm). Those bands are respectively sensitive to chlorophyl concentration, organization of mesophyll, and water content in multitemporal analysis of greenness indices, such as:

- VI=TM4/TM3 (Birth and McVey 1968)
- NDVI= (TM4–TM3)/(TM4+TM3) (Rouse 1974)
- The IR index: II
- The normalized burn ratio, where the TM3 band is substituted with the TM5 band (Hardinsky et al. 1983; Cohen 1991; Marchetti et al. 1995; Key 2003) (Fig. 6.11)

6.6.5
Indices for Danger Assessment

Wildfires occur mainly in dry seasons. This entails the conception of operational methods for near-real-time fire danger assessment using EO. Good

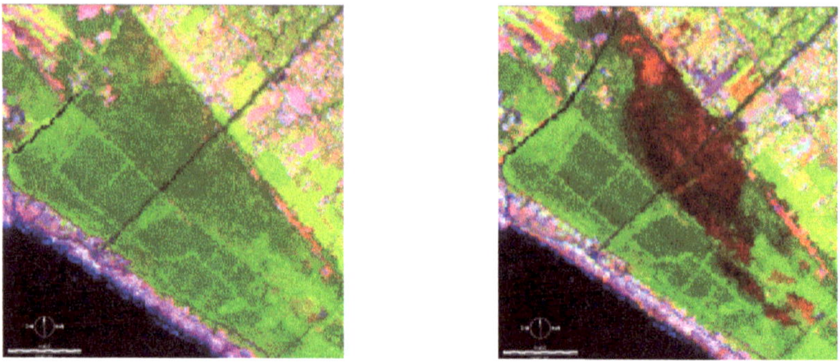

Fig. 6.11. Multitemporal analysis of the normalized-difference vegetation index obtained from Landsat Thematic Mapper images before and after the Castelfusano fire event, close to Rome

knowledge of the fuel status in terms of quantity (load) and quality (moisture, distribution, and size) is mandatory. Several indices can be assessed for estimate of danger. Some authors assess landscape status from surface and near-surface moisture estimation using remote-sensing data (Gillies et al. 1997; Chuvieco et al. 2000; Ceccato et al. 2002a, b; Goward et al. 2002; Sandholt et al. 2002). Other studies are based on vegetation indices (Gonzalez-Alonso et al. 1997; Sannier et al. 2002). Recently the use of thermal properties of land surface has been considered as a possible approach to address moisture status at the surface and near the surface (Goetz 1997; Garouani et al. 2000; Kant and Badarinath 2000; Czajkowski et al. 2002).

References

Achard F, Eva HD, Stibig HJ, Mayaux P, Gallego J, Richards T, Malingreau JP (2002) Determination of deforestation rates of the world's humid tropical forests. Science 297:999–1002

Acharya NG, Bhattarai G, de Gier A, Stein A (2000) Systematic adaptive cluster sampling for the assessment of rare tree species in Nepal. For Ecol Manage 137(1-3): 65–73

Agee JK, Pickford SG (1985) Vegetation and fuel mapping of North Cascades National Park Service Complex. National Park Service Cooperative Park Studies Unit, University of Washington, Seattle

Agresti A (1992) A survey of exact inference for contingency tables. Stat Sci 7(1):131–177

Agresti A, Caffo B (2000) Simple and effective confidence intervals for proportions and differences of proportions result from adding two successes and two failures. Am Stat 54(4):280–288

Akin H, Siemes H (1988) Praktische Geostatistik. Springer, Berlin Heidelberg New York

Ambrosia VG, Brass JA (1988) Thermal analysis of wildfires and effects on global ecosystem cycling. Geocarto Int 3(1):29–39

Ambrosia VG, Brass JA (1989) Remote sensing thermal analysis of the 1988 Yellowstone National Park Wildfires. Association of American Geographers National Meeting, Baltimore, program abstracts, Sect VI, p 3

Amrhein P (1995) Minimax estimation of proportions under random sample-size. J Am Stat Assoc 90(431):1107–1111

Andrews PL, Queen LP (2001) Fire modeling and information system technology. Int J Wildland Fire 10:343–352

Angers C (1989) Note on quick simultaneous confidence intervals for multiple proportions. Am Stat 43:91

Arner SL, Westfall JA, Scott CT (2004) Comparison of annual inventory designs using forest inventory and analysis data. For Sci 50(2):188–203

Aronoff S (1989) Geographical information systems. Management perspective. WDL, Ottawa

Asner GP, Keller M, Silva NM (2004) Spatial and temporal dynamics of forest canopy gaps following selective logging in the eastern Amazon. Global Change Biol 10(5):765–783

Asrar G, Myneni RB, Choudhury BJ (1992) Spatial heterogeneity in vegetation canopies and remote sensing of absorbed photosynthetically active radiation: a modeling study. Remote Sensing Environ 41:85–103

Atkinson PM (1993) The effect of spatial resolution on the experimental variogram of airborne MSS imagery. Int J Remote Sensing 14:1005–1011

Atkinson PM, Lewis P (2000) Geostatistical classification for remote sensing: an introduction. Comput Geosci 26:361–371

Atkinson PM, Tatnall ARL (1997) Neural networks in remote sensing. Int J Remote Sensing 18(4):699–709

Atkinson PM, Dunn R, Harrison AR (1996) Measurement error in reflectance data and its implications for regularizing the variogram. Int J Remote Sensing 17:3735–3750

Auclair D (1983) Mesures des biomasses et des accroissements forestiers. Colloques de l'INRA 19. INRA, Orleans

Avery TE, Burkhart HE (2001) Forest measurements, 5th edn. McGraw-Hill, New York

Azimi-Sadjadi MR, Ghaloum S, Zoughi R (1993) Terrain classification in SAR images using principal components analysis and neural networks. IEEE Trans Geosci Remote Sensing 31(2):511–515

Baatz M, Schäpe A (1999) Object-oriented and multi-scale image analysis in semantic networks. In: Proceedings of the 2nd international symposium on operationalization of remote sensing, ITC, Enschede, 16–20 August 1999

Baatz, M, Heynen M, Hofmann P, Lingenfelder I, Mimier M, Schape A, Weber M, and Willhauck G (2001) eCognition user guide 2.0: object oriented image analysis. Definiens Imaging, Munich

Bachmann P, Köhl M, Päivinen R (eds) (1998) Assessment of biodiversity for improved forest planning. Kluwer, Dordrecht

Banyard SG (1987) Point sampling using constant tallies is biased: a tropical rainforest case study. Commonw For Rev 66:161–163

Barabesi L, Fattorini L (1995) Kernel plant density estimation by a ranked set sampling of point-to plant distances. In: Köhl M, Bachmann P et al (eds) The Monte Verita conference on forest survey designs. Swiss Federal Institute of Forest, Snow and Landscape Research (WSL/FNP), Zurich, Swiss Federal Institute of Technology (ETH), Birmensdorf, pp 71–80

Barbieri MM, Berger JO(2004) Optimal predictive model selection. Ann Stat 32(3):870–897

Baret F, Guyot G (1991) Potentials and limits of vegetation indices for LAI and APAR assessment. Remote Sensing Environ 35:161–173

Barndorff-Nielsen OE, Cox DR (1989) Asymptotic techniques for use in statistics. Chapman and Hall, London

Barnett DT, Stohlgren TJ (2003).A nested-intensity design for surveying plant diversity. Biodivers Conserv 12(2):255–278

Baskerville GL (1965) Estimation of dry weight of tree components and total standing crop in conifer stands. Ecology 46(6):867–869

Batalha MA, Aragaki S, Mantovani W (1998) Chave de identificacao das especies vasulares do cerrado em emas (pirassununga, sp) baseada em caracteres vegetativos. Universidade de Sao Paulo

Beers TW (1962) Components of forest growth. J For 60(4):245–248

Bellhouse DR (1985) Computing methods for variance estimation in complex surveys. J Off Stat 1(3):323–329

Bellhouse DR (1988) Systematic sampling. In: Krishnaiah PR, Rao CR (eds) Sampling. Elsevier, Amsterdam, pp 125–145

Benediktsson JA, Swain PH, Ersoy OK (1990) Neural network approaches versus statistical methods in classification of multisource remote sensing data. IEEE Trans Geosci Remote Sensing 28(4):540–552
Benson ML, Briggs I (1978) Mapping the extent and intensity of major forest fires in Australia using digital analysis of Landsat imagery. In: Proceedings of the international symposium on remotesensing observation and inventory of Earth resources, Freiburg, pp 1965–1980
Bernhardsen T (2002) Geographich information systems. Wiley, New York
Bettinger P, Wing MD (2003) Geographic information systems: applications in forestry and natural resources management. McGraw-Hill, New York
Bettinger P, Bradshaw GA, Weaver GW (1996) Effects of geographic information system vector-raster-vector data conversion on landscape indices. Can J For Res 16:1416–1425
Beyer WH (1968) Handbook of tables for probability and statistics. CRC, Boca Raton
Bickford CA (1959) A test of continuous inventory for national forest management based upon aerial photographs, double sampling and remeasured plots. Proc Soc Am For 143–148
Biolley HE (1921) L'aménagement des forêts par la méthode expérimentale et spécialment la méthode du contrôle. Neuchatel, Paris
Birth GS, McVey G (1968) Measuring the color of growing turf with a reflectance spectrophotometer. Agron J 60:640–643
Bischof H, Schneider W, Pinz AJ (1992) Multispectral classification of landsat-images using neural networks. IEEE Trans Geosci Remote Sensing 30(3):482–490
Bischoff C, Dröschmeier R (eds) (2000) European monitoring for nature conservation. Schriftenr Natursch Landschaftspfl 2
Bishop YMM, Fienberg SE, Holland PW (1975) Discrete multivariate analysis: theory and practice. MIT Press, Cambridge
Bitterlich W (1947) Die Winkelzählmessung. Allg Forst-Holzwirtsch Ztg 58(11/12): 94–96
Bitterlich W (1997) The relascope idea. Commonwealth Agricultural Bureau, Farnham
Böhm M, Löw R, Haag J, Kerwer A, Lüttge U, Rausch T (1993) Evaluation of comparative DNA amplification fingerprinting for rapid species identification within ther genus clusia. Bot Acta 106(5):448–453
Boninsegna JA, Villalba R, Amarilla L, Ocampo J (1989) Studies on tree rings, growth rates and age-size relationships of tropical tree species in Misiones, Argentina. In: Baas P, Vetter RE (ed) Growth rings in tropical trees. Proceedings of the joint session of IUFRO P505 tree ring analyses and IAWA on age and growth rate determination in tropical trees, Sao Paulo, Brazil, 18 May 1988. Int Assoc Wood Anat Bull 10(2):161–169
Boom BM (1989) Use of plant resources by the Chacobo. In: Posey DA, Balee W (eds) Resource management in Amazonia: indigenous and folk strategies. Advances in economic botany 7. New York Botanical Garden, New York, pp 78–96
Boon DR (1970) A critical review of timber survey methods in tropical rainforest. Publications of the International Institute for Aerial Survey and Earth Sciences 58
Borchers DL, Zucchini W, Fewster RM (1998) Mark-recapture models for line transect surveys. Biometrics 54(4):1207–1220
Borman FH, Berlyn G (ed) (1981) Age and growth rate of tropical trees, new directions for research. Bull Sch For Environ Stud Yale Univ 94:20–30

Borner M, Guericke M, Leder B et al (2003) Survey of timber-quality relevant parameters of single trees – standardised methods for young to medium-mature broadleaf species as a basis for scientific studies (Erhebung qualitätsrelevanter Parameter am Einzelbaum. Aufnahmestandards für junge bis mittelalte Laubhölzer als Grundlage für wissenschaftliche Untersuchungen). Forstarchiv 74(6):275–282

Bosdogianni P, Kalviainen H, Petrou M, Kittler J (1997) Robust unmixing of large sets of mixed pixels. Pattern Recogn Lett 18(5):415–424

Bosiak A (1986) Detection and assessment of damage to forest, including that caused by air pollution and fires. In: Sohlberg S, Sokolov VE (eds) Practical application of remote sensing in forestry. Nijhoff, Dordrecht, pp 177–193

Bovio G (1990) La pianificazione antincendi per la difesa del patrimonio boschivo. Acad Ital Sci For IFM 38:431–458

Bowden DC (1979) Multi-level sampling designs for resource inventories. Department of Forest and Wood Sciences, Colorado State University, Rocky Mountain Forest and Range Experimental Station, USDA Forest Service, Ft Collins

Box GEP, Tiao GC (1973) Bayesian inference in statistical analysis. Addison-Wesley, London

Brändli UB (2001) Protection demands of the society. In: Brassel P, Lieschke H (eds) Swiss National Forest Inventory: methods and models of the second assessment. Swiss Federal Reseach Institute WSL, Birmensdorf, pp 232–238

Brändli UB, Ulmer U (1999) Naturschutz und Erholung. In: Brassel P, Brändli U (eds) Schweizerisches Landesforstinventar: Ergebnisse der Zweitaufnahme 1993–1995. Eidgenössiche Forschungsanstalt Forschungsanstalt für Wald, Schnee und Landschaft, Birmensdorf, pp 279–394

Brändli UB, Kaufmann E, Stierlin HR (1995) Survey of biodiversity at the forest margin in the second Swiss NFI. In: Köhl M, Bachmann P et al (eds) The Monte Verita conference on forest survey designs. Swiss Federal Institute of Forest, Snow and Landscape Research (WSL/FNP), Zurich, Swiss Federal Institute of Technology (ETH), Birmensdorf, pp 141–150

Brassel P (1995) Assessment of non-productive forest functions in the Swiss NFI. In: Köhl M, Bachmann P et al (eds) The Monte Verita conference on forest designs. Swiss Federal Institute of Forest, Snow and Landscape Research (WSL/FNP), Zurich, Swiss Federal Institute of Technology (ETH), Birmensdorf, pp 38–46

Brazier JB (1981) Wood quality and properties and utilization of tropical wood – acceptable timber from tomorrow's forests. In: Hillis WE (ed) Proceedings of the XVII IUFRO world congress, Kyoto, Japan, 6–11 September 1981, Division 5, forest products, pp 86–94

Brewer KRW et al (2000) Deriving and estimating an approximate variance for the Horvitz–Thompson estimator using only first order inclusion probabilities. Paper presented at the 2nd International conference on establishment surveys. American Statistical Association, Buffalo

Brewer KRW, Hanif M (1983) Sampling with unequal probability. In: Brillinger D, Fienberg SE, Gani J, Hartigan JA, Krickeberg K (eds) Lecture notes in statistics, vol 15. Springer, Berlin Heidelberg New York, pp 1–164

Brown DG, Pijanowksi BC, Duh JD (2000) Modeling the relationships between land use and land cover on private lands in the Upper Midwest, USA. J Environ Manage 59(4):247–263

Brown JA (2003) Designing an efficient adaptive cluster sample. Environ Ecol Stat 10(1):95–105
Brown JA, Manly BJF (1998) restricted adaptive cluster sampling, Environ Ecol Stat 5(1):49–63
Brown JK (1995) Fire regimes and their relevance to ecosystem management. In: Proceedings of the Society of American Foresters National Convention, Anchorage, 18–22 September 1994. Society of American Foresters, Washington, DC, pp 171–178
Brown S (1997) Estimating biomass and biomass change in tropical forests. FAO forestry paper no 134. FAO, Rome
Bruce D, Max TA (1990) Use of profile equations in tree volume estimation. In: LaBau J, Cunia T (eds) State-of-the-art methodology of forest inventories. U S Dep Agric Gen Techn Rep PNW 263:213–220
Brus DJ, Jansen MJW, De Gruijter JJ (2002) Optimizing two- and three-stage designs for spatial inventories of natural resources by simulated annealing. Environ Ecol Stat 9(1):71–88
Brünig EF (1973) Species richness and stand diversity in relation to site and succession of forests in Sawarak and Brunei (Borneo). Amazoniana 4(3):293–320
Bucur V (1985) Ultrasonic, hardness and X-ray densitometric anaysis of wood ultrasonics 23(6):269–275
Burgan RE, RC Rothermel (1984) BEHAVE: Fire behavior prediction and fuel modeling system – FUEL subsystem. General technical report INT 167. USDA Forest Service
Burk R (1991) Kategoriale Datenanalyse unter komplexem Design. PhD thesis, University of Freiburg
Burke R (2003) Project management. Wiley, New York
Burnham KP, Anderson DR, Laake JL (1980) Estimation of density from line transect sampling of biological populations. Wildl Monogr 72
Burrough PA, McDonnel A (2000) Principles of geographic information systems. Oxford University Press, Oxford
Burschel P, Huss J (2003) Grundriss des Waldbaus. Ulmer, Stuttgart
Cairns MA, Olmsted I, Granados J et al (2003) Composition and aboveground tree biomass of a dry semi-evergreen forest on Mexico's Yucatan Peninsula. For Ecol Manage 186(1–3):125–132
Campbell JB (1996) Introduction to remote sensing, 2nd edn. Taylor and Francis, London
Canada Centre for Remote Sensing (1998) Fundamentals of remote sensing. Natural Resources Canada, Ottawa
Canada Centre for Remote Sensing (2003) Principles of remote sensing. Centre for Remote Imaging, Sensing and Processing (CRISP), National University of Singapore. http:// www.crisp.nus.edu.sg/~research/tutorial/rsmain.htm
Cannell MGR (1982) Untersuchungen zur Wuchsdynamik junger Kiefernbestände. Dissertation, Ludwig Maximilian University, Munich
Cannell MGR, Milne R, Hargreaves KJ, Brown TAW, Cruickshank MM, Bradley RI, Spencer T, Hope D, Billett MF, Adger WN, Subak S (1999) National inventories of terrestrial carbon sources and sinks: the UK experience. Clim Change 42(3):505–530
Cao QV (2004) Predicting parameters of a Weibull function for modeling diameter distribution. For Sci 50:682–685
Carlson TN, Ripley DA (1997) On the relation between NDVI, fractional vegetation cover, and leaf area index. Remote Sensing Environ 62:241–252

Carroll RJ, Ruppert D, Stefanski LA (1995) Measurement error in nonlinear models. Chapman and Hall, London

Casella G, Berger RL (2002) Statistical inference. Duxbury, London

Caetano MS, Mertes LAK, Pereira JMC (1994) Using spectral mixture analysis for fire severity mapping. In: Proceedings of 2nd international conf on forest fire research, Coimbra, pp 667–677

Ceccato P, Gobron N, Flasse S, Pinty B, Tarantola S (2002a) Designing a spectral index to estimate vegetation water content from remote sensing data, part 1. Theoretical approach. Remote Sensing Environ 82:188–197

Ceccato P, Flass S, Grégoire J-M (2002b) Designing a spectral index to estimate vegetation water content from remote sensing data, part 2. Validation and applications. Remote Sensing Environ 82:198–207

Cervantes V, Arriaga V, Meave J et al (1998) Growth analysis of nine multipurpose woody legumes native from southern Mexico. For Ecol Manage 110(1–3):329–341

Chandler C, Cheney P, Thomas P, Trabaud L, Williams D (1983) Fire in forestry, vol 1. Forest fire behavior and effects. Wiley, New York

Chandrasekharan C (1992) Terminology, definition and classification of forest products other than wood. FAO corporate document repository. http://www.fao.org/documents/ show_cdr.asp?url_file=/docrep/V7540e/V7540e28.htm

Chandrasekharan C (1995) Terminology, definition and classification of forest products other than wood. In: Report of the international expert consultation on non-wood forest products, Yogyakarta, Indonesia, 17–27 January 1995. Non-wood forest products no 3. FAO, Rome, pp 345–380

Chandra-Sekar C, Deming WE (1949) On a method of estimating birth and death rates and the extent of registration. J Am Stat Assoc 44(101):21–31

Chavez P, Bauer B (1982) An automatic optimum kernal-size selection technique for edge enhancement. Remote Sensing Environ 12:23–38

Chen D (1998) Measurement errors in line transect surveys. Biometrics 54(3):899–908

Chen D, Cowling A (2001) Measurement errors in line transect surveys where detectability varies with distance and size. Biometrics 57(3):732–742

Chen D, Shao J (2001) Jackknife variance estimation for nearest-neighbor imputation. J Am Stat Assoc 96(453):260–269

Chen D, Stow DA, Tucker L, Daeschner S (2001) Detecting and enumerating new building structures utilizing very-high resolution image data and image processing. Geocarto Int 16:69–82

Cheng Tan L, Ruiz Pérez M, Ibch M (1996) Non-timber forest product database. CIFOR special publication. CIFOR, Jakarta

Cherrill AJ, Fuller RM (1994) The use of classified Landsat-5 thematic mapper imagery in the characterization of landscape composition: a case study in northern England. J Environ Manage 40(4):357–377

Chilès J-P, Delfiner P (1999) Geostatistics: modeling spatial uncertainty. Wiley, New York

Chirici G, Corona P, Marchetti M, Portoghesi L, Travaglini D (2002) Feasibility assessment very high resolution satellite imagery for hierarchical classification of burnt pine stands in a Mediterranean coastal environment. Atti del III congresso SISEF Alberi e foreste per il nuovo millennio, pp 67–72

Chirici G, Corona P, Marchetti M, Travaglini D (2003a) Testing Ikonos and Landsat 7 ETM+ potential for stand-level forest type mapping by soft supervised approach.

In: Corona P, Kohl M, Marchetti M (eds) Advances in forest inventory for sustainable forest management and biodiversity monitoring. Forestry sciences, vol 76. Kluwer, Dordrecht, pp 71–87
Chirici G, Corona P, Travaglini D (2003b) Sperimentazione di tecniche di classificazione object-oriented di immagini QuickBird a fini forestali. Ital For Mont 58(4):231–250
Chong GW, Reich RM, Kalkhan MA, Stohlgren TJ (2001) New approaches for sampling and modeling native and exotic plant species richness. West North Am Nat 61(3):328–335
Chopra K, Kumar P (2004) Forest biodiversity and timber extraction: an analysis of the interaction of market and non-market mechanisms. Ecol Econom 49(2):135–148
Chou Y-H (1997) Exploring spatial analysis in geographich information systems. OnWord, Albany
Chrisman NR (1987) The accuracy of map overlays: a reassessment. Landscape Urban Planning 14:427–439
Christman N (2000) A review of quadrat-based sampling of rare, geographically clustered populations. J Agric Biol Environ Stat 5(2):168–201
Christman N (2001) Exploring geographic information. Wiley, New York
Christman N (2002) Efficiency of some sampling designs for spatially clustered populations. Envirometrics 8:145–166
Chung YS, LE HV (1984) Detection of forest-fire smoke plumes by satellite imagery. Atmos Environ 18(10)2143–2151
Chuvieco E, Congalton RG (1988) Mapping and inventory of forest fires from digital processing of TM data. Geocarto Int 4:41–53
Chuvieco E, Martin MP (1994) Global fire mapping and fire danger estimation using AVHRR images. Photogramm Eng Remote Sensing 60(5)563–570
Chuvieco E, Vaughan P, Riano D, Cocero D (2000) Fire danger and fuel moisture content estimation from remotely sensed data. Proceedings crossing the millennium: integrating spatial technologies and ecological principles for a new age in fire management. US University of Idaho, model. Remote Sensing Environ 79:225–242
Clark I (1979) Practical geostatistics. Applied Sciences, London
Clayton G, Cox TF (1986) Some robust density esimators for spatial point processes. Biometrics 42:753–767
Clement J (1988) Inventorying and monitoring forests. Proceedings of joint meetings of IUFRO subject groups S4.01, S4.02, S4.04 and S6.05 at the 18th IUFRO world congress, Ljubljana, 1986
Cochran WG (1977) Sampling techniques. Wiley, New York
Cohen WB (1991) Response of vegetation indices changes in three measures of leaf water stress. Photogramm Eng Remote Sensing 57 195–202
Congalton RG (1988) Using spatial autocorrelation analysis to explore the errors in maps generated from remotely sensed data. Photogramm Eng Remote Sensing 54(5):587–592
Congalton RG (1991) A review of assessing the accuracy of classifications of remotely sensed data. Remote Sensing Environ 37:35–46
Congdon RG (2001) Bayesian statistical modelling. Wiley, Chichester
Cordero LDP, Kanninen M (2002) Wood specific gravity and aboveground biomass of Bombacopsis quinata plantations in Costa Rica. For Ecol Manage 165:3–9

Cormack RM (1993) Variance of mark-recapture estimates. Biometrics 49:1188–1193
Corona P, Zeide B (eds) (1999) Contested issues of ecosystem management. Food Product, Binghampton
Corona P, Köhl M, Marchetti M (eds) (2003) Advances in assessments for sustainable forest management and biodiversity monitoring. Kluwer, Dordrecht
Correll RL, Cellier KM (1987) Effects of plot size, block size and buffer rows on the precision of forestry trials. Aust For Res 17:11–18
Cotta H (1804) Systematische Anleitung zur Taxation der Waldungen. Sander, Berlin
Coulter L, Stow DA, O'Leary J, Hope A, Longmire P, Peterson S (2000) Comparison of high-spatial resolution imagery for efficient generation of GIS vegetation layers. Photogramm Eng Remote Sensing 66:1329–1335
Cox DD, Cox LH, Ensor KB (1997) Spatial sampling and the environment: some issues and directions. Environ Ecol Stat 4:219–233
Cressie NAC (1991) Statistics for spatial data. Wiley, New York
Cressie N, Chan NH (1989) Spatial modeling of regional variables. J Am Stat Assoc 84:393–401
Crow TR (1978) Common regressions to estimate tree biomass in tropical stands. For Sci 24:110–114
Cruickshank MM, Tomlinson RW, Trew S (2000) Application of CORINE land-cover mapping to estimate carbon stored in the vegetation of Ireland. J Environ Manage 58(4):269–287
Cunia T (1979) On sampling trees for biomass table construction: some statistical comments. In: Frayer WE (ed) Forest resource inventories, workshop proceedings, vol II. Colorado State University, Ft Collins, pp 643–664
Cunia T (1981) Some theory on reliability of volume estimates in a forest inventory sample. For Sci 11(1):115–128
Cuzick J (2001) Event-based analysis times for randomised clinical trials. Biometrika 88(1):245–253
Czajkowski K, Goward S, Shirey D, Walz A (2002) Thermal remote sensing of near surface water vapor. Remote Sensing Environ 79:253–265
Czaplewski RL (1989) Graphical analysis of stem taper in model building. Can J For Res 19:522–524
Czaplewski RL, Brown AS, Walker RC (1989) Profile models for estimating log end diameters in the Rocky Mountain Region. U S Dep Agric For Serv Res Pap RM 284
Dalenius T, Gurney M (1951) The problem of optimal stratification. Skand Akt 34:133–148
da Silva JA (1982) Aplicacao do relascopio de banda larga em inventarios de florestas tropicais. Silvic Sao Paulo 16A(1):601–603
Datta GS, Fay RE, Ghosh M (1991) Hierarchical and empirical Bayes multivariate analysis in small area estimation. Proceedings of the Bureau of the Census 1991. Annual Research Conference. US Bureau of the Census, Washington, DC, pp 63–79
David M (1977) Geostatistical ore reserve estimation. Elsevier, Amsterdam
Davis-Case D (1990) The community's toolbox: the idea, methods and tools for participatory assessments, monitoring and evaluation in community forestry. FAO, Rome
Davison AC, Hinkley DV (1988) Saddlepoints approximations in resampling methods. Biometrika 75(3):417–431

De Beer JH, McDermott M (1989) The economic value of non-timber forest products in South East Asia. The Netherlands Committee for IUCN, Amsterdam
de Carvalho JOP (1981) Distribuicao diametrica de especies comerciais e potenciais em floresta tropical umida natural na Amazonia, Boletim de Pesquisa no 23. Centro de Pesquisa Agropecuaria do Tropico Umido, EMBRAPA, Brasília,
Dees MG (1988) Einfache statistische Analyse bei ordinaler Skalierung der Schadensvariablen und einer Auswahl der Bäume in Probekreisen. Forstliche Biometrie und Informatik, Vorträge anläßlich der 2 Sektionstagung. Mitt FVA Baden-Württemberg Heft 142:85–99
de Gier A (1989) Woody biomass for fuel. ITC publication 9. ITC, Enschede
Dejhan K, Wisetphanichkij S, Kerdyou P, Cheevasuvit F, Mitatha S, Pienvijarnpong C, Soonyeekan C (2000) Flood area assessment with fused multi-spectral multi-sensor by using texture feature analysis and neural network classification. Paper presented at the 21st Asian conference on remote sensing (ACRS 2000), Taipei, Taiwan, 4–8 December 2000
del Valle JI (1986) La ecuacion de crecimiento de von Bertalanffy en la determinacion de la edad y el crecimiento de arboles tropicales. Rev Fac Nac Agron Medellin 39(19):61–74
De Mers MN (1997) Fundamentals of geographic information systems. Wiley, New York
De Mers MN (2000) Fundamentals of geographic information systems, 2nd edn. Wiley, New York
Denison DGT, Holmes CC, Mallick BK, Smith AFM (2002) Bayesian methods for nonlinear classification and regression. Wiley, New York
Deutsch CV (2002) Geostatistical reservoir modelling. Oxford University Press, Oxford
Devall MS, Parresol BR, Wright SJ (1995) Dendrochronological analysis of *Cordia alliodora*, *Pseudobombax septenatum* and *Annona sraguei* in central Panama. Int Assoc Wood Anat J 16:411–424
De Vasconcelos MJP, Gonçalves A, Catry FX, Paúl JU, Barros FG (2002) A working prototype of a dynamic geographical information system. Int J Geogr Inf Sci 16(1):69–91
de Vries PG (1986) Sampling theory for forest inventory. A teach yourself course. Springer, Berlin Heidelberg New York
Diblasi A, Bowman AW (2001) On the use of the variogram in checking for independence in spatial data. Biometrics 57(211):218
Dobbertin M, Biging GS (1996) A simulation study of the effect of scene autocorrelation, training sample size and sampling method on classification accuracy. Can J Remote Sensing 22(4):360–367
Dobson JE (1995) Defining the university consortium for geographic information sciences. GIS World 8(3):44–46
Draper NR, Smith H (1981) Applied regression analyses. Wiley, New York
Dunn RM, Out DO & Wong JLG (1994) Report of the reconnaissance inventory of the high forest and swamp forest areas in Cross River State, Nigeria. Cross River State Forestry Project (ODA Assisted), Calabar
Dutt CBS, Udayalakshmt V, Sadhasivaiah AS (1994) Role of remote sensing in forest management, NRSA, Bangalore. http:// www.GISdevelopment.net
Dykstra DP (1997) Information systems in forestry Unasylva 189:10–15
EAFV (1988) Schweizerisches Landesforstinventar, Ergebnisse der Erstaufnahme 1982–1986 Eidg Anstalt Forstl Versuchswes Ber 304

Eastman JR (1999) Idrisi 32 – guide to GIS and image processing. Clark University, Worcester

EC (1992) Coordination of information on the environment land cover (CORINE). European Commission, Brussels

EC (1997) Study on European forestry information and communication System: report on forest inventory and survey systems. European Commission, Luxembourg

Eckstein D, Ogden J, Jacoby GC, Ash J (1981) Age and growth rate determination in tropical trees: the application of dendrochronological methods. In: Borman FH, Berlyn G (eds) Age and growth rate of tropical trees: new directions for research. Bull Sch For Environ Stud Yale Univ 94:83–106

Edwards I (1991) Quantitative ethnobotanical survey of a hectare of tropical forest near Toraut, Dumogo Bone National Park, Northern Sulawesi, Indonesia. Sulawesi Ethnobotanical Project. Preliminary report

Efford M (2004) Density estimation in live-trapping studies. Oikos 106(3):598–610

Efron B (1982) The jackknife, and other resampling plans. Society of Industrial and Applied Mathematics, Philadelphia

Efron B (1994) Missing data, imputation and the bootstrap. J Am Stat Assoc 89:463–474

Efron B (2004) The estimation of prediction error: covariance penalties and cross-validation. J Am Stat Assoc 99 (467):619–632

Efron B, Tibshirani RJ (1993) An introduction to the bootstrap. Chapman and Hall, Boca Raton

Eichhorn F (1904) Beziehung zwischen Bestandeshöhe und Bestandesmasse. Allg Forst J Z 80:45–49

Ekström M, Sjöstedt de Luna S (2004) Subsampling methods to estimate the variance of sample means based on non-stationary spatial data with varying expected values. J Am Stat Assoc 99(465):82–95

Ella AB, Escobin RP (1993) Taxonomy and wood anatomy of the manggasinoro species (shorea spp.): Dipterocarpaceae. The Philippine Journal of Science, Forest Products Research and Development Institute, Department of Science and Technology College, Laguna, Philippines

Ella AB, Pitarguen (1998) Identification and end-use characterization of unexploited woody vines based on morphological and anatomical feature. In: Abstracts of the 4th pacific regional wood anatomy conference, New horizons in wood anatomy on the threshold of a new miielium, 26–29 October

Elliot MR, Little RJA (2000) A Bayesian approach to combining information from a census, a coverage measurement survey, and demographic analysis. J Am Stat Assoc 95(450):351–362

Enquist BJ, AJ Leffler (2001) Long-term tree ring chronologies from sympatric tropical dry-forest trees: individualistic responses to climatic variation. J Trop Ecol 17:41–60

Environmental Systems Research Institute (1994) Understanding GIS: the ARC/INFO method, version 7 for UNIX and Open VMS. ESRI, Redlands

Fabbro L (2000) Amazonia biodiversity estimation using remote sensing and indigenous taxonomy. ERS-Envisat symposium Looking down to earth in the new millennium SP-461. European Space Agency Publication Division

Fall A, Fall J (2001) A domain specific language for models of landscape dynamics. Ecol Modell 141:1–18

FAO (1995) Forest resources assessment 1990 – global synthesis. FAO forestry paper 124. FAO, Rome
FAO (1998) Guidelines for the management of tropical forests 1. The production of wood. FAO forestry paper 135. FAO, Rome
FAO (2001) Forest resources assessment 2000, main report. FAO forestry paper 140. FAO, Rome
FAO (2003) State-of-the-world's forests. FAO, Rome
FAO/ECE/ILO (1992) Proceedinds of the joint FAO/ECE/ ILO committee seminar on the use of information systems in forestry, Garpenberg, Sweden, 14–18 September 1992. ECE/ FAO Agriculture and Timber Division, Geneva
Faraklioti M, Petrou M (2000) Recovering more classes than avilable bands for sets of mixed pixels in satellite images. Image Vision Comput 18:705–713
Fearnside PM, Graca PMLD, Leal N et al (1999) Tropical forest burning in Brazilian Amazonia: measurement of biomass loading, burning efficiency and charcoal formation at Altamira, Para. For Ecol Manage 123(1):65–79
Felix-Medina MH (2003) Asymptotics in adaptive cluster sampling. Environ Ecol Stat 10(1):61–82
Ferrare RA, Fraser RS, Kaufman YJ (1990) Saatellite measurements of large-scale air pollution: measurements of forest fire smoke. J Geophys Res 95(D7):9911–9925
Fienberg SE, Holland PW (1973) Simultaneous estimation of multinomial cell probabilities. J Am Stat Assoc 68:683–691
Fisher AR, Gorbert AS, Williams CB (1943) The relation between the number of species and the number of individuals in a random sample of an annual population. J Anim Ecol 12:42–58
Flewelling JW, Thomas CE (1984) An improved estimator for merchantable basal area growth based on point samples. For Sci 30:813–821
Flores LA, Martínez LI (2000) Land cover estimation in small areas using ground survey and remote sensing. Remote Sensing Environ 74:240–248
Foody GM (1996) Fuzzy modelling of vegetation from remotely sensed imagery. Ecol Modell 85:3–12
Foody GM (2000) Accuracy of thematic maps derived from remote sensing. Accuracy 2000. In: Heuvelink GBM, Lemmens MJPM (eds) Proceedings of the 4th international symposium on spatial accuracy assessment in natural resources and environmental sciences. Delft University Press, Amsterdam, pp 217–224
Foresman TW (ed) (1998) The history of geographic information systems: perspectives from the pioneers. Prentice Hall, Upper Saddle River
Forman RTT, Godron M (1986) Landscape ecology. Wiley, New York
Franco-Lopez H, Ek AR, Bauer ME (2001) Estimation and mapping of forest stand density, volume, and cover type using the k-nearest neighbors method. Remote Sensing Environ 77:251–274
Franklin J (2001) Remote sensing for sustainable forest management. Lewis, Boca Raton
Franklin J, Woodcock CE, Warbington R (2000) Digital vegetation maps of forest lands in California: integrating satellite imagery, GIS modeling, and field data in support of resource management. Photogramm Eng Remote Sensing 66:1209–1217
French NHF, Kasischke ES, Bourgearu-Chavez LL, Berry D (1995) Mapping the location of wildfires in Alaskan boreal forests using AVHRR imagery. Int J Wildland Fire 5(2):55–62

Fries J, Matérn B (1965) On the use of multivariate methods for the construction of tree taper curves. In: IUFRO 3rd conference of the advisory group on forest statistics section 25, Jouy-en-Josas, France, INRA 3:277–286

Fröhlich C, Mettenleiter M (2004) Terrestrial laser scanning: new perspectives in 3D surveying. In: Thies M, Koch B, Spiecker H, Weinacker H (eds) Laser-scanners for forest and landscape assessment. Int Arch Photogramm Remote Sensing Spatial Inf Sci 36(8/W2):7–13

Frost P (1990) Wood biomass of Brachystegia spiciformis in Zimbabwe, internal report. Forest Research Centre, Highlands, Harare

Fukunaga K (1990) Introduction to statistical pattern recognition, 2nd edn. Academic. Boston

Fuller WA (1987) Measurement error models. Wiley, New York

Furness GO (1976) The dispersal, age-structure and natural enemies of the long-tailed mealy bug, *Pseudococcus longispinus* (Targoni-Tozetti) in relation to sampling and control. Aust J Zool 24:237–247

Gamba P, Houshmand B (2001) An efficient neural classification chain for optical and SAR urban images. Int J Remote Sensing 22:1535–1553

Garouani AEl, Boussema MR, Ennabli M (2000) Utilisation du systeme d'information geographique (SIG) et des donnees de teledetection pour l'estimation de l'evapotranspiration reelle a l'echelle regionale. Int J Remote Sensing 21(15):2811–2830

Gaudard M, Karson M, Linder E, Sinha D (1999) Bayesian spatial prediction. Environ Ecol Stat 6(2):147–171

Gausman HW (1977) Reflectance of leaf components. Remote Sensing Environ 6:1–9

Gertner GZ (1984) The sensitivity of measurement error in stand volume estimations. Can J For Res 16:1120–1123

Gertner GZ, Köhl M (1992) An assessment of some nonsampling errors in a national survey using an error budget. For Sci 38:525–538

Ghosh M, Meeden G (1997) Bayesian methods for finite population sampling. Chapman and Hall, London

Ghosh M, Rao JNK (1994) Small area estimation: an appraisal. Stat Sci 9:55–93

Gholz HL, Curran PJ, Kupiec JA, Smith GM (1997) Assessing leaf area and canopy biochemistry of Florida pine plantations using remote sensing. In: Gholz HL, Nakane K, Shimoda H (eds) The use of remote sensing in the modeling of forest productivity. Kluwer, Dordrecht, pp 3–22

Giles RH, Trani MK (1999) Key elements of landscape pattern measures. Environ Manage 123:477–481

Gillies RR, Carlson TN, Cui J, Kustas WP, Humes KS (1997) A verification of the 'triangle' method for obtaining surface soil water content and energy fluxes from remote measurements of the NDVI and surface radiant temperature. Int J Remote Sensing 18(15):3145–3166

Gjertsen AK, Tomppo E, Tomter S (1999) National forest inventory in Norway: using sample plots, digital maps, and satellite images. In: IEEE 1999 international geoscience and remote sensing symposium, Hamburg, pp 729–731

Glackin DL (1998) International space-based remote sensing overview: 1980–2007. Can J Remote Sensing 24:307–314

Godambe VP, Thompson ME (1988) On single stage unequal probability sampling. In: Krishnaiah PR, Rao CR (eds) Sampling. Elsevier, Amsterdam, pp 111–123

Godoy R, Lubowski R, Markandya A (1993) A method for the economic valuation of non-timber tropical forest products. Econ Bot 47(3):220–233

Goetz SJ (1997) Multi-sensor analysis of NDVI, surface temperature and biophysical variables at a mixed grassland site. Int J Remote Sensing 18(1):71–94

Goelz SJ, Burk TE (1996) Measurement error causes bias in site index equations. Can J For Res 26:1585–1593

Gold RZ (1963) Tests auxiliary to chi-squared tests in a Markov chain. Ann Math Stat 34:56–74

Gomarasca MA (2004) Elementi di geomatica. Associazione Italiana di Telerilevamento, Milan

Gonzales RC, Woods RE (1992) Digital image analysis. Addison-Wesley, Reading

Gonzalez-Alonso F, Cuevas JM, Casanova JL, Caille A, Illera P (1997) A forest fire risk assessment using NOAA-AVHRR images in the Valencia area, eastern Spain. Int J Remote Sensing 18(10):2201–2207

Goodchild MF, Gopal S (1989) The accuracy of spatial databases. Taylor and Francis, London

Goodenough D, Bhogal AS, Fournier R, Hall RJ, Iisaka J, Leckie D, Luther JE, Magnussen S, Niemann O, Strome WM (1998) Earth observation for sustainable development of forests, project description. Natural Resources Cananda, Victoria

Goodman LA (1964) Simultaneous confidence intervalls for contrasts among multinomial proportions. Ann Math Stat 35:716–725

Goodman LA (1965) On simultaneous confidence intervalls for multinomial proportions. Technometrics 7(2):247–254

Goovaerts P (1997) Geostatistics for natural resources evaluation. Oxford University Press, New York

Goovaerts P (1999) Geostatistics for natural resources evaluation. Oxford University Press, Oxford

Gottle A, Sène E-H (1997) Forest functions related to protection and environmental conservation. http://www.fao.org/ documents/show_cdr.asp?url_file=/docrep/w6251e/w6251e06.htm

Goulding C (1979) Cubic spline curves and calculations of volume of sectionally measured trees, N Z J For Sci 9:89–99

Gove JH, Patil GP, Swindel BF, Taillie C (1994) Ecological diversity and forest management. In: Patil GP, Rao CR (eds) Environmental statistics. Handbook of statistics, vol 12. Elsevier, Amsterdam, pp 409–462

Gove JH, Patil GP (1998) Modeling the basal area-size distribution of forest stands: a compatible approach. For Sci 44: 285–297

Goward S, Markham B, Dye D, Dulaney W, Yang J (1991) Normalized difference vegetation index measurements from the advanced very high resolution radiometer. Remote Sensing Environ 35:257–277

Goward S, Xue Y, Czajkowski KP (2002) Evaluating land surface moisture conditions from the remotely sensed temperature/ vegetation index measurements. An exploration with the simplified simple biosphere model. Remote Sensing Environ 79:225–242

Grandell J, Pulliainen J, Hallikainen M (1998) Subpixel land use classification and retrieval of forest stem volume in the Boreal forest zone by employing SSM/I data. Remote Sensing Environ 63:140–154

Graves HS (1906) Forest mensuration. Wiley, New York,
Gray A (2003) Monitoring stand structure in mature coastal Douglas-fir forests: effect of plot size. For Ecol Manage 175(1–3):1–16
Green EJ, Clutter M (2000) Using auxiliary information to estimate stand tables. Can J For Res 30:865–872
Green EJ, Valentine HT (1998) Bayesian analysis of the linear model with heterogenous variance. For Sci 44 (1):134–138
Green EJ, Thomas CE, Strawderman WE (1987) Stein-rule estimation of timber removals by county. For Sci 33(4):1054–1061
Green EJ, Köhl M, Strawderman WE (1992) Improved estimates for cell values in a two way table. Biom Inform Med Biol. 23:24–30
Gregoire TG (1998) Design-based and model-based inference in survey sampling: appreciating the difference. Can J For Res 28:1429–1447
Gregoire TG, Scott ZT (1990) Sampling at stand boundaries: a comparison of the statistical performance among eight methods. In: Burkhart HE, Bonnor GM, Lowe JJ (eds) Research in forest inventory, monitoring, growth and yield. Publication FWS-3-90. School of Forestry and Wildlife Resources, Virginia Polytechnic Institute and State University, Blacksburg, pp 78–85
Gregoire TG, Scott CT (2003) Altered selection probabilities caused by avoiding the edge in field surveys. J Agric Biol Environ Stat 8(1):36–47
Gregoire TG, Valentine HT (1996) Sampling methods to estimate stem length and surface area of tropical tree species. For Ecol Manage 83:229–235
Gregoire TG, Valentine HT (1999) Composite and calibration estimation following 3P sampling. For Sci 45(2):179–185
Gregoire TG, Walters DK (1988) Composite vector estimator by weighting inversely proportional to variance. Can J For Res 18:282–284
Gregoire TG, Valentine HT, Furnival GM (1995) Sampling methods to estimate foliage and other characteristics of individual trees. Ecology 76(4):1181–1194
Grosenbaugh LR (1952) Plotless timber estimates – new, fast, easy. J For 50:32–37
Grover KD, Steven MD, Rondeaux G, Clark JA (2000) Estimating albedo from limited spectral and angular data. Int J Remote Sensing 21(1):155–165
Grundy IM (1995) Regeneration and management of Brachystegia spiciformis and Julbernardia globiflora in Miombo Woodland, Zimbabwe. PhD thesis, University of Oxford
Günther O, Humbolt GO (1998) Environmental information systems. Springer, Berlin Heidelberg New York, p 244
Gurnaud A (1878) Cahier l'aménagement pour l'application de la méthode par contenance exposée sur la forêt des Eperous. Paris
Gustavson EJ, Parker GR (1992) Relationships between landcover proportion and indices of landscape spatial pattern. Landscape Ecol 7:101–110
Guyot G, Riom J (1988) Review of factors affecting remote sensing of forest canopies. Proceedings of the seminar on remote sensing of forest decline attributed to air pollution. International Institute of Applied System Analysis, Laxenburg, pp 8:1–8:26
Haara A, Maltamo M, Tokola T (1997) The k-nearest-neighbour method for estimating basal area diameter distribution. Scand J For Res 12:200–208
Hailey JR, Morris PI (1988) Application of scanning and imaging techniques to assess decay and wood quality in logs and standing trees. Forintex Canada Corporation, Vancouver

Haines Young R, Chopping M (1996) Quantifying landscape structure: a review of landscape indices and their application to forested landscapes. Prog Phys Geogr 20(4):418–445

Haining RP, Arbia G, Griffith DA (2003) Spatial error propagation when computing linear combinations of spectral bands: the case of vegetation indices. Environ Ecol Stat 10:375–396

Hall AV, de Winter M, de Winter B, van Oosterhout SAM (1980) Threatened plants of South Africa. South African National Scientific Programmes report no 45. CSIR, Pretoria

Hall P, Melville G, Welsh AH (2001) Bias correction and bootstrap methods for a spatial sampling scheme. Bernoulli 7(6):829–846

Hann WJ, Bunnel DL (2001) Fire and land management planning and implementation across multiple scales. Int J Wildland Fire 10:389–403

Hann WJ, Jones JL, Keane RE, Hessburg PF, Gravenmier RA (1998) Landscape dynamics. J For 96(10):10–15

Hanselman DH, Quinn TJ, Lunsford C, Heifetz J, Clausen D (2003) Applications in adaptive cluster sampling of Gulf of Alaska rockfish. Fish Bull 101(3):501–513

Hara Y, Atkins RG, Shin RT, Jin AK, Yueh SH, Kwok R (1995) Application of neural networks for sea ice classification in polarimetric SAR images. IEEE Trans Geosci Remote Sensing 33(3):740–748

Hardinsky MA, Klemas V, Smart RM (1983) The influence of soil salinity, growth form and leaf moisture on the spectral radiance of Spartinia alterniflora canopies. Photogramm Eng Remote Sensing 49:77–83

Hardy CC, Schmidt KM, Menakis JM, Samson NR (2001) Spatial data for national fire planning and fuel management. Int J Wildland Fire 10:353–372

Hartig GL (1795) Anweisung zur Taxation der Forste oder zur Bestimmung des Holzertrags der Wälder. Heyer, Gießen

Hartley HO (1966) Systematic sampling with unequal probability and without replacement. J Am Stat Assoc 71:739–748

He HS, Ventura SJ, Mladenoff DJ (2002) Effects in spatial aggregation approaches on classified satellite imagery. Int J Geogr Inf Sci 16(1):93–109

Hedley SL, Buckland ST (2004) Spatial models for line transect sampling. J Agric Biol Environ Stat 9(2):181–199

Heinselman ML (1981) Fire intensity and frequency as factors in the distribution and structure of Northern ecosystems. In: Proceedings of the conference Fire regimes and ecosystem properties, Honolulu, 11–15 December 1978. General technical report WO-26. USDA Forest Service, Washington, DC, pp 7–57

Hemmila T, Sipi M (2004) The effect of site and tree size on the quality of swan timber from small logs of Scots pine. Pap Puu Pap Timber 86(7):502–507

Herfert MR, Lulla KP (1990) Mapping continental-scale biomass burning and smoke palls over the Amazon Basin as observed from the Space Shuttle. Photogramm Eng Remote Sensing 10(56):1367–1373

Herold M, Gardner M, Hadley B, Roberts D (2002) The spectral dimension in urban land cover mapping from high-resolution optical remote sensing data. In: Proceedings of the 3rd symposium on remote sensing of urban areas, Istanbul, 11–13 June

Herzog F, Lausch A, Müller E, Thulke H-H (1999) Das Monitoring von Landschaftsveränderungen mit Landschaftsstrukturmaßen – Fallstudie Espenhain. IÖR-Schr 29:93–107

Hess GR, Bay JM (1997) Generating confidence intervals for composition-based landscape indexes. Landscape Ecol 12(5):309–320
Higgins HG, Phillips FH, Logan AF, Balodis V (1973) Pulping of tropical hardwoods: individual and mixed species, wood and paper properties, resource assessment. Division of Applied Chemistry technological paper no 70. CSIRO, Collingwood
Hirsch KG (2000) Canadian Forest Fire Behavior Prediction (FBP) System: user's guide. Special report no 7. Natural Resources Canada. Canada Forest Service-Northern Center
Hitchcock HC, Hoffer RM (1974) Mapping a recent forest fire with ERTS-1 MSS data. LARS information note 032674. LARS, West Lafayette
Hohenadl W (1936) Die Bestandesmessung. Forstwiss Centralbl 58:51–61, 69–86, 114–127
Holmgren P, Thuresson T (1998) Satellite remote sensing for forestry planning – a review. Scand J For Res 13:90–110
Holmgren P, Joyce S, Nilsson M, Olsson H (2000) Estimating stem volume and basal area in forest compartments by combining satellite image data with field data. Scand J For Res 15(1):103–111
Holmström H (2002) Estimation of single-tree characteristics using the kNN method and plotwise aerial photograph interpretations. For Ecol Manage 167:303–314
Holton JR, Haynes PH, McIntyre ME, Douglass AR, Rood RB, Pfister L (1995) Stratosphere-troposphere exchange. Rev Geophys 33:403–439
Hord MR, Brooner W (1976) Land-use map accuracy criteria. Photogramm Eng Remote Sensing 42:671–677
Hug C, Ullrich A, Grimm A (2004) LiteMapper-5600: a waveform-digitizing LIDAR terrain and vegetation mapping system. In: Thies M, Koch B, Spiecker H, Weinacker H (eds) Laser-scanners for forest and landscape assessment. Int Arch Photogramm Remote Sensing Spatial Inf Sci36(8/W2):24–29
Husby E (1995) Wilderness quality mapping in the Euro-Arctic Barents region. DN-rapport 1995-4. Directorate for Nature Management, Trondheim.
Hush B (1962) Tree weight relationships for white pine in southeastern New Hampshire. Univ N H Agric Exp Stat Tech Bull 106
Hush B, Miller CI, Beers TW (1982) Forest mensuration. Wiley, New York
Hush B, Miller CI, Beers TW (2003a) Forest mensuration, 3rd edn (reprinted). Krieger, Malabar
Hush B, Beers TW, Kershaw JA (2003b) Forest mensuration, 4th edn. Wiley, New York
Hutchinson ID (1985) Rapid evaluation of tropical forests by the relative occurence of botanical families. Commonw For Rev 64(1):43–56
Inglis A (1991) Harvesting local forestry knowledge: A comparison of RRA and conventional surveys. RRA notes 12. Sustainable agricultural programme. IIED, London
IPCC (2004) Good practice guidance for land use, land-use change and forestry. http://www.ipcc-nggip. iges.or.jp/public/gpglulucf/gpglulucf.htm
Isaacson DL, Smith HG, Alexander CJ (1982) Erosion hazard reduction in a wildfire damaged area. In: Remote sensing for resource management. Jo-hannsen CJ, Sanders JL (eds). Soil Conservation Society of America. Ankeny, pp 179–190
Ito Y, Omatu S (1998) A polarimetric SAR data classification method using neural networks. IEEE Int Geosci Remote Sensing 1790:246–251
Itten KI, Nanayakkara SDFC, Humbel R, Bichsel M, Sommer M (1985) Inventory and monitoring of Sri Lankan forests using remote sensing techniques. In: Schmid-Haas

P (ed) Inventorying and monitoring endangered forests. Proceedings of the IUFRO conference, Zurich, 19–24 August 1985, pp 93–98
ITTO (2004) For services rendered: the current status and future potential of markets for the ecosystem services provided by tropical forests. ITTO technical series 21, International Tropical Timber Organization, Yokohama
IUFRO (1959) The standardization of symbols in forest mensuration. University of Maine, Orono
Jackson WJ, Ingeles AW (1998) Participatory techniques for community forestry–a field manual. IUCN, Gland
Jakubauskas M, Lulla KP, Mausel PW (1990) Assessment of vegetation change in a fire-altered forest landscape. Photogramm Eng Remote Sensing 56:371–377
Jensen JR (2000) Remote sensing of the environment: an earth resource perspective. Prentice Hall, Upper Saddle River
Jensen JR, Cowen DJ (1999) Remote sensing of urban/suburban infrastructure and socio-economic attributes. Photogramm Eng Remote Sensing 65:611–622
Jessen RJ (1942) Statistical investigations of a sample survey for obtaining farm facts. Iowa Agric Exp Stat Res Bull 304
Jijia A et al (1989) Detection of forest fires in Da Hinggan Ling region by meteorological satellite. Acta Meteorol Sin 3:562–568
Johnson NL, Kotz S, Kemp AW (1992) Univariate discrete distributions. Wiley, New York
Johnston DC (1982) Theory and application of selected multilevel sampling designs. PhD thesis, Colorado State University, Ft Collins
Jones SG (2004) Non-timber forest products: potential for sustainable forest income. In: Fladder SL (ed) Towards sustainability for Missouri forests. General technical report NC-239 – North Central Station. USDA Forest Service, pp 98–105
Jost A (1993) Geostatistische Analyse des Stichprobenfehlers systematischer Stichproben. Mitt Abt Forstl Biom 93-1
Journel AG, Huijbregts CJ (1978) Mining geostatistics. Academic, London
Jovanovic BD, Levy PS (1997) A look at the rule of three. Am Stat 51(2):137–139
Jung DH, Kim JY, Chun KS (2001) Development of compromise method between zero-line and center-line survey for environmentally friendly forest road route selection. KFRI J For Sci (Seoul) 64:68–77
Kaennel M (1998) Biodiversity: a diversity of definitions. In: Bachmann P, Köhl M, Päivinen R (eds) Assessment of biodiversity for improved forest planning. Kluwer, Dordrecht, pp 71–81
Kalton G, Anderson DW (1986) Sampling rare populations. J R Stat Soc Ser A 149:65–82
Kangas A (1996) On the bias and variance in tree volume predictions due to model and measurement errors. Scand J For Res 11:281–290
Kant Y, Badarinath KVS (2000) Studies on land surface temperature over heterogeneous areas using AVHRR data. Int J Remote Sensing 21(8):1749–1756
Kapos V, Iremonger SE (1998) Achieving global and regional perspectives on forest biodiversity and conservation. In: Bachmann P, Köhl M, Päivinen R (eds) Assessment of biodiversity for improved forest planning. Kluwer, Dordrecht, pp 3–13
Kasischke ES, Christensen NL, Stocks, BJ (1995) Fire, global warming and the carbon balance of boreal forests. Ecol Appl 5:437–451
Kasischke ES, Melack JM, Dobson MC (1997) The use of imaging radars for ecological applications – a review. Remote Sensing Environ 59:141–156

Kathleen M, Bergen M, Dobson C (1999). Integration of remotely sensed radar imagery in modeling and mapping of forest biomass and net primary production. Ecol Modell 122:257–274
Katila M, Tomppo E (2001) Selecting estimation parameters for the Finnish multi-source national forest inventory. Remote Sensing Environ 76:16–32
Katila M, Heikkinen J, Tomppo E (2000) Calibration of small-area estimates for map errors in multi source forest inventory. Can J For Res 30:1329–1339
Kaufman Y, Setzer A, Ward D, Tanre D, Holben B, Menzel P, Pereira M, Rasmussen R (1992) Biomass burning airborne and spaceborne experiment in the Amazonas (BASE-A). J Geophys Res 97(D13):14581–14599Kaufman YJ, Tucker CJ, Fung I et al (1990) Remote sensing of biomass burning in the tropics. In: Goldammer JG (ed) Fires in the tropical biota. Springer, Berlin Heidelberg New York, pp 371–399
Kaufmann E (1991) Volumenermittlung im zweiten Schweizerischen Landesforstinventar. Eidg Anstalt Forstl Versuchswes, internal report
Kauth RJ, Thomas GS (1976) The tasseled cap – a graphic description of the spectral-temporal development of agricultural crops as seen in Landdat. In: Proceedings of the symposium on machine processing of remotely sensed data, Purdue University, West Lafayette, 29 June 29–1 July 1976, LARS, West Lafayette, pp 41–51
Keen EA (1950) The relaskop. Emp For Rev 29(3):253–264
Keller M (2001) Aerial photography. In: Brassel P, Lischke H (eds) Swiss National Forest Inventory: methods and models of the second assessments. WSL, Birmensdorf, pp 45–64
Kendall M, Stuart A, Ord JK (1983) The advanced theory of statistics. MacMillan, Simon and Schuster, New York
Key CH (2003) The normalized burn ratio (NBR): a landsat TM radiometric measure of burn severity. http://nrmsc.usgs.gov/research/ndbr.htm
Kirkman H (1996) Baseline and monitoring methods for seagrass meadows. J Environ Manage 47(2):191–201
Kish L (1965) Survey sampling. Wiley, New York
Kleinn C (1991) Der Fehler von Flächenschätzungen mit Punkterastern und linienförmigen Stichproben. Mitt Abt Forstl Biom Univ Freiburg 91-1
Kleinn C (1996) Ein Vergleich der Effizienz von verschiedenen Clusterformen in forstlichen Grossrauminventuren. Forstwiss Centralbl 115:378–390
Kleinn C, Laamanen R, Malla SB (1996) Integrating the assessment of non-wood forest products into the forest inventory of a large area: experiences from Nepal. In: Domestication and commercialization of non-timber forest products in agroforestry systems. Proceedings of an international conference held in Nairobi. FAO, Rome, pp 23–31
Knight J (2003) Zoologists prime traps for California wildlife survey. Nature 424:987
Köhl M (1990) National inventories and inventories of endangered forests in Europe. In: LaBau VJ, Cunia T (ed) State-of-the-art methodology of forest inventory: a symposium, proceedings, Sytacuse, 30 July–5 August 1989, pp 356–363
Köhl M, Brassel P (2001) Zur Auswirkung der Hangneigungskorrektur auf Schätzwerte im Schweizerischen Landesforstinventar (LFI). Schweiz Z Forstwesen 152(6):215–225
Köhl M, Gertner G (1997) Geostatistics in evaluating forest damage inventories: considerations on methods for describing spatial distributions. For Ecol Manage 95:131–140

Köhl M, Kushwaha SPS (1994) A four-phase sampling method for assessing standing volume using landsat-TM-data aerial photography and field assessments. Commonw For Rev 73:35–41

Köhl M, Lautner M (2001) Erfassung von Waldökosystemen durch Hyperspektraldaten. Photogr Fernerkund Geoinform 2:107–117

Köhl M, Oehmichen K (2003) Comparison of landscape indices under particular consideration of the geometric and geographic moving window concept. In: Corona P, Köhl M, Marchetti M (eds) Advances in assessments for sustainable forest management and biodiversity monitoring. Forestry sciences, vol 76. Kluwer, Dordrecht, pp 231–244

Köhl M, Scott CT (1994) Zur Auswertung von Gruppenstichproben bei extensiven Forstinventuren. Allg Forst-Jagdz 165(5/6):101–106

Köhl M, Zingg A (1995) Eignung von Diversitätsindices bei Langzeituntersuchungen zur Biodiversität. Allg Forst Z 167:76–85

Köhl M, Bachmann P, Brassel P, Preto G (eds) (1995) The Monte Verita conference on forest survey designs. Simplicity versus efficiency and assessment of non-timber ressources. Swiss Federal Institute of Forest, Snow and Landscape Research (WSL/FNP), Zurich, Swiss Federal Institute of Technology (ETH), Birmensdorf

Köhl M, Thiele D, Zingg A (2000) GIS-based geo-statistical analysis of long term experiments in forest research. In: Heuvelink GBM, Lemmens MJPM (eds) Proceedings, spatial accuracy assessment, accuracy 2000. Proceedings of the 4th international symposium on spatial accuracy assessment in natural resources and environmental sciences, Amsterdam, July 2000, pp 373–380

Kohm KA, Franklin JF (1997) Creating a forestry for the 21st century. The science of ecosystem management. Island, Washington, DC

Kollmann FFZ, Côté J (1968) Principles of wood science and technology I: solid wood. Springer, Berlin Heidelberg New York

Komiyama A, Havanond S, Srisawatt W et al (2000) Top/root biomass ratio of a secondary mangrove (Ceriops tagal (Perr.) CB Rob.) forest. For Ecol Manage 139(1–3):127–134

Kotz S, Johnson NL (1988) Taylor-series linearization. Wiley, New York, pp 646–647

Koukal T, Schneider W (2003) Mapping and monitoring tree resources outside the forest in Central America. In: Corona P, Köhl M, Marchetti M (eds) Advances in assessments for sustainable forest management and biodiversity monitoring. Forestry sciences, vol 76. Kluwer, Dordrecht, pp 313–323

Koutsias M, Karteris M (2000) Burned area mapping using logistic regression modeling of a single post-fire Landsat-5 Thematic Mapper image. Int J Remote Sensing 21:673–687

Kozak A, Smith JHG (1966) Critical analysis of multivariate techniques for estimating tree taper suggests that simpler methods are best. For Chron 42:458–463

Kozak A, Munro DD, Smith JHG (1969) Taper functions and their application in forest inventory. For Chron 45:278–283

Krige DG (1951) A statistical approach to some basic mining valuation problems on the Witwatersrand. J Chem Metall Min Soc S Afr 52(6):119–139

Krige DG (1966) Two-dimensional weighted moving average trend surfaces for ore-evaluation. J South Afr Inst Min Metall 66:13–38

Krutzsch H, Lötsch F (1938) Holzvorratsinventur und Leistungsprüfung der naturgemässen Forstwirtschaft. Neumann, Radebeul

Kublin E (1987) Grenzen der klassischen Regressionsmodelle. Mitt FVA Baden-Württemberg Heft 134
Küchler AW (1975) Potential natural vegetation of the conterminous United States, 2nd edn. Map 1:3,168,000. American Geographical Society, New York
Kuswhaha SPS (1990) Forest type mapping and change detection from satellite imagery. J Photogramm Remote Sensing 45:175–181
Kuusela K (1979) Sampling of tree stock by angle gauge in proportion to tree characteristics. Commun Inst For Fenn 95.7
Laclau P (2003) Root biomass and carbon storage of ponderosa pine in a northwest Patagonia plantation. For Ecol Manage 173(1–3):353–360
Lahiri R (2003) On the impact of bootstrap in survey sampling and small-area estimation. Stat Sci 18(2):199–210
Laird NM (1978) Empirical Bayes methods for two-way contingency tables. Biometrics 65:581–590
Lambin EF, Erlich D (1996) The surface temperature-vegetation index space for land cover and land cover change analysis. Int J Remote Sensing 17:463–487
Lancaster VA, Keller-McNulty S (1998) A review of composite sampling methods. J Am Stat Assoc 93(443):1216–1230
Landres PB, Morgan P, Swanson FJ (1999) Overview of the use of natural variability concepts in managing ecological systems. Ecol Appl 9:1179–1188
Langaas S, Kane R (1991) Temporal spectral signatures of fire scars. In: Savanna woodland geoscience and remote sensing symposium 1991. IGARSS '91, vol 3, pp 1157–1160
Lanly JP (1977) Les inventaires de forets tropicales humides pour les decisions en matiere d'investissement industriel. Bois For Trop 171:45–46
Lappi J (1991) Calibration of height and volume equations with random parameters. For Sci 37:781–801
Lappi J (2001) Forest inventory of small areas combining the calibration estimator and a spatial model. Can J For Res 31:1551–1560
Larsson TB, Svensson L, Angelstam P, Balent Gbarbate A, Bijlsma R-J, Boncina A, Bradshaw R, Bücking W, Cancio O, Corona P, Diaci J, Dias S, Ellenberg H, Fernandes FM, Fernandez-Gonzales F, Ferris R, Frank G, Moeller PF, Giller PS, Gustafson L, Halbritter K, Hall S, Hansson L, Innes J, Jactel H, Kaennel-Dobertin M, Klein M, Marchetti M, Mohren F, Niemelä P, O'Halloran J, Rametsteiner E, Refeo F, Scheidegger C, Scotti R, Sjöberg K, Spanos I, Spanos K, Standovar T, Tommaras A, Trakolis D, Uuttera J, Walsh PM, Vanderkerkhove K, Watt AD, van den Meerschaut D (2001) Biodiversity evaluation tools for European forests. A report from the FAIR project "Indicators for monitoring and evaluation of forest biodiversity in Europe", CT97-3575, European Commission RTD Programme. Ecol Bull 50
Leimgruber P, McShea WJ, Schnell GD (2002) Effects of scale and logging on landscape structure in a forest mosaic. Environ Monit Assess 74(2):141–166
Lesser VM, Kalsbeek WD (1999) Nonsampling errors in environmental surveys. J Agric Biol Environ Stat 4(4):473–488
Lesslie R, Maslen M (1995) National wilderness inventory handbook, 2nd edn. Australian Heritage Commission, Australian Government Publishing Service, Canberra
Le Toan T, Ribbes F, Wang LF et al (1997) Rice crop mapping and monitoring using ERS-1 data based on experiment and modeling results. IEEE Trans Geosci Remote Sensing 35(1):41–56

Levy PS, Lemeshow S (1991) Sampling of populations: methods and applications. Wiley, New York

Li HG, Schreuder HT (1985) Adjusting estimates in large two-way tables in surveys. For Sci 31:366–372

Liebermann M, Lieberman D, Vandermeer JH (1988) Age-size relationships and growth behaviour of the palm Welfia georgii. Biotropica 20(4):270–273

Liew SC, Lim OK, Kwoh LK, Lim H (1998) A study of the 1997 forest fires in South East Asia using SPOT quicklook mosaics. Proc Int Geosci Remote Sensing Symp 2:879–881

Linddal M (1995) Assessment of non-wood goods and services – a forest policy appraoch. In: Köhl M, Bachmann P et al (eds) The Monte Verita conference on forest designs. Swiss Federal Institute of Forest, Snow and Landscape Research (WSL/FNP), Zurich, Swiss Federal Institute of Technology (ETH), Birmensdorf, pp 213–227

Little RJ (2004) To model or not to model? Competing modes of inference for finite population sampling. J Am Stat Assoc 99(466):546–556

Little RJA, Rubin DB (1987) Statistical analysis with missing data. Wiley, New York

Lloyd CJ (1999) Analysis of categorical variables. Wiley, New York

Lo NCH, Griffith D, Hunter JR (1997) Using a restricted adaptive cluster sampling to estimate Pacific hake larval abundance. Calif Coop Ocean Fish Invest Rep 38:103–113

Lodhiyal N, Lodhiyal LS (2003) Biomass and net primary productivity of Bhabar Shisham forests in central Himalaya, India. For Ecol Manage 176(1–3):217–235

Loetsch F (1957) Report to the government of Thailand on inventory methods for tropical forests. FAO report no 545. FAO, Rome

Loetsch F, Haller KE (1964) Forest inventory, vol I. BLV, Munich

Loetsch F, Zöhrer F, Haller KE (1973) Forest inventory, vol II. BLV, Munich

Longley PA, Goodchild MF, Maguire DJ, Rhind DW (2001) Geographic information systems and science. Wiley, New York

López Garcia MJ, Caselles V (1991) Mapping burns and natural reforestation using Thematic Mapper data. Geocarto Int 6:31–37

Loveland TR, Sohl TL, Stehman SV, Gallant AL, Sayler KL, Napton DE (2002) A strategy for estimating the rates of recent United States land-cover changes. Photogramm Eng Remote Sensing 68:1091–1099

Lugo AE, Brown S, Chapman J (1988) An analytical review of production rates and stemwood biomass of tropical forest plantations. For Ecol Manage 23(2–3):179–200

Lund HG (ed) (1998) IUFRO guidelines for designing multiple forest resources inventories. IUFRO world series, vol 8. IUFRO, Vienna

Lund HG, Smith WB (1997) The Unites States forest inventory program. In: Nyssönen A, Ahti A (eds) Proceedings of the FAO/ECE meeting of experts on global forest resources assessment. Finn For Res Inst Res Pap 620:331–333

Lund HG, Caballero Deloya M, Villarreal Canton R (1987) Land and resource evaluation for national planning in the topics. An international conference and workshop, Chetumal, Mexica, 25–31 January 1987. General technical report WO-39. USDA Forest Service, Washington, DC

Lunetta RS, Elvidge CD (1999) Remote sensing change detection. Environmental monitoring methods and applications.Ann Arbor Press, Ann Arbor

Lunetta RS, Lyon JG, Guindon L, Elvidge CD (1998) North American landscape characterization dataset development and data fusion issues. Photogramm Eng Remote Sensing 64(8):821–829

Luxmi C, Raturi RD, Rao RV, Dayal R (1995) Wood anatomy of Indian Flacourtiaceae, wood anatomy, botany division. Forest Research Institute, Dehra Dun
Luxmi C et al (1998) Identification of Indien bamboos using culm epidermal features – an overview. In: Proceeding of a national seminar on processing and utilisation of plantation timbers and bamboo, IPIRTI, Bangalore, 23–24 July 1998, pp 66-73
Magnussen S (1986) Diameter distributions in Picea abies described by the Weibull model. Scand J For Res 1:493–502
Magnussen S (1994) A coordinate-free area variance estimator for forest stands with a fuzzy outline. For Sci 42:76–85
Magnussen S (1998) Tree height tarifs and volume estimation errors in New Brunswick. North J Appl For 15(1):7–13
Magnussen S (2000) PPP sampling for total volume with and without prior predictions. In: Proceedings of the Society of American Foresters 1999 convention. Society of American Foresters, Bethesda, pp 83–92
Magnussen S (2001) Fast pre-survey computation of the mean spatial autocorrelation in large plots composed of a regular array of secondary sampling units. Math Modell Sci Comput 13(3–4):204–217
Magnussen S (2003) Stepwise estimators for three-phase sampling of categorical variables. J Appl Stat 30(5):461–475
Magnussen SD, Burgess (1996) Stochastic resampling techniques fpr quantifying error propagations in forest field experiments. Can J For Res 27:630–637
Magnussen S, Köhl M (2002) Polya posterior frequency distributions for stratified double sampling of categorical data. For Sci 48(3):569–581
Magnussen S, Stehman SV, Corona P, Wulder MA (2004) A Pòlya-urn resampling scheme for estimating precision and confidence intervals under one-stage cluster sampling: Application to map classification accuracy and cover-type frequencies. For Sci 50(4):1–13
Magurran AE (1988) Ecological diversity and its measurement. Croom Helm, London
Mahalanobis PC (1946) On large-scale sample surveys. Philos Trans B 231:329–451
Makela A, Makinen H (2003) Generating 3D sawlogs with a process-based growth model. Forest Ecol Manage 184(1–3):337–354
Malingreau JP (1990) The contribution of remote sensing to the global monitoring of fires in tropical and subtropical ecosystems. In: Goldammer JG (ed) Fires in the tropical biota. Springer, Berlin Heidelberg New York, pp 337–370
Mandallaz D (1991) A unified approach to sampling theory for forest inventory based on infinite population and superpopulation models. Chair of forest management and planning. Swiss Federal Institute of Technology, Zurich
Mandallaz D (1993) Geostatistical methods for double sampling schemes: application to combined forest inventories. PhD thesis, Swiss Federal Institute of Technology, Zurich
Mandallaz D (1994) Geostatistical methods for double sampling schemes: application to combined forest inventory. Chair of forest management and planning. Swiss Federal Institute of Technology, Zurich
Mandallaz D (2000) Estimation of the spatial covariance in universal kriging: application to forest inventory. Environ Ecol Stat 7(3):263–284
Mandallaz D (2001) Optimal sampling schemes based on the anticipated variance with lack of fit. Swiss Federal Insitute of Technology, Zurich

Marchetti M, Castagnoli A (1989) Proposed methods for application of new techniques for vegetation inventory, forest damage evaluation, their cartographic representation and permanent monitoring. In: Proceedings of the international conference on global natural resources monitoring and assesment: preparation for the 21st century, FAO-IUFRO, Venice, pp 496–503

Marchetti M, Ricotta C, Volpe F (1995) A qualitative approach to mapping postfire regrowth in mediterranean vegetation with Landsat data. Int J Remote Sensing 16(13):2487–2494

Marchetti M, Oradini A, Mariotti B, Cumer A (1998) Progetto transitalia: telerilevamento dei danni sulle foreste di aree appenniniche ed insulari. Doc Territor 39:20–30

Mariaux A (1981) Past efforts in measuring age and annual growth in tropical trees. In: Borman FH, Berlyn G (eds) Age and growth rate of tropical trees, new directions for research. Bull Sch For Environ Stud Yale Univ 94:20–30

Marques TA (2004) Predicting and correcting bias caused by measurement error in line transect sampling using multiplicative error models. Biometrics 60(3):757–763

Marques FFC, Buckland ST (2003) Incorporating covariates into standard line transect analyses. Biometrics 59(4):924–935

Matérn B (1956) On the geometry of the cross-section of a stem. Medd Stat Skogsforsoeksves Inst 46(11)

Matérn B (1964) A method of estimating the total length of roads by means of a line survey. Stud For Suec 18:68–70

Matérn B (1980) Spatial variation. Springer, Berlin Heidelberg New York

Matheron G (1965) Les variables regionalisèes et leur estimation, Masson, Paris

Matson M, Stephens G, Robinson J (1987) Fire detection using data from the NOAA-N satellites. Int J Remote Sensing 8:961–970

Max TA, Burkhart HE (1976) Segmented polynomial regression applied to taper equations. For Sci 22:283–289

Mayers J, Bass S (2004) Policy that works for forests and people; real prospects for governance and livelihoods. Earthscan, London

McCloy KR (1995) Resource management information systems, CRC, Boca Raton

McCormack A (1998) Guidelines for inventorying non-timber forest products. MSc thesis, University of Oxford

McGarigal K, Marks BJ (1994) Fragstats. Spatial pattern analysis program for quantifying landscpe structure, version 2.0. Oregon State University, Corvallis

MCPFE (2003) Improved pan-European indicators for sustainable forest management. MCPFE Liaison Unit, Vienna

McRoberts RE (2001) Imputation and model-based updating techniques for annual forest inventories. For Sci 47(3):322–330

McRoberts RE, Nelson MD, Wendt DG (2002) Stratified estimation of forest area using satellite imagery, inventory data, and the k-nearest neighbors technique. Remote Sensing Environ 82(2–3):457–468

Meeden G (1999) A noninformative Bayesian approach for two-stage cluster sampling. Sankhya Ser B 133:144

Meier P (1953) Variance of a weighted mean. Biometrics 9:59–73

Merchant JW (1984) Using spatial logic in classification of landsat TM data In: Proceedings of the pecora IX symposium, Sioux Falls, pp 378–385

Mertens K (2003) Using genetic algorithms in sub-pixel mapping. Int J Remote Sensing 24(21):4241–4247

Mesavage C, Grosenbaugh LR (1956) Efficiency of several cruising designs on small tracts in North Arkansas. J For 54:569–576

Miller RG (1974) The jackknife – a review. Biometrika 61:1–15

Miller RG (1981) Simultaneous statistical inference. Springer, Berlin Heidelberg New York

Milne AK (1986) The use of remote sensing in mapping and monitoring vegetational change associated with bushfire events in Eastern Australia. Geocarto Int 1(1):25–34

Minghelli-Roman A, Mangiolini M, Petit M, Polidori L (2001) Spatial resolution improvement of MeRIS images by fusion with TM images. IEEE Trans Geosci Remote Sensing 39:1533–1536

Minick GR, Shain WA (1981) Comparison of satellite imagery and conventional aerial photography in evaluating a large forest fire. In: Proceedings of the 7th interational symposium on machine processing of remotely sensed data, Purdue University, West Lafayette, pp 544–546

Moeur M, Crookston NL, Stage AR (1995) Most similar neighbor: an improved sampling inference procedure for natural resource planning. For Sci 41:337–359

Mollicone D, Achard F, Eva HD, Belward AS, Federici S, Lumicisi A, Rizzo VC, Stibig H-J, Valentini R (2003) Land use change monitoring in the framework of the UNFCCC and its Kyoto protocol: report on current capabilities of satellite remote sensing technology. EUR report 20867 EN. European Communities, Luxembourg

Montagu K, Duttmer K, Barton C, Cowie A (2002) Estimating above-ground biomass carbon of Eucalyptus pilularis across eight contrasting sites – what works best? In: International conference on eucalypt productivity, Hobart, 10–15 November 2002, pp 49–50

Moore N, Tubert-Bitter P, Fourrier A, Begaud B (2003) A simple method to estimate sample sizes for safety equivalence studies using inverse sampling. J Clin Epidemiol 56(5):433–435

Moran EF, Brondizio ES, Tucker JM, da Silva-Forsberg MC, McCracken S, Falesi I (2000) Effects of soil fertility and land-use on forest succession in Amazônia. For Ecol Manage 139(1–3):93–108

Morgan P, Aplet GH, Haufler JB, Humphries HC, Moore MM, Wilson WD (1994) Historical range of variability: a useful tool for evaluating ecosystem change. In: Sampson N, Adams DL (eds) Assessing forest ecosystem health in the inland west. Haworth, New York, pp 87–111

Morgan P, Bunting SC, Black AE, Merrill T, Barrett S (1996) Fire regimes in the Interior Columbia River Basin: past and present. Final report for RJVA-INT-94913, on file at USDA Forest Service, Rocky Mountain Research Station, Fire Sciences Laboratory, Missoula

Morgan BJT (1997) Strip adaptive cluster sampling: probability proportional to size selection of primary units. Biometrics 53:1092–1096

Muchoney DM, Haack B (1994) Change detection for monitoring forest defoliation. Photogramm Eng Remote Sensing 60:1243–1251

Muller-Landau HC (2004) Interspecific and inter-site variation in wood specific gravity of tropical trees. Biotropica 36.1:20–32

Mushove PT (1994) Biomass functions for brachystegia spiciformis in Zimbabwe, internal report. Forest Research Centre, Highlands, Harare

Muttlak HA, Khan A (2002) Adjusted two-stage adaptive cluster sampling. Environ Ecol Stat 9(1):111–120

Myers RH (1986) Classical and modern regression with applications. Duxbury, Boston

Næsset E (1998) Positional accuracy of boundaries between clearcuts and mature forest stands delineated by means of aerial photointerpretation. Can J For Res 28:368–374

Næsset E (1999) Effects of delineation errors in forest stand boundaries on estimated area and timber volumes. Scand J For Res 14(6):558–566

Nelson BW, Mesquita R, Pereira JLG et al (1999) Allometric regressions for improved estimate of secondary forest biomass in the central Amazon. For Ecol Manage 117(1–3):149–167

Nelson RM Jr (2002) An effective wind speed for models of fire spread. Int J Wildland Fire 11:153–161

Nezry E, Rémondiére S, Solaas G, Genovese G (1995) Mapping of next season's crops during the winter using ERS SAR data. EOQ no 50. European Space Agency. http://esapub.esrin.esa.it/eoq/eoq50/nezry50.htm

Nichols P (1991) Social survey methods: A field guide for development workers. Development guidelines no 6. Oxfam, Oxford

Noack D (1971) Evaluation of properties of tropical timbers. J Inst Wood Sci 5(5):17–23

Norris JL III, Pollock KH (1998) Non-parametric MLE for Poisson species abundance models allowing for heterogeneity between species. Environ Ecol Stat 5(4):391–402

Nusser SM, Breidt EJ, Fuller WA (1998) Design and estimation for investigating the dynamics of natural resources. Ecol Appl 8(2):234–245

Nyerges AE, GM Green (2000) The ethnography of landscape: GIS and remote sensing in the study of forest change in West African Guinea savanna. Am Anthropol 102(2):271–289

O'Brien SM, Dunson DB (2004) Bayesian multivariate logistic regression. Biometrics 60(3):739–746

Odum EP (1968) Energy flows in ecosystems: a historic review. Am Zool 8:11–18

Oehmichen K (2001) Vergleich von Landschaftsindizes unter besonderer Berücksichtigung des geometrischen-und geographischen Window-Konzeptes. Diploma, TU Dresden Tharandt

Oleson KW, Sarlin S, Garrison J, Smith S, Privette JL, Emery WJ (1995) Unmixing multiple land-cover type reflectances from coarse spatial resolution satellite data. Remote Sensing Environ 54(2):98–112

Olsen ER, Ramsey RD, Winn DS (1993) A modified fractal dimensions as a measure of landscape diversity. Photogramm Eng Remote Sensing 59(10):1517–1520

Olsen AR, Sedransk J, Edwards D, Gotway CA, Liggett W, Rathburn S, Reckhow KH, Young LJ (1999) Statistical issues for monitoring ecological and natural resources in the United States. Environ Monit Assess 54:1–45

O'Neill RV, Krummel JR, Gardner RH, Sugihara G, Jackson B, Deangelis DL, Milne BT, Turner MG, Zygmunt B, Christensen SW, Dale VH, Graham RL (1988) Indices of landscape pattern. Landscape Ecol 7(1):153–162

O'Regan WG, Arvanitis LG (1966) Cost effectiveness in forest sampling. For Sci 12:406–414

Ord K (1990) Statistical methods for point pattern data. In: Griffith DA (ed) Spatial statistics: past, present, future. Institute of Mathematical Geography, Ann Arbor, pp 31–53

Ott WR (1978) Environmental indices: theory and practice. Ann Arbour Science, Ann Arbour

Page B, Hilty LM (1995) Umweltinformatik. Oldenbourg, Munich

Palmer JG, Murphy, JO (1993) An extended tree-ring chronology (teak) from Java. Proc K Ned Akad Wet-Biol Chem Geol Phys Med Sci 96:27–41

Palmer JR (1975) Towards more reasonable objectives in tropical high forest management for timber production. Commonw For Rev 54(3/4):273–289

Palmer MW (1988) Fractal geometry: a tool for describing spatial patterns of plant communities. Vegetatio 75(1–2):91–102

Pancel L (1984) Agroforstliche Landnutzungsmethode spezieller subhumider Vegetationsformen in Guinea Bissau. Forstarchiv 55(5):186–194

Panchapakesan S, Childs A, Humphrey BH, Balakrishnan N (1998) Inverse sampling procedures to test for homogeneity in a multinomial distribution. Elsevier, Amsterdam,

Panshin AJ, Zeeuw CD (1980) Textbook of wood technology, 4th edn. McGraw-Hill, New York

Panzer KF (1975) Quantification of decay and related volume loss in tropical forest inventories. Mitt Bundesforschungsanst Forst Holzwirtsch 109:117–128

Paola JD, Schowengerdt RA (1995) A detailed comparison of backpropagation neural network and maximum-likelihood classifiers for urban land use classification. IEEE Trans Geosci Remote Sensing 33(4):981–996

Pardé J (1980) Forest biomass. For Abstr 41:343–362

Parresol BR (1999) Assessing tree and stand biomass: a review with examples and critical comparisons. For Sci 45(4):573–593

Patil GP, Taillie C (1979) An overview of diversity. In: Grassle JF, Patil GP, Smith WK, Taillie C (eds) Ecological diversity in theory and practice. International Cooperative Publishing House, Fairland, pp 3–27

Patil GP, Sinha AK, Taillie C (1994) Ranked set sampling. In: Patil GP, Rao CR (eds) Environmental statistics. Handbook of statistics 12. North-Holland, Amsterdam, pp 167–200

Patterson HD (1950) Sampling on successive occasions with partial replacement of units. J R Stat Soc Ser B 12(2):241–255

Payandeh B, Ek AR (1986) Distance methods and density estimators. Can J For Res 16(5):918–924

Payn TW, Clough ME (1988) Differential fertilisation on pine plantations on acid forest soils. South Afr For J 147:16–25

Peet RK (1974) The measurements of species diversity. Ann Res Ecol Syst 5:285–307

Peet RK (1978) Forest vegetation in the Colorado Front Range: patterns of species diversity. Vegetatio 37(2):65–78

Pellemans AHJM, Jordans RWL, Allewijn R (1993) Merging multispectral and panchromatic SPOT images with respect to radiometric properties of the sensor. Photogramm Eng Remote Sensing, 59(1):81–87

Pellico-Netto S (1979) Die Forstinventuren in Brasilien: neue Entwicklungen und ihr Beitrag für eine geregelte Forstwirtschaft. Mitteilungen aus dem Arbeitskreis für Forstliche Biometrie

Pelz DR, Cunia T (1985) Forest inventories in Europe. Mitt Abt Forstl Biom Univ Freiburg 3/85

Pelz DR (1995) Assessment of non-timber functions: state-of-the-art. In: Köhl M, Bachmann P et al (eds) The Monte Verita conference on forest designs. Swiss Federal Institute of Forest, Snow and Landscape Research (WSL/FNP), Zurich, Swiss Federal Institute of Technology (ETH), Birmensdorf, pp 103–109
Peralta R, Hartshorn GS, Lieberman D, Lieberman M (1987) Resena de estudios a largo plazo sobre composicion floristica y dinamica del bosque tropical en La Selva, Costa Rica. Rev Biol Trop Ecol Ecofisiol Plant Bosques Mesoam 35(Suppl 1):23–39
Pereira MC (1999) A comparative evaluation of NOAA/AVHRR vegetation indexes for burned surface detection and mapping. IEEE Trans Geosci Remote Sensing 37(1):217–226
Pereira MC, Setzer AW (1993) Spectral characteristics of fire scars in Landsat-5 TM images of Amazonia. Int J Remote Sensing 14:2061–2078
Perry GLW (1998) Current approaches to modeling the spread of wildland fire: a review. Prog Phys Geogr 22(2):222–245
Peterson DL, Running SW (1989) Applications in forest sceinces and management. In: Asrar G (ed) Theory and applications optical remote sensing. Wiley, New York, pp 429–473
Peterson DJ, Resetar S, Browe J, Diver R (1999) Forest monitoring and remote sensing a survey of accomplishments and opportunities for the future. Prepared for the White House Office of Science and Technology Policy. Science and Technology Policy Institute
Pfeffermann D (2002) Small area estimation – new developments and directions. Int Stat Rev 70(1):125–143
Pfeffermann D, Skinner CJ, Holmes DJ, Goldstein H, Rasbach J (1998) Weighting for unequal selection probabilities in multilevel models. J R Stat Soc Ser B 60(1):23–40
Pielou EC (1970) Introduction to mathematical ecology. Wiley, New York
Pielou EC (1975) Ecological diversity. Wiley, New York
Pielou EC (1977) Statistical ecology. Wiley, New York
Philip MS (1976) The role of forest inventories in the development of tropical moist forests. Commonw For Rev 55(1):57–64
Pollard (1971) On distance estimators of density in randomly distributed forests. Biometrics 27:991–1002
Poole (2002) Bayesian estimation of survival from mark-recapture data. J Agric Biol Environ Stat 7(2):264–276
Pozo D, Olmo FJ, Alados-Arboledas L (1997) Fire detection and growth monitoring using a multitemporale technique on AVHRR mid-infrared and thermal channels. Remote Sensing Environ 60:111–120
Prance GT, Balée W, Boom BM, Carbeuri RL (1987) Quantitative ethnobotany and the case for conservation in Amazonia. Conserv Biol 1(4):296–310
Pretzsch H (2001) Modellierung des Waldwachstums. Parey, Berlin
Prodan M (1965) Holzmesslehre. Sauerländers, Frankfurt
Puffenberger HE (1976) Forest inventories in the tropics: a consideration. J For 74(1):28–38
Pumijumnong N, Eckstein D, Sass U (1995) Tree-ring research on *Tectona grandis* in northern Thailand. Int Assoc Wood Anat J 16:385–392
Puumalainen J, Angelstam P, Banko G, Brandt J, Caldeira M, Estreguil C, Folving S, Garcia del Barrio JM, Keller M, Kennedy P, Köhl M, Marchetti M, Neville P, Plsson

H, Parviainen J, Pretzsch H, Ravn HP, Stahl G, Tomppo E, Uuttera J, Watt A, Winkler B, Wrbka T (2002) Forest biodiversity – assessment approaches for Europe. EUR report 20423 EN. European Commission, Joint Research Centre, Ispra
Quegan S, Le Toan T, Yu JJ (2000) Multitemporal ERS SAR analysis applied to forest mapping. IEEE Trans Geosci Remote Sensing 38:741–753
Quesenberry CP, Hurst DC (1964) Large sample simultaneous confidence intervals for multinomial proportions. Technometrics 6(2):191–195
Rabii HA (1979) An investigation of the utility of Landsat-2 MSS data to the fire-danger rating area, and forest fuel analysis within Crater Lake National Park, Oregon. PhD dissertation, Oregon State University
Rai SN (1975) Rate of diameter increment of Cryptocaria wightiana in the tropical rain forests of Western Ghats-India, Sivan. Eester Ranger College, Kurseong, pp 54–59
Rai SN (1978) Rate of growth of Dalbergia latifolia and Xylia dolabriformis. Malays For 41(3):241–252
Rai SN (1983) Basal area and volume increment in tropical rain forests of India. Indian For 109(4):198–211
Rai SN, Sarma CR (1987) Diameter increment of iron wood tree (Mesua ferrea Linn.) in humid tropics of India. J Trop For 3:324–329
Rajora OP, Zsuffa L (1991) Screening populus deltoids marsh. Selections by allozymes to assure species identity. Scand J F Res 6:471–478
Ramirez-Maldonado H (1988) On the relevance of geostatistical theory and methods to forest inventory problems. PhD thesis, University of Georgia
Ramsey FL, Harrison K (2004) A closer look at detectability. Environ Ecol Stat 11(1):73–84
Ramsey, Gates CE, Patil GP (1988) On transect sampling to assess wildlife populations and marine resources. Elsevier, Amsterdam, pp 515–532
Rao CR (1988) Variance estimation in sample surveys. In: Krishnaiah PK, Rao CR (eds) Handbook of statistics. Elsevier, Amsterdam, pp 427–447
Rao CR (2003) Small area estimation. Wiley, Hoboken
Rao CR, Wu CFJ (1988) Resampling inference with complex survey data. J Am Stat Assoc 83:231–241
Rao CR, Wu C-FJ, Yue K (1992) Some recent work on resampling methods for complex surveys. Surv Methods 18:209–217
Rashed T, Weeks JR, Gadalla MS, Hill AG (2001) Revealing the anatomy of cities through spectral mixture analysis of multispectral satellite imagery: a case study of the greater Cairo region, Egypt. Geocarto Int 16:5–15
Razafimpanilo H, Frouin R, Iacobellis, SF; Somerville RCJ (1995) Methodology for estimating burned area from AVHRR reflectance data. Remote Sensing Environ 54:273–289
Reed DD, Green EJ (1984) Compatible stem taper and volume ratio equations. For Sci 30:977–990
Reed WJ, McKelvey KS (2002) Power-law behaviour and parametric models for the size-distribution of forest fires. Ecol Modell 150:239–254
Reid WV, Miller KR (1989) Keeping options alive: the scientific basis for conserving biodiversity. World Resources Institute, Washington
Rencher AC (2002) Methods of multivariate analysis. Wiley-Interscience, New York
Richards GD (1990) An elliptical growth model of forest fire fronts and its numerical solution. Int J Num Methods Eng 30:1163–1179

Richards GD (1995) A general mathematical framework for modeling two-dimensional wildland fire spread. Int J Wildland Fire 5(2):63–72
Richards JA (1984) Thematic mapping from multitemporal image data using the principal components transformation. Remote Sensing Environ 46:16–35
Richards PW (1996) The tropical rain forest, 2nd edn. Cambridge University Press, Cambridge
Ricotta C, Avena GC, Olsen ER (1998) Geographic windows in the fractal analysis of local landscape complexity. Abstr Bot 22:143–147
Ricotta C, Avena G, de Palma A (1999) Mapping and monitoring net primary productivity with AVHRR NDVI time-series: statistical equivalence of cumulative vegetation indices. Photogramm Eng Remote Sensing 54:325–331
Ricotta C, Marchetti M, Marignani M, Carranza ML (2003a) Misure di incertezza per classificazioni sfumate: alcune idee per la quantificazione. Ital For Mont 58(4):279–288
Ricotta C, Marchetti M, Corona P, Chirici G, Innamorati S (2003b) LADY: software for assessing local landscape diversity profiles of raster land cover maps using geographic windows. Environ Model Software 18:373–378
Riggan PJ, Lockwood RN, Jacks PJ, Colver CG, Weirich F, DeBano LF, Brass JA (1994) Effects of fire severity on nitrate mobilization in watersheds subject to chronic atmospheric deposition. Environ Sci Technol 28:369–375
Ringvall A, Ståhl G (1999) Field aspects of line intersect sampling for assessing coarse woody debris. For Ecol Manage 119:163–170
Ripley BD (1981) Spatial statistics. Wiley, New York
Ritz C (2004) Goodness-of-fit tests for mixed models. Scand J Stat 31(3):443–458
Robinson JM (1991) Fire from space: global fire evaluation using infrared remote sensing. Int J Remote Sensing 12:3–24
Roesch F (1993) Adaptive cluster sampling for forest inventories. For Sci 39:655–669
Rogan J, Franklin J, Roberts DA (2002) A comparison of methods for monitoring multi-temporal vegetation change using Thematic Mapper imagery. Remote Sensing Environ 80:143–156
Rogan J, Miller J, Stow D, Franklin J, Levien L, Fischer C (2003) Land cover change mapping in California using classification trees with Landsat TM and ancillary data. Photogramm Eng Remote Sensing 69(7):793–804
Rouse JW Jr (1974) Monitoring vegetation systems in the Great Plains with ERST, Washington, DC, Goddard Space Flight Center, earth resources technology satellite-1 symposium, pp 309–317
Ronchetti E, Field C, Blanchard W (1997) Robust linear model selection by cross-validation. J Am Stat Assoc 92:1017–1023
Roy PS, Kaul RN, Sharma Roy MR, Garbyal SS (1985) Forest type stratication and delineation of shifting cultivation areas in the eastern part of Arunchai Pradesh using Landsat MSS data. Int J Remote Sensing 6(3/4):411–418
Royall RM (1998) A review of optimal designs in survey sampling. Can J Stat 12(1):53–65
Royall RM (2001) Model robust confidence intervals using maximum likelihood estimators. Int Stat Rev 54(2):221–226
Rubin DB (1981) The Bayesian bootstrap. Ann Stat 9(1):130–134
Rubin DB (1987) Multiple imputation for nonrersponse in surveys, Wiley, New York
Ruschel AR, Nodari ES, Guerra MP et al (2003) Uses and valuation evolution of the timber species from the Brazilian semi-evergreen forest of Alto-Uruguai, SC. Cien Flor 13(1):153–166

Saborowski J (1990) Schätzung von Varianzen und Konfidenzintervallen aus mehrstufigen Stichproben am Beispiel von Waldschadensinventuren. Schriften aus der Forstlichen Fakultät der Universität Göttingen und der niedersächsischen forstlichen Versuchsanstalt, vol 99

Saborowski J, Smelko S (1998) Evaluation of inventories based on sample plots of variable size. Allg Forst Jagdz 169(4):71–75

Saho J, Sitter RR (1996) Bootstrap for imputed survey data. J Am Stat Assoc 91:1278–1288

Sai VS, Mishra M (1986) Comparison of some indices of species diversity in the estimation of the actual diversity in a tropical forest: a case study. Trop Ecol 27:195–201

Salazar LA(1982) Remote sensing techniques aid in preattack planning for fire management. USDA research paper PSW-162

Salehi M (1999) Rao–Blackwell versions of the Horvitz–Thompson and Hansen–Hurwitz in adaptive cluster sampling. Environ Ecol Stat 6(2):183–195

Salehi M (2003) Comparison between Hansen-Hurwitz and Horvitz-Thompson estimators for adaptive cluster sampling. Environ Ecol Stat 10(1):115–127

Salehi SM, Seber GAF (1997) Two-stage adaptive cluster sampling. Biometrics 53:959–970

Salang AT, Sugawa T (1997) Identification of dipterocarp timbers of Sarawak. In: Proceedings of the TRTTC-JICA Research '97, Wisma Sumber Alam, Kuching, Malaysia, 18–20 November 1997. Timber Research and Technical Training Centre (TRTTC), Kuching, pp 110–127

Salick J, Mejia A, Anderson T (1995) Non-timber forest products integrated with natural forest management, Rio San Juan, Nicaragua. Ecol Appl 5(4):878–895

Santner TJ, Duffy DE (1989) The statistical analysis of discrete data. Springer, Berlin Heidelberg New York

Särndal CE (1996) Efficient estimators with simple variance in unequal probability sampling. J Am Stat Assoc 91:1289–1300

Särndal CE, Swensson B, Wretman J (1992) Model assisted survey sampling. Springer, Berlin Heidelberg New York

Samra JS, Gill MS, Bhatia VK (1989) Spatial stochastic modelling of growth and forest resource evaluation. For Sci 35(3):663–676

Sandholt I, Rasmussen K, Anderson J (2002) A simple interpretation of the surface temperature/vegetation index space for assessment of surface moisture status. Remote Sensing Environ 79:213–224

Sannier CAD, Taylor JC, du Plessis W (2002) Real-time monitoring of vegetation biomass with NOAA-AVHRR in Etosha National Park, Namibia, for fire risk assessment. Int J Remote Sensing 23(1):71–89

Santander Flores C, Albertin W (1974) Performance of Dalbergia retusa Hemsl. in the humid tropics, Turrialba, Costa Rica. Turrialba 24(1):76–83

Schade J (1980) Ein mehrphasiges Stichprobensystem für forstliche Grossrauminventuren gestützt auf Landsat MSS Daten, Luftbilder 1:50,000 und ergänzende terrestrische Messungen. Dissertation, University of Freiburg

Schafer JL, Scheinker N (2000) Inference with imputed conditional means. J Am Stat Assoc 95(449):144–154

Schmid P (1967) Die Weiterentwicklung der Leistungskontrolle in der Schweiz. Wiss Zeit Tech Univ Dresden 16/2

Schmid P (1969) Stichproben am Waldrand. Eidg Anstalt Forstl Versuchswes Ber 45(3):234–303
Schmid P (1971) Continuous forest inventory in Switzerland Eidg Anstalt Forstl Versuchswes Mimeogr
Schmid-Haas P (1983) Information on drain by forest inventories. In: Forest inventories for improved management, vol 17. Helsingin Yliopiston Metsänarvioimistieteen Laitos, Tiedonantoja, pp 50–61
Schmid-Haas P, Werner J, Baumann E (1978) Kontrollstichproben: Aufnahmeinstruktion, 2nd edn. Eidg Anstalt Forstl Versuchswes Ber 186
Schmidt KM, Menakis JP, Hardy CC, Hann WJ, Bunnell DL (2002) Development of coarse-scale spatial data for wildland fire and fuel management. General technical report RMRS-GTR-87. USDA Forest Service, Rocky Mountain Research Station, Ft Collins
Schmoldt DL, Rauscher HM (1996) Building knowledge-based systems For natural resource. Kluwer, Dordrecht
Schnadt K, Katzenbeiaer R (2004) Unique airborne fiber scanner technique for application-oriented LIDAR products. In: Thies M, Koch B, Spiecker H, Weinacker H (eds) Laser-scanners for forest and landscape assessment. Int Arch Photogramm Remote Sensing Spatial Inf Sci 36(8/W2):19–23
Schowengerdt RA (1997) Remote sensing: models and methods for image processing, 2nd edn. Academic, San Diego
Schreuder HT, Williams MS (2000) Reliability of confidence intervals calculated by bootstrap and classical methods using the FIA 1-ha plot design. General technical report RMRS-GTR-57:1-6. Rocky Mountain Research Station, USDA Forest Service
Schreuder HT, Sedransk J, Ware KD (1968) 3-P sampling and some alternatives I. For Sci 14(4):429–453
Schreuder HT, Sedransk J, Ware KD, Hamilton DA (1971) 3-P sampling and some alternatives, II. For Sci 17(1):103–118
Schreuder HT, Banyard SG, Brink GE (1987) Comparison of three sampling methods in estimating stand parameters for a tropical forest. For Ecol Manage 21(1–2):119–127
Schreuder HT, Gregoire TG, Wood GB (1992) Sampling methods for multiresource forest inventory. Wiley, New York
Schreuder HT, Gregoire TG, Wood GB (1993) Sampling methods for multiresource forest inventory. Wiley, New York
Scott CT (1981) Simplified estimators for sampling on multiple occasions. University of Minnesota Department of Forest Research, staff paper no 23
Scott CT (1984) A new look at sampling with partial replacement. For Sci 30:157–166
Scott CT (1986) An evaluation of sampling with partial replacement. In: Use of auxilliary information in natural resource inventories. Society of American Foresters, Blacksburg, 1–2 October 1985, SAF publ no 86-01:74–79
Scott CT (1992) Multivariate density estimation: theory, practice and visualization. Wiley, New York
Scott CT (1998) Sampling methods for estimating change in forest resources. Ecol Appl 8(2):228–233
Scott CT, Köhl M (1994) Sampling with partial replacement and stratification. For Sci 40(1) 30–46

Seber GAF (1982) The estimation of animal abundance and related parameters. Griffin, London
Seber GAF (1986) A review of estimating animal abundance. Biometrics 42:267–292
Sellers PJ (1987) Canopy reflectance, photosynthesis, and transpiration II: the role of biophysics in the linearity of their interdependence. Remote Sensing Environ 21:143–183.
Serna CB (1990) Rattan resource supply situation and management. In: Torreta NK, Belen EH (eds) Rattan. Proceedings of the national symposium/workshop on rattan, Cebu City, 1–3 June 1988. Philippine Council for Agriculture, Forestry and Natural Resources Research and Development book series no 99, Los Banos, Laguna
Serpico SB, Roli F (1995) Classification of multi-sensor remote-sensing images by structured neural networks. IEEE Trans Geosci Remote Sensing 33(3):562–578
Seto KC, Woodkock CE, Song C, Huang X, Lu J, Kaufmann RK (2002) Monitoring land-use change in the Pearl River Delta using LANDSAT TM. Int J Remote Sensing 23:1985–2004
Settle JJ, Drake NA (1993) Linear mixing and the estimation of ground proportions. Int J Remote Sensing 14:1159–1177
Setzer AW, Pereira MC (1991) Amazonia biomass burnings in 1987 and an estimate of their tro-pospheric emissions. Ambio 20:19–23
Shannon CE, Weaver W (1949) The mathematical theory of communication. University of Illinois Press, Urbana
Shao J (1996) Bootstrap model selection. J Am Stat Assoc 91:655–665
Shao J (2003) Impact of the bootstrap on sample surveys. Stat Sci 18(2):191–198
Shao J, Chen Y (1998) Balanced repeated replication for stratified multistage survey data under imputation. J Am Stat Assoc 93(442):819–831
Shao J, Steel P (1999) Variance estimation for survey data with composite imputation and non-neglible sample fractions. J Am Stat Assoc 94(445):254–265
Sharma M, Zhang SY (2004) Variable-exponent taper equations for jack pine, black spruce, and balsam fir in eastern Canada. For Ecol Manage 198(1–3):39–53
Shasby MV, Burgan RE, Johnson GR (1981) Broad area forest fuels and topography mapping using digital Landsat and terrain data. In: Proceedings of the 7th international symposium on machine processing of remotely Sensed data, Purdue University, West Lafayette, pp 529–538
Shen XT, Huang HC, Ye J (2004) Inference after model selection. J Am Stat Assoc 99(467):751–762
Shenk TM, White GC, Burnham KP (1998) Sampling-variance effects on detecting density dependence from temporal trends in natural populations. Ecol Monogr 68(3):445–463
Sherman M (1996) Variance estimation for statistics computed from spatial lattice data. J R Stat Soc Ser B 58(3):509–523
Shieh G (2001) Sample size calculations for logistic and Poisson regression models. Biometrika 88(4):1193–1199
Shimabukuro YE, Smith JA (1991) The least-squares mixing models to generate fraction images derived from remote sensing multispectral data. IEEE Trans Geosci Remote Sensing 29(1):16–20
Shiver BD, Borders BE (1996) Sampling techniques for forest resource inventory. Wiley, New York

Siljestrom P, Moreno A (1996) Monitoring burnt area by principal component analysis of multitemporal TM data. Int J Remote Sensing 16:1577–1587
Silverman BW (1986) Density estimation for statistics and data analysis. Chapman and Hall, London
Sim HC, Appanah S, Lu WM (eds) (2004) Forests for poverty reduction: can community forestry make money? FAO Regional Office for Asia and the Pacific, Bangkok
Simpson EH (1949) Measurement of diversity. Nature 163:688
Sims DA, Gamon JA (2002) Relationships between leaf pigment content and spectral reflectance across a wide range of species, leaf structures and development stages, Remote Sensing Environ 81(2–3):337–354
Sims DA, Gamon JA (2002) Relationships between leaf pigment content and spectral reflectance across a wide range of species, leaf structures and development stages. Remote Sensing Environ 81(2–3):337–354
Sims DA, Gamon JA (2003) Estimation of vegetation water content and photosynthetic tissue area from spectral reflectance: a comparison of indices based on liquid water and chlorophyll absorption features. Remote Sensing Environ 84(4):526–537
Singh AC, Stukel DM, Pfeffermann D (1998) Bayesian versus frequentist measures of error in small area estimation. J R Stat Soc Ser B 60:377–396
Singh KD (1981) Inventory techniques and mensurational methods to determine the age and growth rate of tropical trees. In: Borman FH, Berlyn G (eds) Age and growth rate of tropical trees, new directions for research. Bull Sch For Environ Stud Yale Univ 94:121–127
Singh KD (1989) Studies on the volume and yield of tropical forest stands. FAO forestry paper 51(1). FAO, Rome
Sison CP, Glaz J (1995) Simultaneous confidence intervals and sample size determination for multinomial proportions. J Am Stat Assoc 90:366–369
Skidmore AK, Turner BJ (1992) Map accuracy assessment using line intersect sampling. Photogramm Eng Remote Sensing 58(10):1453–1457
Sliggers J, Kakabeeke W (2005) Clearing the air: 25 years of the convention on long-range transboundary air pollution. United Nations, New York
Smith AP (1979) Buttressing of tropical trees in relation to bark thickness in Dominica BW I. Biotropica 11(2):159–160
Smith DR, Villella RF, Lemarie DP (2003) Application of adaptive cluster sampling to low-density populations of freshwater mussels. Environ Ecol Stat 10(1):7–15
Smith FH (1938) An empirical law describing heterogeneity in the yields of agricultural crops. J Agric Sci 28:1–23
Snedecor GW, Cochran WG (1971) Statistical methods, 6th edn. Iowa State Univ Press, Ames
Sohn Y, McCoy RM (1997) Mapping desert shrub rangeland using spectra unmixing and modeling spectra mixture with TM data. Photogramm Eng Remote Sensing 63(6):707–716
Spencer RD, Czaplewski RL (1998) National forest inventory in the USA: an outline of the procedure. Aust For 60(1):56–66
Staepelaere RL, Ginsburger PL (1978) Utilisation of bleached sulfate tropical hardwood pulp. In: Auchter RJ (ed) Proceedings of the conference on improved utilization of tropical forests, 21–26 May 1978, pp 451–485
Ståhl G (1997) Transect relascope sampling for assessing coarse woody debris: the case of a pi/2 relascope angle. Scand J For Res 12 (4):375–381

Ståhl G (1998) The transect relascope, an instrument for the quantification of coarse woody debris. For Sci 44(1) 201–215
Ståhl G, Lämås T (1998) Assessment of coarse woody debris, a comparison of probability sampling methods. In: Bachmann P, Köhl M, Päivinnen R (eds) Assessment of forest biodiversity for improved forest management. Kluwer, Dordrecht, pp 241–248
Ståhl G, Ringvall A, Lamas T (2000) Guided transect sampling for assessing sparse populations. For Sci 46(1):108–115
Staudte RG, Sheather SJ (1990) Robust estimation and testing. Wiley, New York
Stehman SV (1992) Comparison of systematic and random sampling for estimating the accuracy of maps generated from remotely sensed data. Photogramm Eng Remote Sensing 58(9):1343–1350
Stein A, Ettema C (2003) An overview of spatial sampling procedures and experimental design of spatial studies for ecosystem comparisons. Agric Ecosyst Environ 94(1):31–47
Sterba H (1980) Stem-curves: a review of literature. For Abstr 41:141–145
Stevens DL, Olsen AR (2004) Spatially balanced sampling of natural resources. J Am Stat Assoc 99(465):262–278
Steward JL, Dunsdon AJ, Hellin JJ, Hughes CE (1992) Wood biomass estimation of central American dry zone species. Tropical forestry papers 26. Oxford Forestry Institute, University of Oxford, Oxford
Stierlin HR (1979) Die Erfassung der Walderschliessung. Eidg Anstalt Forstl Versuchswes Ber 204
Stierlin HR, Zinggeler J (2001) Terrestrial inventory. In: Brassel P, Lischke H (eds) Swiss National Forest Inventory: methods and models of the second assessment. Swiss Federal Research Institute (WSL), Birmensdorf, pp 65–87
Stork N, Davies J (1996) Biodiversity inventories. In: Biodiversity assessment. A guide to good practice. Field manual 1. Data and specimen collection of plants, fungi and mircoorganisms. HMSO, London
Story M, Congalton RG (1986) Accuracy assessment: a user's perspective. Photogramm Eng Remote Sensing 52(3): 397-399
Stott CB, Semmes G (1962) Our changing inventory methods and the CFI system in North America. In: Proceedings of the 5th world forest congress, Seattle, pp 451–454
Stott CB, Ryan EJ (1939) A permanent sample technique adapted to commercial timber stands. J For 37:347–349
Stow DA, Chen DM (2002) Sensitivity of multitemporal NOAA AVHRR data of an urbanizing region to land-use/land-cover changes and misregistration. Remote Sensing Environ 80:297–307
Strand LC (1958) Sampling for volume along a line. Nor Skogst 51
Stroppiana D, Pinnock S, Pereira JMC, Gregoire J-M (2002) Radiometric analysis of SPOT-VEGETATION images for burnt area detection in Northern Australia. Remote Sensing Environ 82:21–37
Stuiver M, de Rebello AL, White JC, Broecker W (1981) Isotopic indicators of age/growth in tropical trees. In: Borman FH, Berlyn G (eds) Age and growth rate of tropical trees, new directions for research. Bull Sch For Environ Stud Yale Univ 94:75–82
Stussi N, Liew SC, Kwoh LK, Lim H (1997) Landcover classification using ERS-SAR/INSAR data over tropical areas. In: Proceedings of the 1997 international geoscience remote sensing symposium, pp 813–815

Sudman S, Sirken MG, Corvan CD (1988) Sampling rare and elusive populations. Science 240:991–995

Sukhatme PV, Sukhatme BV, Sukhatme,S, Asok C (1984) Sampling theory of surveys with applications. Iowa State University Press, Ames

Sutter R (1990) Increasing efficiency in volume estimation by the combination of aerial and terrestrial samples in the Swiss NFI. In: Sylvander R (ed) Proceedings of the IUFRO/SNS workshop, Umea, 26–28 February 1990. Swedish University of Agricultural Sciences, report 4

Svarrer K, Olsen,CS (2005) The economic value of non-timber forest products – a case study form Malaysia. J Sustain For 20(1):17–41

Swindel BF, Conde LF, Smith JE (1984) Species diversity: concept measurement, and response to clearcutting and site preparation. For Ecol Manage 8:11–22

Tanaka S, Kimura H, Suga Y (1983) Preparation of a 125,000 Landsat map of assessment of burnt area on Etajima Island. Int J Remote Sensing 4:17–31

Taras MA, Clark A (1977) Above ground biomass of longleaf pine in a natural sawtimber stand in southern Alabama. USDA Forest Service, Southeast Forest Experimental Station research paper SE-162

Tardif G (1965) Some considerations concerning the establishment of optimum plot size in forest survey. For Chron 41:93–102

Thiel K-H, Wehr A (2004) Performance capabilities of laser scanners: an overview and measurement principle analysis. In: Thies M, Koch B, Spiecker H, Weinacker H (eds) Laser-scanners for forest and landscape assessment. Int Arch Photogramm Remote Sensing Spatial Inf Sci 36(8/W2):14–18

Thill JC (1999) Spatial multicriteria decision making and analysis: a geographic information sciences approach. Ashgate, Aldershot

Thompson SK (1990) Adaptive cluster sampling. J Am Stat Assoc 85:1050–1059

Thompson SK (1991) Stratified adaptive cluster sampling. Biometrika 78(2):389–397

Thompson SK (1992) Sampling. Wiley, New York

Todd SW, Hoffer RM, Milchunas DG (1998) Biomass estimation on grazed and ungrazed rangelands using spectral indices. Int J Remote Sensing 19(3):427–438

Tomppo E (1997) Biodiversity monitoring in finnish forest inventories. In: Bachmann P, Kuuslea K, Uuttera J (eds) Assessment of biodiversity for improved forest management, vol 6. Kluwer, Dordrecht, pp 87–94

Tomppo E, Halme M (2004) Using coarse scale forest variables as ancillary information and weighting of variables in kNN estimation: a genetic algorithm approach. Remote Sensing Environ 92(1):1–20

Torelli N (1987) End-use orientated utilization of tropical forests – testing relevant biological, physical, chemical and mechanical properties. Funkcne integrovane obhospodarovanie lesov a komplexne vyuzitie dreva medzinarodna vedecka konferencia, Zvolen. Zb Ref Sekc 4:251–256

Touber L, Smaling EMA, Andriesse W, Kakkeling RTA (1989) Inventory and evaluation of tropical forest lands: guidelines for a common methodology. Tropenbos technical series 4. Tropenbos Foundation, Wageningen

Trachsler H, Kölbl O, Meyer B, Mahrer F (1980) Stichprobenweise Auswertung von Luftaufnahmen für die Erneuerung der Arealstatistik. Bundesamt für Statistik und Bundesamt für Raumplanung, Bern

Travaglini D, Mason F, Lopresti M, Lombardi F, Marchetti M, Chirici G, Corona P (2005) Aspects of biological diversity in the CONECOFOR plots. IV: Deadwood

surveying experiments in Alpine and Mediterranean forest ecosystems. 3rd report of TF on I&C evaluation of CONECOFOR. University of Florence, CFS:105–142
Treitz PM, Howarth, PJ, Gong P (1992) Application of satellite and GIS technologies for land-cover and land-use mapping at the rural–urban fringe: a case study. Photogramm Eng Remote Sensing 58:439–448
Treitz P, Howarth P (1999) Hyperspectral remote sensing for estimating biophysical parameters of forest ecosystems. Prog Phys Geogr 23(3)359–390
Trotter CM, Dymond JR, Goulding CJ (1997) Estimation of timber volume in a coniferous plantation forest using landsat TM. Int J Remote Sensing 18(10):2209–2223
Tucker CJ (1979) Red and photographic infrared linear combinations for monitoring vegetation. Remote Sensing Environ 8(2):127–150
Tucker CJ (1980) Remote sensing of leaf water content in the near infrared. Remote Sensing Environ 10:23–32
Tuominen S, Fish S, Poso S (2003) Combining remote sensing, data from earlier inventories, and geostatistical interpolation in multisource forest inventory. Can J For Res Rev Can Rech For 33(4):624–634
Turner MG, Gardner RH (eds) (1991) Quantiative methods in landscape ecology: the analysis and interpretation of landscape heterogeneity. Ecological studies, vol 82. Springer, Berlin Heidelberg New York
Tyystjarvi E, Karunen J, Lemmetyinen H et al (1998) Chlorophyll fluorescence can be used to identify plant species. Photosynthesis: mechanisms an effects, vol V. Proceedings of the XIth international congrss on photosynthesis, Budapest, Hungary, 17–22 August 1998, pp 3857–3860
Uebelhör K (1988) Praktische Erfahrungen mit Winkelzählprobe und Breitbandrelaskop im tropischen Regenwald. Forstarchiv 59(2):47–52
Ulaby FT, Pierce L, Oh Y, Sarabandi K, Dobson MC (1991) Scene classification and radar polarimetry for mapping of soil and vegetation parameters. Paper presented at the 6th international symposium on physical measurements and signatures in remote sensing, Val D'Isere, France, 17–21 January
UN-ECE (1998) Manual on methods and criteria for harmonized sampling, assessment, monitoring and analysis of the effects of pollution on forests, 4th edn. IPC, Hamburg
UN-ECE/FAO (2000) Forest resources of Europe, CIS, North America, Australia, Japan and New Zealand. United Nations publication no E.99.II.E.36. UN-ECE, Geneva
UN-ECE (2004) The condition of forests in Europe (2004). Executive report. UN, Geneva
UNESCO (1973) International classification and mapping of vegetation. United Nations Educational, Scientific and Cultural Organisation, Paris
Upton GJG, Fingleton B (1985) Spatial data analysis by example, vol I. Point patterns and quantitative data. Wiley, Chichester
Uriarte M, Canham CD, Thompson J, Zimmerman JK (2004) A neighborhood analysis of tree growth and survival in a hurricane-driven tropical forest. Ecol Monogr 74(4):591–614
Ustin SL, Smith MO, Jacquemoud S, Verstraete M, Govaerts Y (1999) Geobotany: vegetation mapping for earth sciences. In: Renca AN (ed) Remote sensing for the earth sciences, 3rd edn. Manual of remote sensing, vol 3. Wiley, New York, pp 707
Valentine HT, Hilton SJ (1977) Sampling oak foliage by the randomized-branch method. Can J For Res 7:295–298

Valentine HT, Tritton LM, Frurnival GM (1984) Subsampling trees for biomass, volume or mineral content. For Sci 30:673–681
Valentine HT, Gregoire TG, Frurnival GM (1992) Estimation of the aggregate bole volume of a population of trees by stratified, two-stage, probability sampling. Statistician 41:499–507
Valkenburg JLCH (1997) Non-timber forest products of East Kalimantan. Potentials for sustainable forest use. Tropenbos series 16. Tropenbos Foundation, Wageningen
Valliant R, Dorfman AH, Royall RM (2000) Finite population sampling and inference. A prediction approach. Wiley, New York
Vanclay J (1994) Modelling forest growth and yield: applications to mixed tropical forests. CAB, Wallingford
Vanclay JK (1996) Estimating sustainable timber production from tropical forests, CIFOR working paper no 11
Vanclay JK (1998) Towards more rigorous assessment of biodiversity. In: Bachmann P, Köhl M, Päivinen R (eds) Assessment of biodiversity for improved forest planning. Kluwer, Dordrecht, pp 211–232
van Deusen PC (1996) Incorporating predictions into an annual forest inventory. Can J For Res 26:1709–1713
van Deusen PC (1997) Annual forest inventory statistical concepts with emphasis on multiple imputations. Can J For Res 27:379–384
van Deusen PC, Baldwin VC Jr (1993) Sampling and predicting tree dry weight. Can J For Res 23:1826–1829
Van Hees WWS (2002) A comparison of two estimates of standard error for a ratio-of-means estimator for a mapped-plot sample design in southeast Alaska. Pacific North West Research Station notes PNW-532:1–9. USDA Forest Service, Portland
van Lieshout MNM, Baddeley AJ (2002) Extrapolating and interpolating spatial patterns. In: Lawson AB, Denison DGT (eds) Spatial cluster modelling. Chapman and Hall/CRC, Boca Raton, pp 61–86
Vantomme P (2003) What are 'non-wood forest products' for FAO? Int For Rev 5(2):162
van Wagner CE (1987) Development and Structure of the Canadian forest fire weather index system. Forestry technical report 35. Canadian Foresty Service. Ottawa
Velle K (1995) Estimation of standing stock of woody biomass in areas where little or no baseline data are available. A study based on field measurements in Uganda. Doctoral thesis, Agricultural University of Norway, Ås
Venette RC, Moon RD, Hutchison WD (2002) Strategies and statistics of sampling for rare individuals. Annu Rev Entomol 47:143–174
Vetter RE, Botosso RC (1989) Remarks on age and growth rate determination of Amazonian trees. Int Assoc Wood Anat J 10:133–145
Vikhamar D, Solberg R (2003) Subpixel mapping of snow cover in forests by optical remote sensing. Remote Sensing Environ 84(1):69–82
Vogelmann JE, Howard SM (1998) Regional characterization of land cover using multiple sources of data. Photogramm Eng Remote Sensing 64(1):45–57
Volk H, Schirmer C (2004) Leitfaden zur Kartierung der Schutz-und erholungsfunktionen des Waldes, Projektgruppe Forstliche Landespflege. Sauerländers's Verlag, Frankfurt
von Gadow K (1987) Untersuchung zur Konstruktion von Wuchsmodellen für schnellwüchsige Plantagenbauarten. Forstl Forschungsber Münch 77

von Gadow K, Pukkala T, Tomè M (eds) (2000) Sustainable forest management. Kluwer, Dordrecht

von Gadow K, Nagel J, Saborowski J (eds) (2002) Continuous cover forestry – assessment, analysis, scenarios. Kluwer, Dordrecht

von Neumann J, Kent R, Bellison HR, Hart BI (1941) The mean square successive difference. Ann Math Stat 50:226–251

von Segebaden G (1964) Studies of cross-country transport distances and road net extension. Stud For Suec 18

von Segebaden G, Strömnes R, Winer HI (1967) Proposal for international system of terrain classification. IUFRO, 14th Congress Munich, Sect 31–32(8):756-764

Walters DK, Burkhart HE, Reynolds MRJ, Gregoire TG (1991) A Kalman filter approach to localizing height-age equations. For Sci 37(6):1526–1537

Walz U (1999) Charakterisierung der Landschaftsstruktur mit Methoden der Satelliten-Fernerkundung und der Geoinformatik. Dissertation, TU Dresden, Dresden

Wand MP (1997) Data-based choice of histogram bin width. Am Stat 51:59–64

Wang F (1990) Fuzzy supervised classification of remote sensing images. IEEE Trans Geosci Remote Sensing 28(2):194–201

Wang F, Fuller WA (2003) The mean squared error of small area predictors constructed with estimated area variances. J Am Stat Assoc 98(463):716–723

Wan Razali M (1988) Modelling the tree growth in mixed tropical forests. Use of diameter and basal area increment. J Trop For Sci 1/2:114–121

Ware KD (1960) Optimum regression sampling design for sampling of forest populations on successive occasions. PhD dissertation, Yale University

Ware KD, Cunia T (1962) Continuous forest inventory with partial replacement of samples. For Sci Monogr 3

Weaver PL, Birdsey A (1986) Tree succession and management opportunities in coffee shade stands. Turrialba 36(1):47–58

Webster R, Olivier MA (1985) Utilisation exploratoire de la geostatistique pour la cartographie du sol dans la foret de Wyre (GB). Sci Terre Inf Geol 24:171–173

Webster R, Olivier MA (1990) Statistical methods in soil and land resource survey. Oxford University Press, Oxford

Wegmuller U, Werner CL (1995) SAR interferometric signatures of forest. IEEE Trans Geosci Remote Sensing 33:1153–1161

Whitmore TC (1989) Changes over twenty-one years in the Kolombangara rain forests. J Ecol 77(2):469–483

Wiant HV, Wood GB, Miles JA (1989) Estimating the volume of a radiata pine stand using importance sampling. Aust For 52:286–292

Wilkinson GG (1996) A review of current issues in the integration of GIS and remote sensing data. Int J Geograph Inf Syst 19:85–101

Williams DL (1989) The radiative transfer characteristics of Spruce (Picea spp.): implications relative to the canopy microclimate. PhD thesis, University of Maryland

Williams DL (1996) Adaptive optimization of renewable natural resources: solution algorithms and a computer program. Ecol Modell 93:101–111

Williams DL (2001) Nonuniform random sampling: an alternative method of variance reduction for forest surveys. Can J For Res 31:2080–2088

Williams DL, Eriksson M (2002) Comparing the two paradigms for fixed-area sampling in large-scale inventories. For Ecol Manage 168:135-148

Wilson R, Hirsch S, Madden R (1971) Airborne infrared forest fire detection system: final report. USDA Forest Service research paper INT-93
Wintle BA, McCarthy MA, Parris KM, Burgman MA (2004) Precision and bias of methods for estimating point survey detection probabilities. Ecol Appl 14(3):703–712
Wiersum KF (1999) Understanding diversity in NTFP management: a neglected issue in NTFP research. NTFP research in the Tropenbos programme: results and perspectives. Tropenbos Foundation, Wageningen
Wolter MK (1985) Introduction to variance estimation. Springer, Berlin Heidelberg New York
Wong JLG (1998) Non-timber forest products from the reserved forests of Ghana. Consultancy report 11. Forest Sector Development Project, Accra
Wong JLG (2000) The biometrics of non-timber forest product resource assessment: a review of current methodology. Internal report. Wild Resources, Gwynedd
Woodcock CE, Strahler AH (1987) The factor of scale in remote sensing. Remote Sensing Environ 21:333–339
Woodcock CE, Collins JB, Gopal S, Jakabhazy VD, Macomber S, Ryherd S, Harward VJ, Levitan J, Wu Y, Warbington R (1994) Mapping forest vegetation using landsat TM imagery and a canopy reflectance model. Remote Sensing Environ 50:240–254
Wopereis MC, Gascuel-Odoux C, Bourrie G, Soignet G (1988) Spatial variability of heavy metals in soil on a one-hectare scale. Soil Sci 146(2):113–118
Worbes M (1989) Growth rings, increment and age of trees in inundation forests, savannas and amountain forest in the Neotropics. Int Assoc Wood Anat J 10:109–122
Worbes M (1992) Occurrence of seasonal climate and tree-ring research in the tropics. Lundqua Rep 34:338–342
Worbes M, Junk WJ (1989) Dating tropical trees by means of C-14 from bomb tests. Ecology 70:503–507
Wu CB (2003) Optimal calibration estimators in survey sampling. Biometrika 90:937–951
Wulder MA, Franklin SE (eds) (2003) Remote sensing of forest environments: concepts and case studies, Springer, Berlin Heidelberg New York
Wulder MA, Franklin SE, Lavigne MB (1996) High spatial resolution optical image texture for improved estimation of forest stand leaf area index. Can J Remote Sensing 22:441–449
Wullschläger E (1982) Die Erfassung der Waldfunktionen. Eidg Anstalt Forstl Versuchswes Ber 238
Wyatt NL (1991) A methodology for the evaluation of non-timber forest resources. Case study: the forest reserves of southern Ghana. MSc thesis, Cranfield Institute of Technology
Yates F (1949) Sampling methods for censuses and surveys. Griffith, London
Yates F (1981) Sampling methods for censuses and surveys, 4th edn. Macmillan, New York
Young HE (1976) A summary and analysis of weight table studies. OSLO biomass studies. University of Maine, Orono, pp 251–282
Young HE, Strand L, Altenberger R (1964) Preliminary fresh and dry weight tables for seven tree species in Maine. Maine Agric St Orono Tech Bull 12
Yuan M (1997) Use of knowledge acquisition to build wildfire representation in geographical information systems. Int J Geograph Inf Sci 11(8):723–745

Yuan D, Elvidge CD (1998) NALC land cover change detection pilot study: Washington DC area experiments. Remote Sensing Environ 66:166–178

Zeide B (1980) Plot size optimization. For Sci 26:251–257

Zell J, Hanewinkel M, Seeling U (2004) Financial optimisation of target diameter harvest of European beech (Fagus sylvatica) considering the risk of decrease of timber quality due to red heartwood. For Policy Econ 6(6):579–593

Zent S (1996) Behavioural orientations towards ethnobotanical quantification. In: Alexiades MN (ed) Selected guidelines for ethnobotanical research: a field manual. New York Botanical Garden, New York, pp 199–239

Zhan X, Sohlberg RA, Townshend JRG, DiMiceli C, Carroll ML, Eastman JC, Hansen MC, DeFries RS. (2002) Detection of land cover changes using MODIS 250 m data. Remote Sensing Environ 83:336–350

Zhang L-C, Chambers RL (2004) Small area estimates for cross classifications. J R Stat Soc Ser B 66(2):479–496

Zhang L-C, Davidian M (2001) Linear mixed models with flexible distributions of random effects for longitudinal data. Biometrics 57:795–802

Zhao YL, Wall MM (2004) Investigating the use of the variogram for lattice data. J Comput Graph Stat 13(3):719–738

Zhu J, Morgan GD (2004) Comparison of spatial variables over subregions using a block bootstrap. J Agric Biol Environ Stat 9(1):91–104

Zhukov B,Oertel D, Lanzl F, Reinhäckel G (1999) Unmixing-based multisensor multiresolution image fusion. IEEE Trans Geosci Remote Sensing 37:1212–1226

Zingg A (1988) Schweizerisches Landesforstinventar, Anleitung für die Erstaufnahme 1982–1986. Eidg Anstalt Forstl Versuchswes Ber 304

Zinggeler J, Paschedag I, Ulmer U (1999) Erschliessung und Bewirtschaftung. In: Brassel P, Brändli UB (eds) Schweizerisches Landesforstinventar: Ergebnisse der Zweitaufnahmen 1993–1995, Eidgenössiche Forschungsanstalt für Wald, Schnee und Landschaft, Birmensdorf, pp 233–254

Index

K

L

M

N

O

MIX
Papier aus verantwortungsvollen Quellen
Paper from responsible sources
FSC® C105338

If you have any concerns about our products,
you can contact us on
ProductSafety@springernature.com

In case Publisher is established outside the EU,
the EU authorized representative is:
Springer Nature Customer Service Center GmbH
Europaplatz 3, 69115 Heidelberg, Germany

Printed by Libri Plureos GmbH
in Hamburg, Germany